# HYPERTROPHIC RESPONSE in SMOOTH MUSCLE


Editors

**Charles L. Seidel, Ph.D.**
Associate Professor
Department of Medicine
Baylor College of Medicine
Houston, Texas


**Norman W. Weisbrodt**
Professor
Department of Physiology and Cell Biology
University of Texas Medical School
Houston, Texas

CRC Press, Inc.
Boca Raton, Florida

Library of Congress Cataloging-in-Publication Data

Hypertrophic response in smooth muscle.

Includes bibliographies and index.
1. Smooth muscle--Hypertrophy.  I. Seidel, Charles L.
II. Weisbrodt, Norman W.  [DNLM: 1. Hypertrophy--
etiology.  2. Muscle, Smooth--pathology.  WE 500 H998]
QP321.H97  1987        599'.01'852        86-34314
ISBN 0-8493-6299-7



International Standard Book Number 0-8493-6299-7

Library of Congress Card Number 86-34314
Printed in the United States

# PREFACE

The hypertrophic response of smooth muscle has not been as extensively studied as that of striated muscle. The purpose of this book is to describe the current state of knowledge concerning smooth muscle hypertrophy, thereby identifying areas for future research. As the reader will quickly see, there are many gaps in our understanding of the tissue changes that occur and the possible stimuli that initiate these changes; however, these deficiencies will serve as a guide to future experiments.

The book is divided into two broad sections. The first section (Chapters 1 to 5) describes in detail the hypertrophic response of vascular, intestinal, and uteran smooth muscle while the second section (Chapters 6 to 8) describes various stimuli that may be responsible for the hypertrophic response.

# THE EDITORS

Charles L. Seidel is Associate Professor in the Departments of Medicine, Physiology, and Molecular Biophysics at Baylor College of Medicine, Houston, Texas.

Dr. Seidel received his Ph.D. degree in Physiology from the University of Michigan in 1970. After spending one year as a Fulbright-Hays lecturer in Afghanistan, he did postdoctoral work in the Physiology Departments of the University of Michigan and the University of Virginia.

Dr. Seidel is a member of the American Physiological Society, a Fellow of the Cardiovascular Section of the Society, and a member of the Society's Committee on Committees. He is also a member of the Biophysical Society.

Dr. Seidel's primary research interest is hypertrophy of vascular smooth muscle.

Norman W. Weisbrodt is a Professor in the Departments of Physiology and Cell Biology, and Pharmacology at The University of Texas Medical School, Houston, Texas.

Dr. Weisbrodt received his B.S. degree in Pharmacy from the University of Cincinnati in 1965, and his Ph.D. in Pharmacology from the University of Michigan in 1970. After spending two years as a postdoctoral fellow in the Gastroenterology Division of the Department of Medicine at the University of Iowa, he moved to the University of Texas at Houston where he has advanced from Instructor to Professor.

Dr. Weisbrodt is a member of the American Physiological Society, in which he has served as Chairman of the Gastrointestinal Section and as a representative to the Section Advisory Committee; the American Motility Society, in which he has served as Vice-president; and the American Gastroenterological Association. He presently serves as an Associate Editor of the American Journal of Physiology: Gastrointestinal and Liver Physiology.

Dr. Weisbrodt's primary research interest is in gastrointestinal motility and smooth muscle physiology.

# CONTRIBUTORS


Rosemary D. Bevan, M.D.
Associate Professor
Department of Pharmacology
College of Medicine
University of Vermont
Burlington, Vermont

Gordon R. Campbell, Ph.D.
Department of Anatomy
University of Melbourne
Parkville, Victoria, Australia

Julie H. Campbell, Ph.D.
Cell Biology Laboratory
Baker Medical Research Institute
Prahran, Victoria, Australia

Giorgio Gabella, M.D.
Reader in Cytology
Department of Anatomy
University College London
London, England

John R. Guyton, M.D.
Assistant Professor
Department of Medicine and Cell
 Biology
Baylor College of Medicine
Houston, Texas

Gary K. Owens, Ph.D.
Associate Professor
Department of Physiology
University of Virginia School of
 Medicine
Charlottesville, Virginia

Barbara M. Sanborn, Ph.D.
Professor
Departments of Biochemistry and
 Molecular Biology; Obstetrics,
 Gynecology and Reproductive Sciences
University of Texas Medical School
Houston, Texas

Charles L. Seidel, Ph.D.
Associate Professor
Departments of Medicine and
 Physiology and Molecular Biophysics
Baylor College of Medicine
Houston, Texas

Norman W. Weisbrodt
Professor
Department of Physiology and Cell
 Biology
University of Texas Medical School
Houston, Texas

# TABLE OF CONTENTS

## INTERMUSCLE COMPARISON OF HYPERTROPHIC RESPONSE



## HYPERTROPHIC STIMULI

Chapter 1

# MECHANISMS OF SMOOTH MUSCLE GROWTH: CELLULAR HYPERTROPHY VS. HYPERPLASIA

Gary K. Owens

## TABLE OF CONTENTS

# I. INTRODUCTION

A wide variety of functional, structural, and biochemical alterations have been reported in hypertrophic smooth muscle tissues.[1-5] The intent of this chapter is (1) to review studies which have demonstrated that the growth response (i.e., growth in cell size or cellular hypertrophy vs. growth in cell number or hyperplasia) of smooth muscle cells is quite different in different hypertrophic models; and (2) to present evidence suggesting that the diversity of changes that occur in hypertrophic smooth muscle tissues may relate, in part, to the cellular mechanisms of growth.

# II. SMOOTH MUSCLE HYPERTROPHY: ROLE OF CELLULAR HYPERTROPHY VS. HYPERPLASIA

## A. Hypertrophy of Arterial Smooth Muscle

Accelerated vascular smooth muscle cell growth has been identified as having a fundamental role in the formation of atherosclerotic lesions,[6] and is a characteristic feature in arteries of hypertensive patients[7] and animals.[8-11] While a clear clinical association between hypertension and atherosclerosis implies similarities in smooth muscle cell growth mechanisms, there appear to be fundamental differences in the growth behavior of smooth muscle cells in these two diseases. Atherosclerosis in humans or in experimental animal models is characterized by intimal migration and proliferation of smooth muscle cells, and is associated with profound alterations in vessel wall structure and cellular morphology (Figure 1).[6,12,13] In contrast, accelerated smooth muscle growth in hypertension is, in large part, restricted to the media of blood vessels, and vessel wall and cellular morphology are relatively normal (Figure 2).[10,11]

Whereas studies of experimental injury models of atherosclerosis (e.g., balloon embolectomy induced de-endothelialization) have clearly demonstrated the predominant role of smooth muscle cell proliferation in intimal lesion formation,[14,15] there has been considerable controversy in hypertension regarding the relative role of smooth muscle cell hypertrophy vs. hyperplasia in medial hypertrophy. Increased cell number has been inferred from observations of increased numbers of vascular smooth muscle cells labeled with $^3$H-thymidine,[16,17] and from an increase in vessel DNA content.[17,18] Our studies,[9,19,20] however, demonstrated that vascular smooth muscle cell hypertrophy, not hyperplasia, was responsible for the increased mass of smooth muscle in thoracic aortas of spontaneously hypertensive rats (SHR) and Goldblatt hypertensive rats (2-kidney, 1-clip) compared to normotensive controls. Most important, our studies showed that smooth muscle cell hypertrophy was accompanied by an increase in DNA ploidy (Figures 3, 4, 5), and that the increased DNA synthesis and content in these hypertensive models could be accounted for by development of smooth muscle cell polyploidism (Figure 6). These studies established that measurements of DNA synthesis and content alone could not distinguish hypertrophic vs. hyperplastic smooth muscle cell growth.

In contrast to observations in SHR and Goldblatt hypertensive rats, recent studies[21] in this laboratory have demonstrated that the principal smooth muscle cell growth response following induction of acute hypertension by partial ligation of the abdominal aorta between the renal arteries, is cellular hyperplasia. These results are consistent with those of Bevan and co-workers,[17,18] who observed an increased frequency of mitotic smooth muscle cells following aortic coarctation-induced hypertension in rabbits.

Taken together, the studies cited above demonstrate that the cellular growth response of vascular smooth muscle cells is quite different under different growth stimulating conditions (Figure 7). Whereas smooth muscle cell growth in experimental

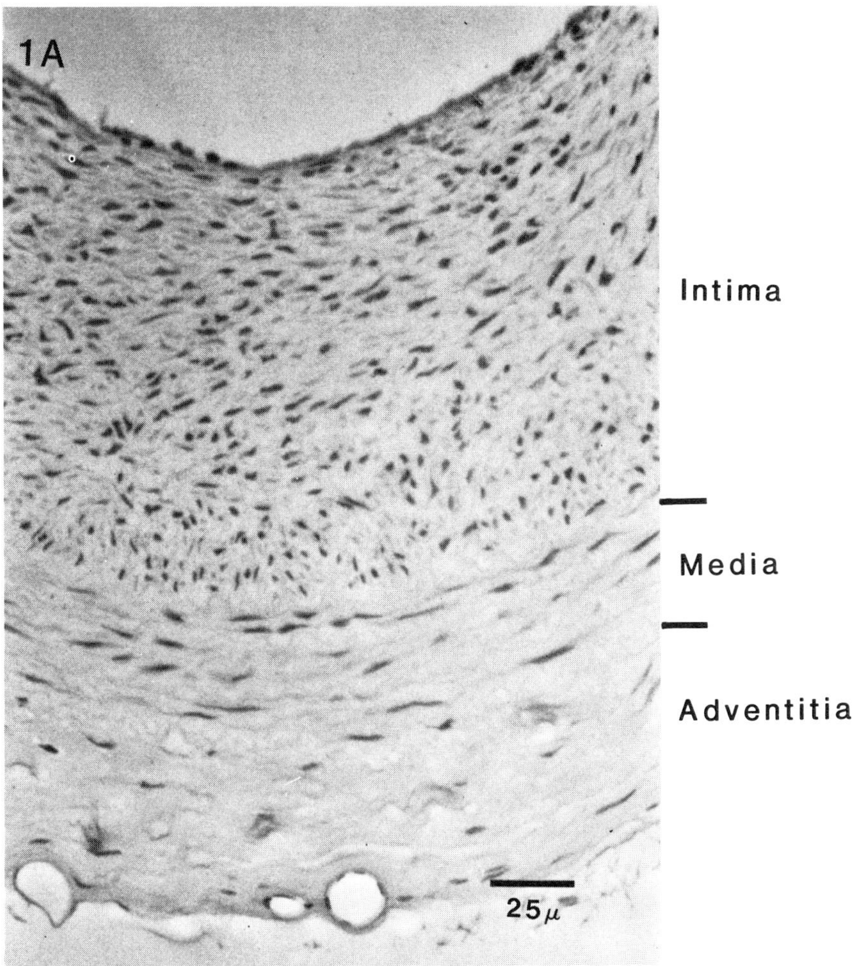


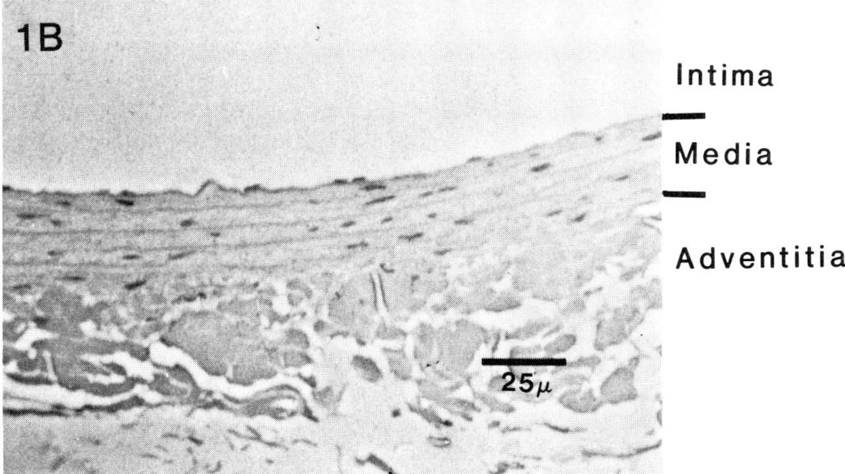


FIGURE 1.   (1A). Light micrograph (hematoxylin stained) of a rat carotid artery 2 weeks follow-ing balloon cathetor induced de-endothelialization. Note the extensive intimal lesion consisting of randomly oriented vascular smooth muscle cells. (1B). Light micrograph of an unballooned (con-trol) carotid.

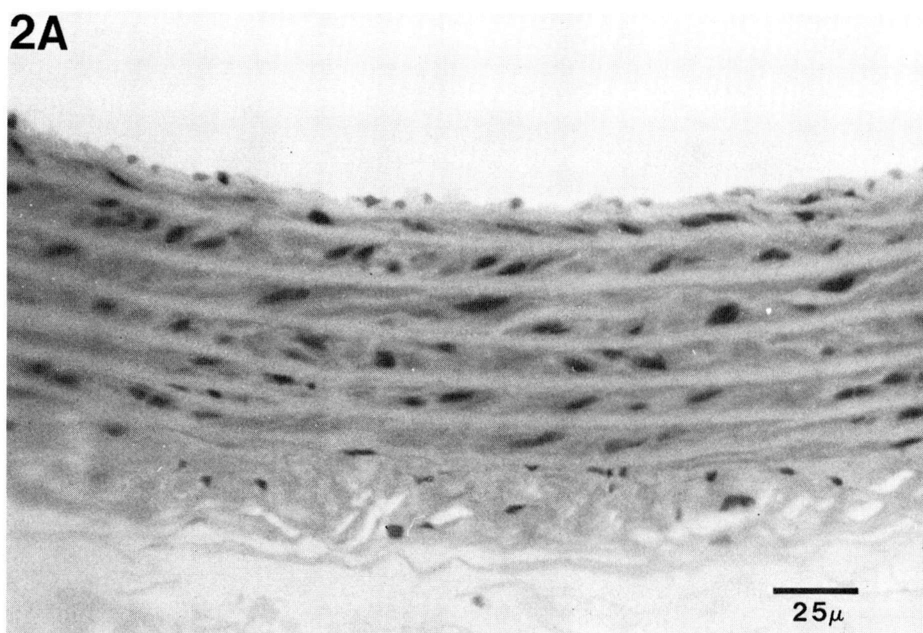


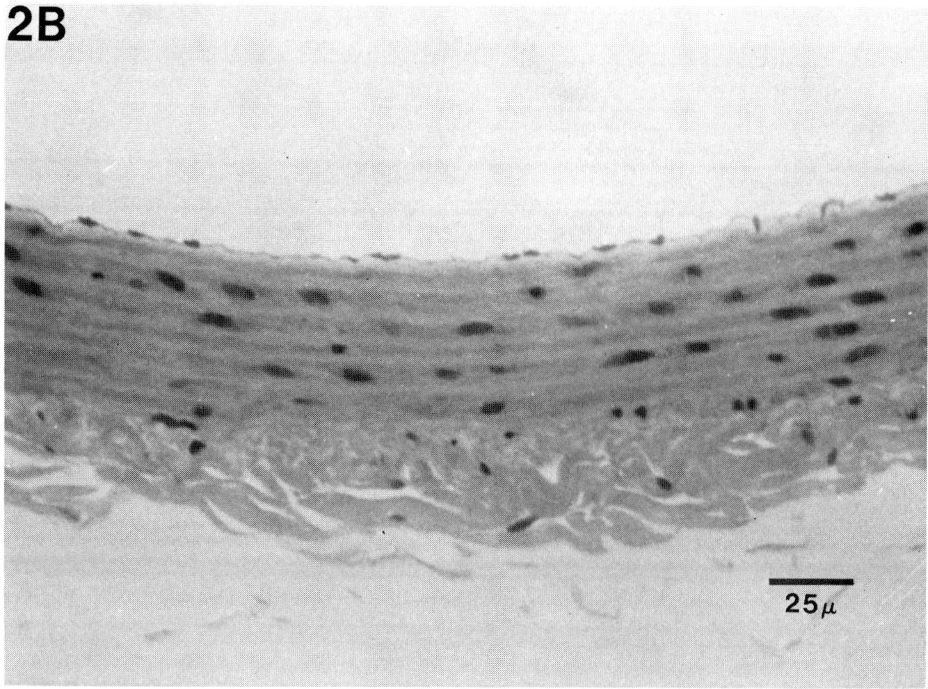


FIGURE 2. Light micrograph (hematoxylin and eosin stained) of the thoracic aorta of a 5-month-old spontaneously hypertensive (2A) and normotensive Wistar-Kyoto (2B) rat. Micrographs are the same magnification, samples were taken from the same aortic location in each animal. In contrast to Figure 1, smooth muscle hypertrophy in the spontaneously hypertensive rat occurs in the media of the blood vessel, and there appears to be no gross disruption of cellular orientation or vessel structure.

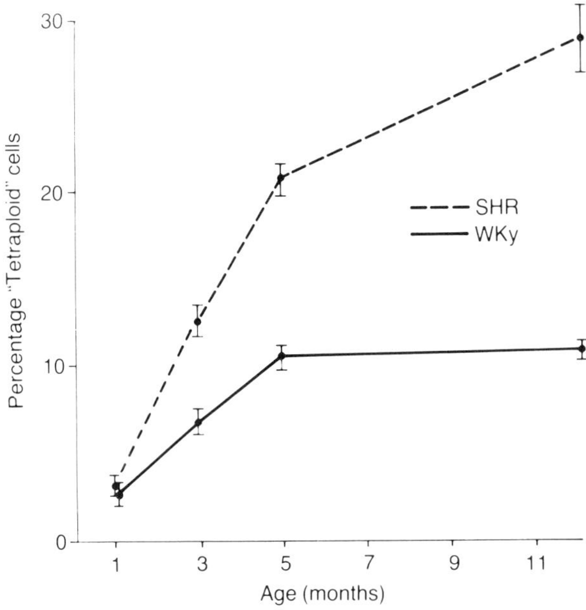



FIGURE 3. Frequency of tetraploid aortic smooth muscle cells in spontaneously hypertensive (SHR) and normotensive Wistar-Kyoto (WKy) rats as determined by flow microfluorimetry. Freshly isolated vascular smooth muscle cells were stained with diamidinophenylindole as described by Owens et al.[19] for flow cytometric determinations. SHR was significantly greater than WKy (P <0.001) at 3 months of age and older. Means ± SE are shown. (From Owens, G. K. and Schwartz, S. M., Alterations in vascular smooth muscle mass in the spontaneously hypertensive rat, *Circ. Res.*, 51, 280, 1982. With permission of the American Heart Association.)

models of atherogenesis[14,15] or following induction of acute hypertension[17,21] occurs principally by cellular proliferation, growth in chronic hypertensive models such as the SHR or Goldblatt is due to smooth muscle cell hypertrophy and hyperploidy without detectable hyperplasia.[9,19,20]

## B. Hypertrophy of Other Smooth Muscle Tissues

There is some evidence suggesting that observations in arterial smooth muscle may extend to other smooth muscle tissues as well. For example, hypertrophy of human uterine smooth muscle during pregnancy is associated with the appearance of polyploid smooth muscle cells.[22] Increases in nuclear volume, suggestive of a change in cell ploidy, have also been reported in hypertrophic smooth muscle in the ligated portal-anterior mesenteric vein,[5] and in urinary bladder,[23] colon,[24] and intestinal smooth muscle.[25]

The relative importance of smooth muscle cell hypertrophy vs. hyperplasia in the hypertrophic response of other (nonarterial) smooth muscle tissues also varies, although, in general, less is known regarding the effects of different hypertrophic stimuli. Brent and Stephens[23] studied hypertrophy of the rabbit urinary bladder induced by outflow obstruction. In adult animals, the muscular thickening was mainly due to an initial 5-fold increase in cell volume, together with a later 3-fold increase in cell number. In young rabbits, the muscle cell number first doubled and then underwent a 6-fold increase in cell volume. Brent[24] reported similar observations in the hypertrophic

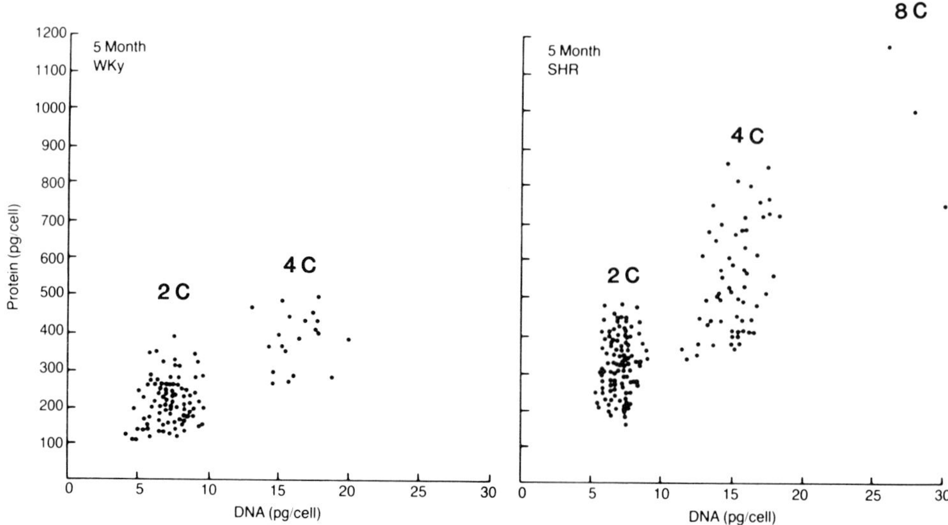


FIGURE 4.    Simultaneous microdensitometric determinations of the protein and DNA content of individual aortic smooth muscle cells from a 5-month SHR and WKy rat. Each data point represents the combined measurements on a single cell. Cells were obtained by enzymatic dissociation of medial preparations of thoracic aortas and protein and DNA determinations done as described by Owens and Schwartz.[9] Distinct populations diploid (2C) and tetraploid (4C) and octaploid (8C) cells were observed in SHR, whereas only diploid and tetraploid cells were found in WKy.

colon. Uvelius et al.[4] observed a 50% decrease in the number of muscle cells per square millimeter cross-sectional area, and a marked increase in cell size following partial ligation of the portal vein, suggesting a predominant role for cellular hypertrophy rather than hyperplasia, in this hypertrophic model. Gabella[3] found a 3- to 4-fold increase in average smooth muscle volume following experimental stenosis of the guinea pig intestine. He also observed an increased frequency of smooth muscle cells in mitosis in hypertrophic muscles, suggesting a role for cell proliferation in this model as well. However, some caution is warranted with this interpretation since: (1) increased cell replication does not necessarily denote increased cell number (i.e., it may reflect increased cell turnover), and (2) it is possible that appearance of mitoses was associated with formation of multinucleated cells[26] rather than cell division (i.e., karyokinesis without cytokinesis).

In summary, there appears to be a variable role of cellular hypertrophy vs. hyperplasia in smooth muscle hypertrophic responses. Now, let us consider the possible importance of distinguishing hypertrophic from hyperplastic cellular growth when assessing structural-functional alterations associated with hypertrophy of smooth muscle tissues, or when exploring possible mediating factors for accelerated smooth muscle growth. With this objective in mind, this discussion is necessarily limited to studies in which the role of cellular hypertrophy and hyperplasia has clearly been defined.

## III. RELATIONSHIP OF SMOOTH MUSCLE CELL GROWTH TO CELL FUNCTION

### A. Structural-Functional Alterations Associated with Smooth Muscle Cell Hyperplasia
#### 1. Acute Changes

Chamley-Campbell et al.[27,28] presented evidence showing that smooth muscle cells do not replicate in cell culture until they have lost much of their contractile apparatus.

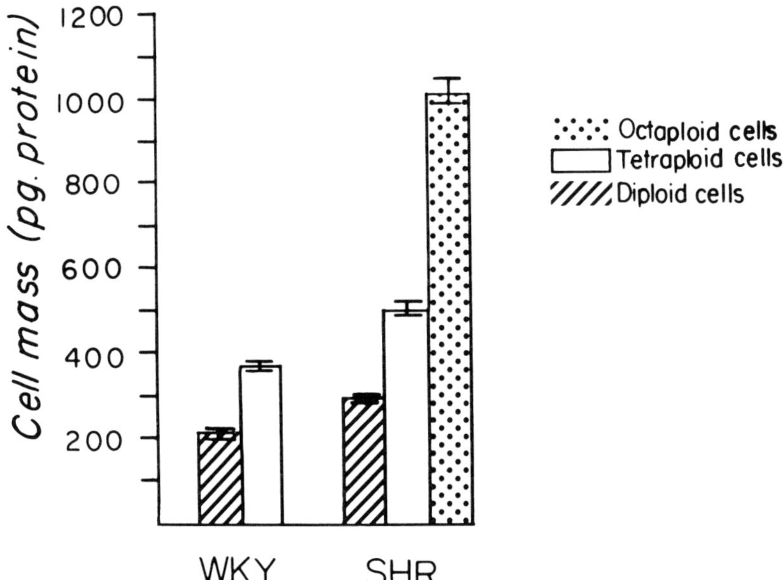



FIGURE 5.   Summary of microdensitometric determinations of protein and DNA content on individual aortic smooth muscle cells from 5-month SHR (n=9) and WKy (n=8) rats. Protein contents (mean ± SE) of diploid, tetraploid, and octaploid smooth muscle cells are shown. Approximately 200 cells were analyzed from each rat. Note that polyploid cells were hypertrophied compared to diploid cells in both rat strains, although each ploidy class in SHR was hypertrophied compared with those from normotensive WKy. (From Owens, G. K. and Schwartz, S. M., Alterations in vascular smooth muscle mass in the spontaneously hypertensive rat, *Cir. Res.*, 51, 280, 1982. With permission of the American Heart Association.)

Based on this and other data, the Campbells suggested that smooth muscle cells normally exist in a nonproliferative, contractile state, and that the cells must first "dedifferentiate" before proliferating. However, numerous morphologic studies of smooth muscle tissues in vivo have demonstrated that mitosis can occur in muscle cells which appear to have a full complement of myofilaments and cellular organelles, characteristic of differentiated smooth muscle cells.[25,29,30] While these observations clearly indicate that total loss of cytodifferentiation is not a prerequisite for smooth muscle proliferation, they are not inconsistent with the idea that smooth muscle cells undergo acute alterations in expression of cell specific characteristics during cell proliferation.

There is substantial biochemical evidence demonstrating that initiation and cessation of smooth muscle cell proliferation is associated with major changes in expression of cell specific proteins. Saborio et al.[31] studied the variant forms of gizzard actin present during chick embryogenesis and found a continuous increase in smooth muscle-specific $\gamma$-isoactin and a decrease in $\beta$-nonmuscle isoactin with embryonic development, and showed that the differential expression of actins was controlled at the transcription level. Studies in our laboratory[32] have shown that similar changes occur during developmental growth of vascular smooth muscle cells in vivo, i.e., decreases in smooth muscle cell replication coincided with a switch in actin expression from the nonmuscle forms (i.e., $\beta$-nonmuscle and $\gamma$-nonmuscle actin) to the smooth muscle cell specific forms ($\alpha$- and $\gamma$-smooth muscle). We have also examined changes in isoactin content and synthesis that occur in association with growth in cultured vascular smooth muscle cells.[32] Results showed that: (1) the onset of cell growth in primary culture was preceded by a dramatic decrease in $\alpha$-smooth muscle actin synthesis relative to $\beta$-nonmus-

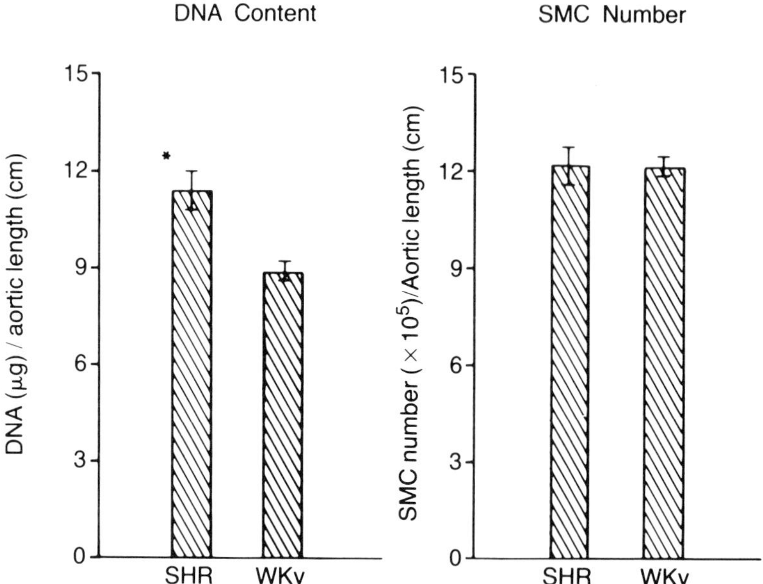


FIGURE 6.   Aortic medial DNA content and smooth muscle cell (SMC) number for 5-month SHR and WKy rats. Cell number was calculated by dividing DNA content by the mean DNA/cell determined by Feulgen-DNA microdensitometry. Whereas DNA content was significantly greater (P<0.001, t-test) in SHR than WKy, no differences were evident in cell number. Results are mean ± SE. (From Owens, G. K., Rabinovitch, P. S., and Schwartz, S. M., Smooth muscle cell hypertrophy versus hyperplasia in hypertension, 1981, *Proc. Natl. Acad. Sci., USA,* 78, 7759, 1981. With permission.)

cle actin; (2) $\alpha$-smooth actin synthesis was low or absent during rapid growth whereas $\beta$-nonmuscle actin synthesis was high; and (3) density arrest of growth was associated with increased $\alpha$-smooth muscle actin synthesis, although the ratio of $\alpha$-actin synthesis to $\beta$-nonmuscle actin synthesis was still much lower than found in vivo. Observations consistent with these have been reported by others.[33-35] Strauch and Rubenstein[35] recently demonstrated that growth arrest of confluent $BC_3H_1$ cells (a presumptive smooth muscle cell line isolated from mouse brain tumors by Schubert and co-workers)[36] in serum-free medium was associated with increased $\alpha$-smooth muscle actin synthesis. $BC_3H_1$ cells also undergo increased expression of muscle type creatine phosphokinase upon growth arrest.[37] Studies by Gown et al.[38] and Gabbiani et al.[34] demonstrated that smooth muscle cells in myointimal lesions of ballooned-injured rats undergo a switch in actin isotype expression from the muscle to the nonmuscle forms, although these studies did not distinguish whether changes were directly related to onset of cell growth, as opposed to other factors associated with intimal migration of cells, (e.g., altered cell to cell contacts, changes in extracellular milieu, etc.)

## 2. Chronic Changes

Whereas the studies cited above show that smooth muscle cell proliferation is associated with acute changes in cell function, cell replication per se, would not be expected to result in a permanent change in individual cell function. That is, each daughter cell should eventually be an exact copy of the parent cell. Thus, following cessation of cell proliferation, and reestablishment of cell to cell contacts disrupted during cell division (e.g., electrochemical and mechanical connections), tissue function would be expected to return to normal.

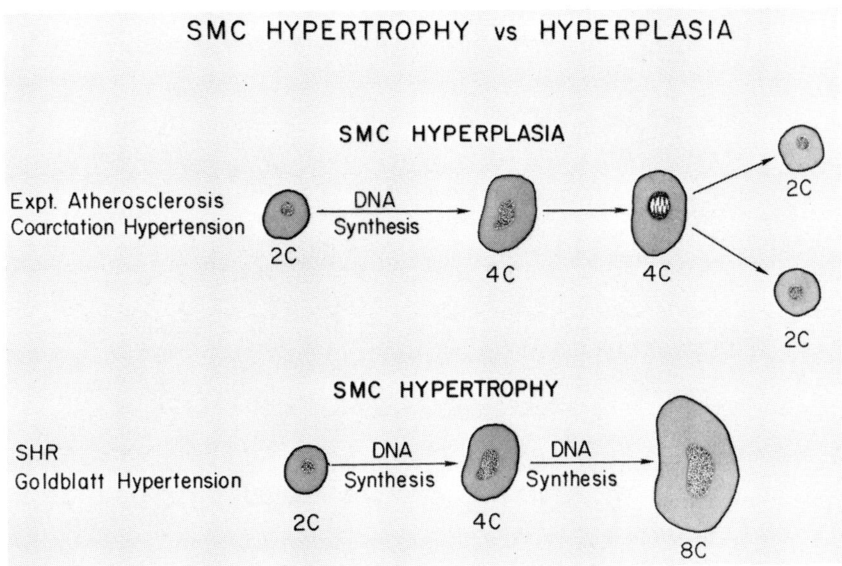



FIGURE 7. Growth response of vascular smooth muscle cells to different hypertrophic stimuli. Note that both cellular hypertrophy and hyperplasia are characterized by increased DNA synthesis.

## B. Structural Functional Alterations Associated with Smooth Muscle Cell Hypertrophy and Hyperploidy

### 1. Acute Changes

In contrast to the acute phenotypic alterations reported for proliferating vascular smooth muscle cells, there is some tentative evidence suggesting that hypertrophic cell growth may not be associated with acute loss of differentiated function. Whereas smooth muscle cells have been reported to undergo a number of functional alterations in hypertension,[1,2,39] such changes do not appear to represent loss of cellular differentiation. Our studies[19] and those of Seidel[40] demonstrated that the actin and myosin content, as well as the distribution of actin variants (Table 1), were not altered in hypertrophic vascular smooth muscle cells from spontaneously hypertensive rats (SHR). These data, however, were based on analysis of cells which were already hypertrophied, and it is unclear whether cells may have been altered during the hypertrophic process, and subsequently reacquired differentiated characteristics. Thus, while on a teleological basis it is interesting to speculate that hypertrophic growth of smooth muscle cells represents a mechanism whereby smooth muscle mass can increase without temporary loss of differentiated function (i.e., without loss of contractile elements, disruption of mechanical and electrochemical connections, etc.), experimental evidence is presently lacking. It is worth noting, however, that there is precedent for this idea in that both cardiac and skeletal myocytes respond to "hypertrophic" stimuli by increasing cell mass without loss of differentiation.[41-43] Furthermore, it is interesting that, like smooth muscle cells, significant numbers of normal adult cardiac myocytes are polyploid and that the incidence and degree of polyploism increases in myocardial hypertrophy.[44]

### 2. Chronic Changes

Whereas smooth muscle hypertrophy may not be associated with acute cellular alterations, there is some evidence suggesting that it is associated with chronic alterations in cell function and/or structure. A change in the surface area to volume ratio as cells

Table 1

CONTRACTILE PROTEIN CONTENTS OF
VASCULAR SMOOTH MUSCLE CELLS FROM
SHR AND NORMOTENSIVE WKy RATS

|  | SHR[a] | WKy[b] |
|---|---|---|
| Actin (% of total cellular protein) | 38% | 35% |
| Actin variants (% of total actin) | | |
| α-smooth muscle | 74.8% | 71.6% |
| β-nonmuscle | 20.3% | 21.7% |
| γ-nonmuscle + γ-smooth muscle | 4.9% | 4.8% |
| Myosin (% of total cellular protein) | 7.1% | 8.1% |

*Notes:* Aortic smooth muscle cells were enzymatically dissociated
from medial preparations of thoracic aortas from SHR
and WKy rats.[9] Actin and myosin contents were determined by densitometry of SDS polyacrylamide gels.[59]
Measurements of actin variants were done by densitome-
tric evaluation of Coomassie Blue stained gels following
two-dimensional gel electrophoresis.[60]

[a]   SHR, Spontaneously Hypertensive Rats.
[b]   WKy, Wistar Kyoto rats.

enlarge may alter cell function through a variety of mechanisms, including alterations
in receptor number and/or density. Both increases[1,2] and decreases[45] in agonist sensitivity have been reported in association with smooth muscle cell hypertrophy, although
the specific relationship of these changes to cellular hypertrophy per se remains to be
determined.

Data of Berner et al.[5] suggest that smooth muscle cell hypertrophy is associated with
major changes in cell structure. Using morphometric techniques, Berner et al.[5] found
a decrease in the relative number of myosin filaments, along with a 250% increase in
the intermediate filaments in hypertrophic venous smooth muscle of the ligated portal
vein, a model characterized by smooth muscle cell hypertrophy, not hyperplasia.[4] Consistent with these observations, Johansson[45] and Uvelius et al.[4] reported decreased
stress generating capacity, a reduced frequency of spontaneous contractions, and a
lower sensitivity to exogenous norepinephrine in this same model. Gabella[3] also observed a remarkable increase in intermediate filaments in hypertrophic smooth muscle
of the guinea pig intestine; and observed slight decreases in force generating capacity
per cross-sectional area, although he did not observe changes in thin and thick myofilaments.

The studies cited above suggest that development of smooth muscle cell hypertrophy
is associated with major qualitative changes in cellular proteins. In contrast, however,
our data[19] (Table 1) and those of Seidel[40] demonstrated the actin and myosin content
increased proportionately with other cellular proteins in hypertrophic-aortic smooth
muscle cells from SHR. Consistent with this, Arner and Uveluis[46] reported that
whereas maximal stress generating capacity was greater in SHR due to an increase in
total smooth muscle mass, the force development per unit cross-sectional area was
unaltered relative to normotensive controls.

The significance of a change in cellular ploidy on smooth muscle cell function is
unclear. It is not known, for example, whether the excess DNA in these cells is functional. Studies in hepatocytes, however, have shown that development of polyploidism
during hypertrophy of these cells is associated with proportionate increases in protein

synthetic capability.[47] An interesting recent observation in this laboratory[48] was that whereas changes in smooth muscle cell mass were reversed by antihypertensive drug treatment of SHR, changes in cellular ploidy were not. Similarly, Van der Heijden and James[22] observed that the polyploidy that occurs in human uterine smooth muscle during pregnancy does not reverse postpartum, despite regression of uterine hypertrophy. Thus, changes in smooth muscle cell ploidy appear to persist despite regression of tissue hypertrophy. The potential significance of this observation remains to be determined.

## IV. GROWTH INITIATING MECHANISMS FOR SMOOTH MUSCLE CELL HYPERTROPHY VS. HYPERPLASIA

A critical question which needs to be addressed is — why does a smooth muscle cell undergo cell replication in response to one set of stimuli, but cellular hypertrophy and hyperploidy to another (Figure 7)? It appears that development of polyploidism is not simply due to an inherent loss in the capacity of cells to divide, since smooth muscle cells in a variety of tissues can clearly be induced to divide. This implies that there must be fundamental differences in the underlying mechanisms responsible for these two growth processes. My intent here is not to attempt to review the extensive literature relating to control of smooth muscle cell growth (see References 6 and 49 for a review on this subject), but instead to focus on evidence relating to how the initiating factors for smooth muscle cell hypertrophy and hyperplasia may be different.

### A. Initiating Mechanisms for Smooth Muscle Cell Hypertrophy and Hyperploidy

One hypothesis is that hypertrophic smooth muscle cell growth represents a response to an increased work load or wall stress. There is precedence for this idea in that both skeletal muscle and cardiac muscle respond to an increased work load by increasing contractile cell mass without cell division.[41,42] While undoubtedly increased tissue stress (or distension) plays an important role in mediation of growth in a variety of models of smooth muscle hypertrophy, evidence that stress induces cellular hypertrophy rather than hyperplasia is largely circumstantial. We have consistently observed a high correlation between the severity and duration of hypertension, and the frequency of polyploid smooth muscle cells.[9,19,20] Furthermore, recent studies in our laboratory demonstrated that lowering of blood pressure in SHR by antihypertensive drug treatment prevented further development of smooth muscle cell hypertrophy and hyperploidy, but did not prevent hyperplasia associated with normal growth (i.e., increases in cell number that occur in both hypertensive and normotensive rats) or with long-term hypertension (there is a moderate increase in medial smooth muscle cell number that occurs in older SHR but not Wistar Kyoto rats (WKy).[48,50] These findings support a direct role for elevated blood pressure in mediation of smooth muscle cell hypertrophy and hyperploidy, but suggest that factors other than blood pressure per se are responsible for smooth muscle cell hyperplasia.

### B. Initiating Mechanisms for Smooth Muscle Cell Hyperplasia

Our studies[21] and those of Bevan[17,18] also support an indirect role for increased blood pressure in mediation of smooth muscle cell proliferation, following aortic-coarctation-induced hypertension. These studies showed that the increased growth of cells was restricted to vessels proximal to the coarctation site where blood pressure was elevated. However, an important difference between this hypertensive model and the SHR or Goldblatt models examined,[9,19,20] was the rapidity and severity of the increase in blood pressure. Following coarctation, blood pressure increased rapidly to high levels within

3 days of surgery,[21] whereas development of hypertension in the SHR and two-kidney one-clip Goldblatt hypertension was much more gradual, developing over a period of weeks or months.[9,20,51] The rapid increase in blood pressure following coarctation-induced hypertension appears to be associated with considerably more endothelial injury than is observed in other hypertensive models. De Chastonay et al.[52] reported that aortic-coarctation-induced hypertension caused a much more rapid and a greater increase in endothelial replication than did uninephrectomy + Na-rich diet or uninephrectomy + 0.9 NaCl drinking water + deoxycorticosterone. In our studies,[21] endothelial DNA replication was increased nearly 20-fold in coarctation rats compared to controls, while cell density and luminal surface area were not altered, implying increased endothelial cell turnover. In contrast, aortic endothelial replication is not increased in SHR compared to Wistar-Kyoto controls,[53] or in chronic Goldblatt (two-kidney one-clip) hypertension,[54] although it is transiently increased in acute Goldblatt hypertension.[55] Consistent with these data, there is considerable morphologic evidence demonstrating that acute increases in blood pressure are associated with endothelial injury or dysfunction.[56-58]

In view of evidence implicating endothelial injury or dysfunction in initiation of smooth muscle cell proliferation in experimental atherogenesis,[6] it is interesting to speculate that the proliferative response of cells following coarctation hypertension may relate to the increased endothelial injury that occurs in this model. Observations by Fernandez and Crane,[16] and Bevan[17,18] that a maximal growth response of cells occurred at sites of the most marked exudation of plasma supports this idea. Furthermore, observations that fibroblasts and epithelial cells adjacent to the site of injury also underwent increased replication, argue for a role of growth factors in this proliferative response, rather than wall tension directly, since it seems unlikely that tension would stimulate replication of nonmuscle cells adjacent to an affected artery. Injury-induced cell proliferation may be attributed to the generation of growth factors derived from platelets or monocytes adherent to the injured wall, generation of growth factors by the endothelial cells themselves, or to increased influx of blood borne mitogenic factors into the vessel wall.[6,49]

In summary, it is interesting to speculate that hypertrophic growth of vascular smooth muscle cells represents a direct response to increased blood pressure (or wall stress), whereas cellular proliferation represents a response mediated by classic cellular mitogens. While this possibility has not been adequately tested, it is clear that identification of the mechanisms responsible for stimulating smooth muscle growth in various hypertrophic models will require a clear distinction regarding the cellular nature of that growth.

## V. CONCLUSIONS

1.  Smooth muscle cells are clearly capable of growth by either cellular hypertrophy or hyperplasia.
2.  Smooth muscle cell hypertrophy is associated with an increase in DNA ploidy, and thus measurements of DNA synthesis and content alone cannot distinguish hypertrophic vs. hyperplastic cell growth.
3.  There is some evidence suggesting that smooth muscle cell hypertrophy is associated with quite different alterations in cellular function than is cellular proliferation. Thus, the diversity of alterations that occur in hypertrophic smooth muscle tissues may relate, in part, to the cellular mechanism of growth.
4.  Whereas it is clear that different hypertrophic stimuli elicit different smooth muscle cell growth responses, the mechanisms responsible for cells responding by

cellular hypertrophy (and hyperploidy) in one case and cellular proliferation in another are not clear.

5.  Many questions remain unanswered. Almost nothing is known, for example, regarding the molecular signals for smooth muscle hypertrophy. If cellular hypertrophy is a response to increased stress or work load, how is this transduced into a signal for cellular growth? How are hypertrophy and hyperplasia different in this regard? Can one predict whether a diploid cell stimulated to grow will divide or will become a tetraploid cell?

# REFERENCES


1. Hermsmeyer, K., Electrogenesis of increased norepinephrine sensitivity of arterial vascular muscle in hypertension, *Circ. Res.*, 38, 362, 1976.
2. Holloway, E. T. and Bohr, D. F., Reactivity of vascular smooth muscle in hypertensive rats, *Circ. Res.*, 23, 678, 1973.
3. Gabella, G., Hypertrophic smooth muscle IV. myofilaments, intermediate filaments and some mechanical properties, *Cell Tissue Res.*, 201, 277, 1979.
4. Uvelius, B., Arner, A., and Johansson, B., Structural and mechanical alterations in hypertrophic venous smooth muscle, *Acta Physiol. Scand.*, 112, 463, 1981.
5. Berner, P. F., Somlyo, A. V., and Somlyo, A. P., Hypertrophy-induced increases of intermediate filaments in vascular smooth muscle, *J. Cell Biol.*, 88, 96, 1981.
6. Ross, R., Atherosclerosis: A problem of the biology of arterial wall cells and their interactions with blood components, *Arteriosclerosis*, 1, 293, 1981.
7. Furuyama, M., Histometrical investigations of arteries in reference to arterial hypertension, *Tohoku J. Exp. Med.*, 76, 388, 1962.
8. Wolinsky, H., Response of the rat aortic media to hypertension, *Circ. Res.*, 26, 507, 1970.
9. Owens, G. K. and Schwartz, S. M., Alterations in vascular smooth muscle mass in the spontaneously hypertensive rat, *Circ. Res.*, 51, 280, 1982.
10. Lee, R. M., Forrest, J. B., Garfield, R. E., and Daniel, E., Ultrastructural changes in mesenteric arteries from spontaneously hypertensive rats, *Blood Vessels*, 20, 72, 1983.
11. Wiener, J., Loud, A. V., Giacomelli, F., and Anversa, P., Morphometric analysis of hypertension-induced hypertrophy of rat thoracic aorta, *Am. J. Pathol.*, 88, 619, 1977.
12. Thomas, W. A., Jones, R., Scott, R. F., Morrison, E., Goodale, F., and Insai, H., Production of early atherosclerotic lesions in rats characterized by proliferation of "Modified Smooth Muscle Cells", *Exp. Molec. Pathol.*, 1, 40, 1963.
13. Lee, K. T., Lee, K. J., Lee, S. K., Imai, H., and O'Neal, R. M., Poorly differentiated subendothelial cells in swine aortas, *Exp. Molec. Pathol.*, 13, 118, 1970.
14. Clowes, A. W., Reidy, M. A., and Clowes, M. M., Kinetics of cellular proliferation after arterial injury: smooth muscle growth in the absence of endothelium, *Lab. Invest.*, 49, 327, 1983.
15. Thomas, W. A., and Kim, D. N., Biology of disease: atherosclerosis as a hyperplastic and/or neoplastic disease, *Lab. Invest.*, 48, 245, 1983.
16. Fernandez, D., and Crane, W. A. J., New cell formation in rats with accelerated hypertension due to partial aortic constriction, *J. Pathol.*, 100, 307, 1969.
17. Bevan, R. D., An autoradiographic and pathological study of cellular proliferation in rabbit arteries correlated with an increase in arterial pressure, *Blood Vessels*, 13, 100, 1976.
18. Bevan, R. D., Eggena, P., Hume, W. R., Marthens, F. V., and Bevan, J. A., Transient and persistent changes in rabbit blood vessels associated with maintained elevation in arterial pressure, *Hypertension*, 2, 63, 1980.
19. Owens, G. K., Rabinovitch, P. S., and Schwartz, S. M., Smooth muscle cell hypertrophy versus hyperplasia in hypertension, *Proc. Natl. Acad. Sci. USA*, 78, 7759, 1981.
20. Owens, G. K. and Schwartz, S. M., Vascular smooth muscle cell hypertrophy and hyperploidy in the Goldblatt hypertensive rat, *Circ. Res.*, 53, 491, 1983.
21. Owens, G. K.and Reidy, M. A., Hyperplastic growth response of vascular smooth muscle cells following induction of acute hypertension in rats by aortic coarctation, *Circ. Res.*, 57, 695, 1985.
22. Van der Heijden, F. L. and James, J., Polyploidy in the human myometrium, *Mikrosk Anat. Forsch Leipzig*, 89, 18, 1975.

23. Brent, L. and Stephens, F. D., The response of smooth muscle cells in the urinary bladder to outflow obstruction, *Invest. Urol.,* 12, 494, 1975.
24. Brent, L., The response of smooth muscle cells in the rabbit colon to anal stenosis: a preliminary report, *Pathol.,* 5, 209, 1973.
25. Gabella, G., Hypertrophic smooth muscle I. Size and shape of cells, occurrences of mitoses, *Cell Tissue Res.,* 201, 63, 1979.
26. Haust, M. D., The nature of bi- and trinuclear cells in atherosclerotic lesions in man, *Atherosclerosis,* 36, 365, 980.
27. Chamley-Campbell, J., Campbell, G. R., and Ross, R., The smooth muscle cell in culture, *Physiol. Rev.,* 59, 1, 1979.
28. Chamley-Campbell, J. H. and Campbell, G. R., What controls smooth muscle phenotype, *Atherosclerosis,* 40, 347, 1981.
29. Kamio, A., Huang, W. Y., Imai, H., and Kummerow, F. A., Mitotic structures of aortic smooth muscle cells in swine and in culture; paired cisternae, *J. Electron Micro.,* 26, 29, 1977.
30. Cobb, J. L. S. and Bennett, T., An ultrastructural study of mitotic division in differentiated gastric smooth cells, *Z. Zellforsch,* 108, 177, 1970.
31. Saborio, J. L., Segura, M., Flores, M., Garcia, R., and Palmer, E., Differential expression of gizzard actin genes during chick embryogenesis, *J. Biol. Chem.,* 254, 11119, 1979.
32. Owens, G. K., Loeb, A., Gordon, D., and Thompson, M. M., Expression of smooth muscle-specific $\alpha$-isoactin in cultured vascular smooth muscle cells: relationship between growth and cytodifferentiation, *J. Cell Biol.,* 102, 343, 1986.
33. Franke, W. W., Schmid, E., Vandekerskhove, S., and Weber, K., A permanently proliferating rat vascular smooth muscle cell with maintained expression of smooth muscle characteristics, including actin of the vascular smooth muscle type, *J. Cell Biol.,* 87, 594, 1980.
34. Gabbiani, G., Kocher, O., Bloom, W. S., Vandekerkhove, J., and Weber, K., Actin expression in smooth muscle cells of rat aortic intimal thickening, human atheromatous plaque, and cultured rat aortic media, *J. Clin. Invest.,* 73, 148, 1984.
35. Strauch, A. R. and Rubenstein, P. A., Induction of vascular smooth muscle $\alpha$-isoactin expression in BC$_3$Hl cells, *J. Biol. Chem.,* 259, 3152, 1984.
36. Schubert, D., Harris, A. J., Devine, C. E., and Heinemann, S., Characterization of a unique muscle cell line, *J. Cell Biol.,* 61, 398, 1974.
37. Olson, E. N., Caldwell, K. L., Gordon, J. I., and Glaser, L., Regulation of creatine phosphokinase expression during differentiation of BC$_3$Hl cells, *J. Biol. Chem.,* 258, 2644, 1983.
38. Gown, A. M., Gordon, D., and Vogel, A. M., Analysis of smooth muscle cell-specific cytoskeletal antigens with monoclonal antibodies, *Fed. Proc.,* 42, 502, 1983.
39. Jones, A. W., Altered ion transport in vascular smooth muscle from spontaneously hypertensive rats: influences of aldosterone, norepinephrine and angiotensin, *Circ. Res.,* 33, 563, 1973.
40. Seidel, C. L., Aortic actomyosin content of maturing normal and spontaneously hypertensive rats, *Am. J. Physiol.,* 237, 434, 1979.
41. Carlson, B. M., The regeneration of skeletal muscle — a review, *Am. J. Anat.,* 137, 119, 1973.
42. Rumyantsev, P. P., Interrelations of the proliferation and differentiation processes during cardiac myogenesis and regeneration, *Int. Rev. Cytol.,* 51, 187, 1977.
43. Baserga, R., Introduction to cell growth: growth in size and DNA replication, in *Tissue Growth Factors,* Baserga, R., Ed., Springer-Verlag, New York, 1981, 1.
44. Sandritter, W. and Scomazzoni, G., Deoxyribonucleic acid content (Feulgen photometry) and dry weight (interference microscopy) of normal and hypertrophic heart muscle fibers, *Nature,* 202, 100, 1964.
45. Johansson, B., Structural and functional changes in rat portal veins after experimental portal hypertension, *Acta Physiol. Scand.,* 98, 381, 1976.
46. Arner, A. and Uvelius, B., Force-velocity characteristics and active tension in relation to content and orientation of smooth muscle cells in aortas from normotensive and spontaneously hypertensive rats, *Circ. Res.,* 50, 812, 1982.
47. Rumeur, E. L., Beaumont, C., Guillouzo, C., Rissel, M., Bourel, M., and Guillouzo, A., All normal rat hepatocytes produce albumin at a rate related to their degree of ploidy, *Biochem. Biophys. Res. Commun.,* 101, 1038, 1981.
48. Owens, G. K., Differential effects of antihypertensive drug therapy on vascular smooth muscle cell hypertrophy, hyperploidy, and hyperplasia in the spontaneously hypertensive rat, *Fed. Proc.,* 42, 1173, 1983.
49. Schwartz, S. M. and Ross, R., Cellular proliferation in atherosclerosis and hypertension, *Prog. Cardiovasc. Dis.,* 26, 355, 1984.
50. Owens, G. K., unpublished observations, 1984.

51. Limas, C., Westrum, B., and Limas, C., The evolution of vascular changes in the spontaneously hypertensive rat, *Am. J. Pathol.*, 98, 357, 1980.

52. De Chastonay, C., Gabbiani, G., Elemer, G., and Huttner, I., Remodeling of the rat aortic endothelial layer during experimental hypertension, *Lab. Invest.*, 48, 45-52, 1983.

53. Schwartz, S. M., and Lombardi, D. M., Effect of chronic hypertension and antihypertensive therapy on endothelial, cell replication in the spontaneously hypertensive rat, *Lab. Invest.*, 47, 510, 1982.

54. Schwartz, S. M. and Standaert, D. M., Endothelial cell turnover in rats: effects of angiotensin, age, and chronic hypertension, in *Atherosclerosis Reviews*, Gotto, A. M., Jr. and Paoletti, R., Eds., Raven Press, New York, 1982, 109.

55. Schwartz, S. M. and Benditt, E. P., Aortic endothelial cell replication, effects of age and hypertension in the rat, *Circ. Res.*, 41, 248, 1977.

56. Giese, J., Acute hypertensive vascular disease: 2. studies on vascular reaction and permeability changes by means of vital microscopy and colloidal tracer technique, *Acta Path. Microbiol. Scand.*, 62, 497, 1964.

57. Goldby, F. S. and Beilin, L. J., How an acute rise in arterial pressure damages arterioles, electron microscopic changes during angiotensin infusion, *Cardiovasc. Res.*, 6, 659, 1972.

58. Huttner, I., More, R. H., and Rona, G., Fine structural evidence of specific mechanism for increased permeability in experimental hypertension, *Am. J. Pathol.*, 61, 395, 1970.

59. Murphy, R. A., Herlihy, J. T., and Megerman, J., Force-generating capacity and contractile protein content of arterial smooth muscle, *J. Gen. Physiol.*, 64, 691, 1974.

60. O'Farrell, P. H., High resolution two-dimensional electrophoresis of proteins, *J. Biol. Chem.* 250, 4007, 1975.

Chapter 2

# HYPERTROPHY OF VASCULAR SMOOTH MUSCLE*

Charles L. Seidel

## TABLE OF CONTENTS



* This manuscript is supported by grants HL23815 and HL34280.

## I. INTRODUCTION

The purpose of this chapter is to compare the changes that the vascular wall undergoes in various examples of hypertrophy in order to delineate a "basic" response of the vascular smooth muscle cell to which variations might be added, depending upon the particular initiating stimuli for wall hypertrophy.

In this chapter a broad definition of hypertrophy will be used: it will be defined as an increase in vessel wall thickness. This can occur because of an increase in the size of the smooth muscle cell (true cellular hypertrophy), an increase in smooth muscle ceil number (hyperplasia), an increase in the amount of extracellular connective material or a combination of all three. The latter is included because it is now generally accepted that the smooth muscle cell is responsible for the synthesis and secretion of the extracellular proteins, collagen, and elastin,[1] and therefore changes in the amount of these proteins in the extracellular space represent an adaptive response of the smooth muscle cell to a stimulus.

Where possible, attempts will be made to correlate the morphological and biochemical changes associated with wall hypertrophy with functional changes. However, an extensive review of the various pharmacological changes occurring with vessel wall hypertrophy, especially those associated with systemic arterial hypertension, is beyond the scope of this review. Also, since a major objective of this review is to delineate a "basic" hypertrophic response, a wide range of hypertrophic models will be discussed. Therefore, a particular parameter that has been well characterized only in a particular model of hypertrophy (e.g., Na flux in systemic arterial hypertension) will not be covered, since comparisons with other models cannot be made.

Because an attempt is made to combine morphological and biochemical observations with functional observations, information presented centers on in vitro preparations. The response of entire vascular beds to various physiological stimuli in the presence of wall hypertrophy is of course very relevant; however, this review is concerned with specific changes of the vascular smooth muscle cell which cannot always be derived from studies of an entire vascular bed.

Three models of vascular hypertrophy will be discussed. The first is strain-induced hypertrophy. Several models have been grouped together under this heading because the primary stimulus for hypertrophy is an increase in wall strain. Included in this group are studies involving vessel coarctation, venous tissue grafted into the arterial circulation, stretched in vitro preparations of vessel wall, and stretched cultures of smooth muscle cells. The second model to be discussed is hypoxic-induced pulmonary hypertension. The pulmonary artery atrophies during the postnatal period as the pulmonary artery pressure decreases, and therefore information about the normal atrophy process of smooth muscle can be obtained. It will be seen that the hypertrophic response of the pulmonary vasculature to hypoxia will vary, depending upon whether it is induced during the postnatal period or during adulthood. The third model to be described is systemic arterial hypertension. In the discussion of this model, an attempt will be made to contrast the vascular responses to genetic, DOCA-salt and renal hypertension as well as to differentiate between the responses of microvessels and large conduit vessels.

However, before beginning this discussion, a description of normal vascular smooth muscle development is necessary. Since the morphological, biochemical, and functional characteristics of the vessel wall change during the life of the animal, any hypertrophic stimulus will be applied to this changing base line. It is therefore not surprising to find that the characteristics of the hypertrophic response vary with the time of onset.

## II. NORMAL DEVELOPMENT

### A. Morphological Changes

The great majority of work concerning the normal developmental changes of the vasculature has been done on the rat aorta. The 1967 and 1970 morphological studies of Cliff[2,3] formed the basis of future work. During the first 12 weeks after birth,[2] wall thickness increases due to an increase in the amount of extracellular protein; the smooth muscle cell changes its orientation, becoming more obliquely arranged; and the cells become more widely separated because of the increase in extracellular material, and acquire a more irregular outline with a consequent increase in surface to volume ratio. The number of lamella remain the same but become thicker. As the rat ages beyond 12 weeks,[3] the amount of connective material continues to increase, the muscle cells increase in size, but not as fast as the increase in extracellular material, so that the number of cells per unit area decreases.

Since the early descriptive work of Cliff, morphological studies have become more quantitative, and direct chemical analyses of specific wall components have been performed. Olivetti et al.[4] examined the morphology of aorta from rats 1 to 11 days of age. During this period, doubling of the wall thickness was due to the combined effect of a 1.8 fold increase in muscle cell volume, a 2.6 fold increase in the number of muscle cells, and an 8 and 10 fold increase in the amount of elastic laminae and collagen, respectively. Because the increase in extracellular material was greater than the increase in the number or size of the muscle cells, the actual volume percent of the wall comprised of muscle cells decreased. All subcellular components of the muscle cell increased during this time, with the greatest increases being in rough and smooth endoplasmic reticulum (2.3 and 2.7 fold, respectively). The increase in rough endoplasmic reticulum corresponded to the increase in production of extracellular proteins.

Gerrity and Cliff[5] also applied the technique of quantitative morphology to the developing rat aorta from birth to 12 weeks. They observed that the percentage of media volume occupied by smooth muscle cells decreased from over 60% at birth to under 30% at 12 weeks of age. At the same time, the volume occupied by connective material increased from over 30% at birth to 70% at 12 weeks of age. They also observed that the volume percentage of myofilaments increased from 5% at birth to a maximum of approximately 20% at 8 weeks.

To obtain a more accurate estimate of the changes in smooth muscle cell volume with development, Bucher and co-workers[6,7] measured the volumes of isolated muscle cells from aorta of rats between 3 and 18 weeks of age. It was observed that during this time the volume doubled (487 to 1070 $\mu m^3$) with the greatest increase occurring during the first 7 weeks, which indicates that the smooth muscle cell continues to enlarge after birth. It was also observed that there were two populations of cells with different volumes. The population with the larger volume (1.4 to 1.5 times larger than the mean cell volume) was approximately 7% of the total number of isolated cells and its relative amount remained constant during development. The function of these two cell types is unknown; however, it has been determined by antibody staining of isolated rat aortic smooth muscle cells that some (60 to 70%) stain for the intermediate filament protein vimentin, while others (25 to 30%) stain for both vimentin and desmin.[8] It has been suggested[9] that those cells rich in vimentin are the smooth muscle cells that migrate into the intima and contribute to intimal thickening following endothelial damage. Whether the two cell populations described by Bucher correspond to these immulogically distinct cells is unknown.

As described above, the work of Olivetti et al.[4] indicated that smooth muscle hyperplasia was occurring in the rat aorta during the first 11 days after birth. To determine if hyperplasia continued during development, other investigators measured aortic DNA content as an indication of muscle cell number.[7,10-13] The absolute amount of DNA increases rapidly during the first 8 weeks after birth and then increases slowly during the remainder of the first year.[11] However, if the increasing aortic length is taken into consideration, the amount of DNA per unit length doubles between 3 and 7 weeks after birth, reaching a peak at 7 weeks and then declining.[12,13] This suggests that the increase in media thickness occurring during the first 7 weeks is due to muscle cell hyperplasia as well as hypertrophy.

The use of DNA content as an indication of cell number assumes that each cell contains the same amount of DNA and that this relationship does not change during development. Bucher, et al.[7] found this to be true; however, Owens and Schwartz[14] have suggested that the percentage of tetraploid cells increases during the development of rat aorta, reaching a peak of approximately 10% of the cells at 12 weeks of age. Total DNA measurements made on vessels from mature animals would overestimate the true number of cells by approximately this percentage; however, measurements made on vessels from developing animals where polyploidy is small (10%) and changes in DNA content are large (2 to 3 times), would indicate hyperplasia. As will be discussed briefly later in this chapter and in more detail by Dr. Owens in a separate chapter in this volume, polyploidy becomes significant when one attempts to differentiate between hypertrophy and hyperplasia during the response of the vessel wall to systemic arterial hypertension.

## B. Protein Changes

Direct chemical quantitation of the various protein constituents of the vessel wall during development, support the morphological observations. The work of Looker and Berry,[11] Wolinsky,[15,16] Berry and Greenwald,[17] and Gerrity and Cliff[5] indicates that the absolute amounts of collagen and elastin per rat aorta continue to increase to the longest time studied (64 weeks of age); however, the amount of these two proteins relative to the vessel dry weight (or total protein content) does not increase with the same time course. The proportion of elastin increases until approximately 5 weeks after birth,[5,11] and then remains constant for as long as 64 weeks. On the other hand, the proportion of collagen continues to increase even at 64 weeks. In addition, Looker and Berry[11] observed that there were differences along the length of the aorta in the proportion of collagen and elastin. At any age, the proportion (% of dry fat free weight) of collagen was greater in the abdominal aorta, while the proportion of elastin was greater in the thoracic aorta. McCloskey and Cleary[18] also observed a similar change in the proportion of collagen and elastin along the length of adult rabbit aorta. These observations suggest that the net production of collagen and elastin by the smooth muscle cell does not proceed in parallel during development and that their relative production varies along the vessel length. Therefore, it may be expected that they would not only change differently during wall hypertrophy but that the change may vary with the vessel location.

Wolinsky measured the noncollagenous alkali soluble protein fraction of the developing rat aortic wall and observed that the amount relative to dry weight did not change between 10 and 64 weeks of age.[15,16,19] This protein fraction represents, in part, the contractile proteins, and together with the morphological observations of Gerrity and Cliff[5] discussed above (see Morphological Characteristics), would suggest that the net production of actin and myosin relative to the total protein content of the vascular wall increases during the first 10 weeks after birth in the rat aorta, and then remains constant.

To test this directly, we measured the content of actin and myosin in aorta from rats between 3 days and 43 weeks of age.[12,13,20] These studies indicate that the absolute amount of actin plus myosin continues to increase up to 43 weeks; however, the amount relative to total protein changes at different stages of development. During the first 4 weeks after birth, the net production of actin and myosin increases in parallel with the increased net production of total protein, averaging 4% of the total protein. Between 4 and 7 weeks, the net production of actin and myosin is greater than that of total protein, so the percentage of contractile protein increases to over 15% of the total protein. After 7 weeks, the production of contractile proteins again parallels that of total protein. At no time during development is the weight ratio of actin to myosin different. When the amount of actomyosin per cell (based on DNA) is determined, it changes over the same time course as actomyosin per total protein. Between the 3rd and 7th week after birth the cellular actomyosin content increases 3 times. As indicated previously, this is also a period of hyperplasia.

Recently, Kocher et al.[21] have extended these observations to the quantitation of the actin variants, tropomyosin, vimentin, and desmin in aorta from fetal to 12 week old rats. They confirmed our previous observation that actin per muscle cell increased during early postnatal life, and demonstrated that there is a shift in the actin variant pattern from the beta to the alpha form during maturation. The cellular content of tropomyosin, vimentin, and desmin also increases with maturation.

These studies indicate that in the rat aorta certain periods of development are characterized by hyperplasia and increased cell size, which includes an increase in contractile and intermediate filament protein content. It is not known what stimulates this simultaneous increase in cell size and number.

## C. Functional Changes

The maximum contractile response of vessel preparations changes during development. Cohen and Berkowitz[22] observed that the maximum contractile response of rat aorta to both norepinephrine and potassium chloride increases in magnitude until approximately 9 to 13 weeks of age and then declines over the remainder of a year. This is not due to a shift in the active length-tension curve or to a specific change in drug receptor interaction. They also observed that the ability of isoproterenol, but not nitroglycerin or adenosine, to relax decreases with age.[23] Up to 13 weeks of age, this decreasing ability of isoproterenol to induce relaxation is due to a reduction in the effectiveness of cyclic AMP, rather than to a reduction in the ability of isoproterenol to stimulate adenylate cyclase.[23] However, by 24 weeks of age the ability of isoproterenol to stimulate cyclic AMP formation is also reduced.[25] The decreasing responsiveness of arteries to isoproterenol with increasing age does not appear to occur in venous tissue, since rat portal and mesenteric veins maintain their responsiveness with increasing age.[26,27]

These studies have been extended to neonatal animals,[20] where it was observed that the maximum response to several contractile agents increases between 3 days and 1 month after birth. In addition, Seidel and Allen[20] observed that aorta from rats less than 1 week of age do not respond to either norepinephrine or isoproterenol, even though they contracted in response to potassium chloride. This suggests that in the rat aorta, the development of functional adrenergic receptors does not occur until several days after birth. This is in contrast to other species such as sheep, where functional adrenergic receptors have been demonstrated in the fetus.[28]

The explanation for these changes in maximum response with development have not been completely explained. The changes between 4 and 7 weeks of age can be explained in part by the increasing proportion of actomyosin;[13] however, the observed increase at younger ages or the decline after 13 weeks of age cannot be explained. Recent work by Kocher et al.[21] indicates that aorta from young rats possess more nonmuscle actin, while those from older animals possess more muscle type actin. It is possible that these actin forms interact differently with myosin, resulting in changes in force generation. The reduction in maximum contractile response of aorta from older animals could be due to a structural reorganization of muscle cells or contractile filaments within the muscle cell, a change in the chemomechanical transduction process, or in the availability of calcium. None of these possibilities have been adequately tested.

Variable changes in the ED50 for various contractile agents have been observed to occur during development. Aortas from rats less than 15 days of age were observed to be less sensitive to potassium and norepinephrine than aortas from older animals.[20] As the rat continues to age between 6 weeks and a year, the sensitivity of the aorta again tends to decrease, but not significantly.[22] Cohen and Berkowitz[22] observed that the ability of serotonin and norepinephrine to contract in the absence of extracellular calcium was reduced in aorta from 36- to 52-week-old rats compared with aorta from 6- to 8-week-old rats. Since contraction in response to these agents is thought to involve the release of an intracellular or a more tightly bound pool of calcium, these results suggest that either the amount of calcium stored within this pool or its releasability is reduced during maturation. Such a reduction in the availability of calcium may also be involved in the general decrease in maximum contractile response seen with maturation.

## D. Intervessel Comparisons

As indicated above, most observations on the developmental changes in the vasculature have been made using aorta from rats; however, observations have also been made in other animals and with other vessels.

### 1. Protein Changes

Cox[29] examined carotid arteries from rats at 8, 48, and 96 weeks of age and observed that the amount of collagen per dry weight was not elevated until 96 weeks of age and that the amount of elastin per dry weight continuously decreased from 8 weeks. This is in contrast to the aorta as described above where the relative amount of collagen increases continuously with maturation[16,19] and the relative amount of elastin peaks at 5 weeks.[11]

Warshaw[30,31] examined isolated rat mesenteric arteries less than 250 $\mu$m in diameter and observed that during development (6 to 50 weeks of age) there is no change in media thickness; however, there is an increase in the cross-sectional area of the muscle cells. Miller et al.[32] have determined the dimensional characteristics of arterioles in the intestine of rats between 6 to 8 and 10 to 12 weeks of age. They observed that at any given age, the length and width of the smooth muscle cell does not vary with arteriolor diameter (30 to 50 $\mu$m); however for any given diameter the cell length and width increases with maturation. Therefore, as in larger vessels, smooth muscle hypertrophy also occurs in the microvasculature during development.

Differences between aorta from male and female rats have been noted during development. Wolinsky[16] observed that the absolute amount of collagen, elastin, and noncollagenous alkali soluble protein increases with age in the male; however, in the female, only collagen increases, which causes the relative amount of elastin and alkali soluble protein to decrease with age.

In the rat aorta from Wistar and Wistar Kyoto rats a difference in the amount of actomyosin per media weight was observed.[12] Aorta from Wistar rats have a larger amount of actomyosin because of a greater number of smooth muscle cells (based on DNA content) containing the same amount of actomyosin as cells in aorta from Wistar Kyoto rats.

Looker and Berry[11] observed changes in collagen and elastin content of developing rat aorta; however, they noted that the change with development occurred quicker in the abdominal portion than in the thoracic portion.

McClosky and Cleary[18] observed that the amount of collagen and elastin relative to the dry weight of rabbit aorta continued to increase up to 20 weeks of age and then declined. This observation is in contrast to that in the rat aorta where the relative amount of elastin does not increase after 5 weeks of age.

### 2. Functional Changes

Duckles et al.[33] observed no change in sensitivity of femoral or renal arteries or renal vein to norepinephrine in rats 24 to 108 weeks of age. In addition, the maximum contractile response of these vessels to norepinephrine, potassium chloride, or nerve stimulation also does not change with age. These observations suggest that not all vessels from the rat undergo a decrease in maximum contractile ability after 13 weeks of age like the aorta.

Cox et al.[34] determined the maximum contractile responses of iliac, renal, carotid, and mesenteric arteries and aorta from developing dogs (0 to 20 weeks). The maximum response to norepinephrine or KCl relative to wall cross-sectional area increases with age in all vessels. In addition, the ability of the vessel to respond maximally to norepinephrine increases during development, so that in the adult animal, the maximum response to norepinephrine and KCl are almost identical. This differential change in maximum response to these agents suggests that their mechanism of excitation-contraction coupling is maturing at different rates. In contrast to arterial tissue from rat, aorta from dog exhibited an increase in sensitivity to norepinephrine, phenylephrine, and tyramine during development.[35] The reason for these interspecies and intervessel differences in sensivity change is not known.

### E. Summary (See Table 1)

Using the data from rats as a general model for large vessel maturation, the following pattern emerges. During the early postnatal period the smooth muscle cell increases in size and number. Associated with this increase in size is an increase in cellular content of both contractile and noncontractile proteins, and smooth and rough endoplasmic reticulum. In addition, the net production of elastin and collagen increases, with the latter continuing on through the first year. The increase in wall thickness is due to an increase in muscle cell size and number plus an increase in the amount of extracellular material. Mechanically there is (1) an increase in the maximum contractile response which peaks and either declines or remains constant and (2) a reduction in the effectiveness of beta adrenergic receptor-induced relaxation of arterial tissue without a change in nonadrenergic mediated dilation.

## III. STRAIN INDUCED HYPERTROPHY

Three experimental models are included as examples of strain induced hypertrophy because they represent examples where a single perturbation (elevated strain) is the primary stimulus acting on the vascular smooth muscle cell. They are (1) vessel coarctation, (2) venous to arterial autologous grafts, and (3) in vitro preparations of either stretched vessel wall or isolated vascular smooth muscle cells.

### A. Coarctation

The coarctation model has been used primarily on the abdominal aorta above the renal arteries and on the portal vein. This model is thought to represent primarily the

Table 1

DEVELOPMENTAL CHANGES OF THE VASCULATURE

| | Rat aorta | Rabbit aorta | Rat carotid |
|---|---|---|---|
| Morphological changes | ↑ Wall thickness, SMC hypertrophy and hyperplasia.[2-5]<br>↑ SMC content of rough and smooth ER.[2-5]<br>↓ % of media volume made up of SMCs.[2-5]<br>↑ % of media volume made up of connective material.[2-5] | — | — |
| Chemical changes | Absolute amounts of E and C increase; % E peaks at 5 wks, % C continues to ↑. Relative amounts of each change along aortic length.[10,11,15,16]<br>Cellular AM content increases between 3rd and 7th week.[12,13,20]<br>Increase in alpha actin variant and decrease in $\beta$ actin variant.[21] | Changes in E and C similar to rat except % E peaks at 20 weeks.[18] | % C remains constant, % E ↓ after 8 weeks of age.[29] |
| Functional changes | Max contractile force ↑ up to 13 weeks. Effectiveness of $\beta$ agonists ↓.[20,22-26] | — | — |

*Abbreviations:* SMC, smooth muscle cell; ER, endoplasmic reticulum; E, elastin; C, collagen; AM, actomyosin.

response of the vessel to elevated transmural pressure (strain) rather than to increased levels of circulating pressor agents or to increased neural activity. However, if the coarctation is placed between the two renal arteries, there is an elevation in the plasma renin activity and presumably an elevation in angiotensin levels[36] which complicate the interpretation of results from this model.

Over the last decade, the Bevans have extensively studied the morphological and contractile responses of vessels on either side of an aortic coarctation in the adult rabbit.[37-40] Within 2 weeks, vessels above the coarctation and therefore exposed to an elevated transmural pressure exhibit an increase in wall thickness and, based on morphological appearance, large (but not small) diameter vessels exhibit an increase in collagen and elastin content.[37,38]

Using ³H-thymidine uptake as a measure of DNA turnover and therefore hyperplasia, they observed an increase in labelling that peaked at 2 weeks.[38] The extent of labelling vaires inversely with the vessel caliber; smaller vessels above the coarctation are labelled more than the larger arteries (aorta, carotid).

The incorporation of ³H-proline and lysine into connective and nonconnective tissue proteins was also determined.[39] Vessels above the coarctation had greater labelling of collagen, elastin, and nonconnective protein pools.

To determine if these vascular wall changes continued beyond 2 weeks, similar parameters were examined after 8 weeks of coarctation.[40] Wall thickness is still elevated above control (it was not indicated whether it was elevated further from 2 weeks);

however, [3]H-thymidine, proline, and lysine incorporation is not elevated. This suggests that by 8 weeks the wall completes its adaptation to the elevated transmural pressure.

The contractile ability of vessels above the aortic coarctation were also examined.[37] There is no change in sensitivity ($ED_{50}$) or maximum contractile response to contractile agonists. It was concluded that coarctation induces wall hypertrophy in vessels above the coarctation which is due to both hyperplasia and hypertrophy of the smooth muscle cells; however, neither the sensitivity nor the maximum contractile response of the muscle cells is altered.

Olivetti[41] has determined the morphological changes occurring above a coarctation of the aorta in 4-week-old rats. After 8 days the lumen diameter of the aorta is unchanged; however, the wall thickness increases 39%. The percentage of the wall composed of smooth muscle cells, collagen, or elastin is unchanged, indicating that they increase in proportion to the increase in wall thickness. Muscle cell volume increases 57%, elastic laminae 30%, and collagen 136%. Within the smooth muscle cell only the relative volume of the rough endoplasmic reticulum is elevated while all other components increase in proportion to the increase in cell volume, with an absolute increase in nuclear volume of 27%, myofilament volume of 47%, and mitochondrial volume of 62%. There is no indication of an increase in the number of smooth muscle cells, and it was argued that the increased nuclear area could explain why other investigators had suggested hyperplasia based on [3]H-thymidine labelling. The authors concluded that within 8 days the aortic wall of the rat responds to an elevation in transmural pressure with smooth muscle cell hypertrophy which includes an increased net production of extracellular connective tissue protein as well as intracellular contractile protein.

Owens and Reidy[42] have also morphologically examined the coarcted rat aorta; however, their study suggests that the increase in wall thickness is due to smooth muscle hyperplasia, not hypertrophy. The difference between the conclusion from this work and that of Olivetti et al.[41] may be due to the fact that Owens and Reidy: (1) used older rats (20 weeks old), (2) produced a greater degree of coarctation, and (3) placed the constriction between rather than above the renal arteries. Such differences in results emphasize that one must be cautious about generalizing the response of the smooth muscle cell to hypertrophic stimuli.

The portal vein in rats and rabbits has also been used as a coarctation model. Johansson[43] and Uvelius et al.[44] have examined the morphological and mechanical properties of the portion of the rat portal vein exposed to the high pressure induced by coarctation for up to 15 days. Within a week, the portal vein exhibits an increase in smooth muscle cell area as well as in the area made up of connective tissue. There is no indication of smooth muscle hyperplasia.

As in arteries subjected to the high pressure of coarctation, there is no change in the sensitivity of the portal vein to norepinephrine;[43] however, the vein loses its spontaneous contractile activity[44,45] which appears to be due to an increase in the resting membrane potential.[45] When the hypertrophied vein is depolarized with 15 m$M$ potassium chloride, normal action potentials and spontaneous contractile activity return, suggesting that the membrane is capable of generating action potentials but because of hyperpolarization, the resting potential is too far from threshold to intiate spontaneous activity. The reason for this hyperpolarization is unknown.

As in arteries subjected to the high pressure of coarctation, the portal vein exhibits (after 15 days) an increase in the absolute amount of maximum force developed; however, when expressed relative to total wall, media, or cell cross-sectional area, maximum stress is significantly reduced.[43,44] Arteries on the other hand, show no change when force is expressed relative to wall area.[37] After only 5 days of coarctation, the cell cross-sectional area doubles but the force per cell area decreases by 40%, suggesting that the muscle cell increases in area but that there is not a parallel increase in its contractile protein content.[44] This is different than that observed in arteries after 8 days of coarctation[41] and suggests that venous and arterial tissue respond differently to an elevation in transmural pressure.

Arner and Uvelius[46] examined the metabolic cost of tension maintenance of rat portal veins after 5 days of coarctation-induced high pressure. They observed that the slope of the relationship between ATP consumption and developed force was less steep for hypertrophied portal veins, suggesting a lower metabolic tension cost. Since this change occurs soon (5 days) after coarctation, it probably does not represent the synthesis of contractile proteins with altered enzyme activity. Similar metabolic measurements have also been made on arterial tissue from animals with systemic arterial hypertension (see Systemic Arterial Hypertension) and no changes in metabolic cost of tension maintenance are observed. This suggests either that arterial and venous tissues change differently during wall hypertrophy, or that all models of hypertrophy do not produce the same vascular changes.

Berner et al.[47] produced portal vein hypertrophy in rabbits by coarctation and after 2 weeks observed an increase in wall thickness and smooth muscle cell cross-sectional area as observed previously[43,44] in rat portal veins. However, when the number of thin, intermediate, and thick filaments is determined, the number of thin and intermediate filaments increases in hypertrophied veins, but the number of thick do not, resulting in a significant increase in the thin to thick filament ratio. These data suggest that the net production of actin, intermediate filament protein, and myosin do not change similarly during vascular smooth muscle hypertrophy, and that changes in net production of some of these proteins can occur within 2 weeks after an increase in pressure. Junker et al.,[48] using the same rabbit coarctation model as Berner et al.,[47] determined the intracellular distribution and concentration of various ions by electronprobe analysis. There are no changes in the cytoplasmic, nuclear, or mitochondrial concentrations of the measured elements following wall hypertrophy.

## B. Autologous Vein Grafts

The use of venous tissue as a replacement for damaged arterial tissue is a common clinical procedure to improve arterial perfusion. Such autologous vein grafts are used both in the coronary and systemic circulations and are subjected to the elevated pressure, shear stress, and $PO_2$ of the arterial circulation. The graft undergoes wall hypertrophy[114] which may progress to occlusion, rather than reaching a new steady state when tension per wall area becomes normal as occurs in arterial tissue with systemic hypertension.[15,16,19] The cause of the hypertrophy is unknown, but is due in part to the increase in transmural pressure to which the venous tissue is subjected.

We have studied the response of the intact saphenous vein when it is grafted into the arterial circulation.[49] Within 1 week, there is a significant increase in the absolute amount of total protein but not in the amount of actin or myosin in grafted saphenous veins. At 1 month, the absolute amount of collagen is significantly elevated and continues to increase at 8 weeks; however, the content of actin and myosin does not become significantly elevated until 8 weeks after grafting. At all times the actin to myosin weight ratio does not change, which is different than that predicted from morphological studies of hypertrophied rabbit portal veins.[47]

The maximum contractile response of grafted saphenous veins is decreased when expressed relative to wall cross-sectional area or myosin content, suggesting that the force generating ability of myosin may be compromised. This reduction in force generating ability is observed 1 week after grafting and is maintained for 8 weeks (the duration of study). The explanation for this decrease is unknown, but may involve structural reorganization of the contractile system, a change in the force generating ability of the contractile proteins, or a change in excitation-contraction coupling. As pointed out above, the coarcted portal vein also underwent a decrease in maximum force generation,[43,44] while arteries showed no change in force.[37]

## C. In Vitro Preparations

To obtain more direct information about the effect of elevated transmural pressure (strain) on smooth muscle cells, in vitro preparations have been used. Hume[50] mounted rabbit ear artery segments on a spring loaded frame and incubated these preparations in sterile culture medium for up to 9 days. By adjusting the tension of the spring, various levels of strain could be produced. The uptake of $^3$H-thymidine or proline was determined and it was observed that an elevation in their uptake is not detected until 5 days after the application of the strain. The amount of uptake increases with the amount of applied strain.

Leung et al.[51,52] performed similar uptake experiments using cultured rabbit aortic smooth muscle cells grown on elastin membranes and subjected to rhythmical strain. The muscle cells were derived from aortic explants and had been passed once before seeding on the elastin membrane. The membrane was stretched 10% of its initial length at a rate of 52 times/min, conditions that mimic the changes in aortic wall dimension produced by the pulse pressure.[53] Within 8 hr of the beginning of rhythmic strain, an increase is observed in incorporation of proline into both the total protein pool and collagen. Strain does not increase the incorporation of thymidine or the total DNA content, suggesting that hyperplasia is not occurring. In a similar preparation, Sotti-urai et al.[54] observed that cyclical stretching increases the number of rough endoplasmic reticulum profiles while retarding the loss of myofilaments that occurred in stationary cultures.

Recently we have performed similar studies using primary cultures of venous smooth muscle cells grown on a polyurethane membrane and subjected to a strain comparable to arterial conditions.[55] These studies indicate that after three days of strain, there is an increase in the total protein content of the cells, but no change in the actin or myosin contents, suggesting that an elevation of strain first induces an increase in the net production of noncontractile protein.

## D. Summary (Table 2)

The majority of evidence suggests that the response of vascular smooth muscle exposed primarily to an increase in strain is to undergo hypertrophy, which involves a rapid (within 1 week) increase in the net production of noncontractile protein followed later (weeks) by an increase in the net production of contractile proteins. The maximum contractile ability of arterial tissue is not altered; however, in venous tissue, maximum force relative to wall or cell cross-sectional area or to myosin content is reduced. Insufficient data exists at the present time to be able to explain the changes in contractile ability of venous tissue.

## IV. HYPOXIA INDUCED PULMONARY VASCULAR HYPERTENSION

### A. Normal Developmental Changes

Due to the elevated pulmonary vascular resistance in the fetus, both the pulmonary arterial pressure and wall thickness are greater than in the adult. In contrast to the systemic arterial vasculature (see Normal Development above) which undergoes an increase in wall thickness during neonatal development, the pulmonary arterial vasculature undergoes a decrease in wall thickness and in muscularity. Haworth and

## Table 2
### STRAIN INDUCED CHANGES IN THE VASCULATURE

| | Arteries | Veins | Isolated cells |
|---|---|---|---|
| Morphological changes | ↑ Wall thickness; SMC hypertrophy in young animals; hyperplasia in older.[37-42] | ↑ Wall thickness; SMC hypertrophy.[43,44,47] | — |
| | ↑ Connective material.[37-41] | ↑ Thin and intermedial filament number.[47] | |
| Chemical changes | ↑ A² labelling of vessel proteins.[39] | ↑ Connective tissue content.[49] | ↑ A² labelling of collagen.[51,52] No change in cell content of A or M, but noncontractile protein increased.[55] |
| Functional changes | No changes in $ED_{50}$ or in maximum force/cross-sectional area.[37] | No change in $ED_{50}$;↓ maximal force/cross-sectional area.[43,44,49] ↓ Metabolic cost of tension maintenance.[46] | — |

*Abbreviations:* SMC, smooth muscle cell; A², amino acid; A, actin; M, myosin.

Hislop[56,57] examined the morphology of the pulmonary vasculature in developing pigs from birth to 24 weeks, and observed that medial thickness decreases in all arteries examined; however, the decrease in thickness occurs sooner (within 24 hr after birth) in arteries with diameters less than 250 μm. In larger vessels, a decrease in medial thickness is not observed until several days after birth. The rapid decrease in small artery thickness is probably due to the physical effect of the rapid fall in intrapleural pressure that occurs upon lung inflation at birth and the subsequent increase in small vessel diameter. The slower change in larger vessels probably represents actual changes in wall composition due to the effect of a sustained reduction in transmural pressure. Meyrick and Reid[58] studied the morphology of the hilar pulmonary artery in 8- to 28-day-old rats. They observed a fluctuating change in vessel wall thickness which was paralleled by similar changes in muscle cell thickness. Over the period studied the volume density of collagen and elastin increased, with collagen increasing more. Because of this increase in connective material there is a decrease in the percentage of the media made up of smooth muscle cells (from 80% to 60%).

In support of these morphological observations, Leung et al.[59] chemically determined collagen and elastin contents of pulmonary arteries in developing rabbits and observed that the percentage of the dry weight composed of elastin and collagen increases during the first 2 weeks after birth; however, these increases are much less than those seen in the aorta. It is suggested that this difference is the response of the aortic smooth muscle cells to the higher aortic transmural pressure; however, it is interesting that net connective tissue protein production occurs, even though pulmonary artery pressure decreases during development.

The contractile properties of the pulmonary vasculature also change with development. Nuwayhid et al.[60] observed in fetal lambs between 60 days and term that pulmonary blood pressure does not change when beta adrenergic agonists are infused; however, vasoconstriction and vasodilation are obtained following infusion of norepinephrine and acetylcholine, respectively. In contrast to the fetal pulmonary vasculature, the adult pulmonary arterial vasculature responds in the anticipated way to a wide variety of pharmacological agents.[61,62] In addition, in the rat it undergoes a loss of responsiveness to beta adrenergic agents with age; however, this loss is not as great as is in the aorta.[63] This difference has been suggested to be due to a predominance of beta 1 receptors in the pulmonary arterial vasculature which may be more resistant to changes with age than the beta two receptors of the systemic vessels.[63]

## B. Hypoxia Induced Hypertension

A detailed description of the proposed mechanisms by which alveolar hypoxia results in hypertension is beyond the scope of this review and the interested reader is referred to two excellent reviews by Fishman.[64,65] In general the hypertension does not require either division of the autonomic nervous system and therefore is primarily a response of the smooth muscle cell to hypoxia. It is not known whether the vasoconstriction is a direct response of the smooth muscle cell to hypoxia or is a secondary response to a vasoactive agent released from another cell type within the lung; however, it is the property of both large[66] and small[67] pulmonary arteries.

Since the pulmonary vasculature changes during development, it is not surprising that the response of the pulmonary vasculature to hypoxia induced hypertension is different in neonates and adults. A description of the response of the adult pulmonary system will be presented first followed by that of the neonate.

Rabinovitch et al.[68] examined the morphological characteristics of pulmonary arteries from adult rats exposed to hypoxia for up to 2 weeks and observed that within 2 days the alveolar duct arteries, which are not normally muscularized, became more muscular, and after 3 days, the ratio of wall thickness to vessel diameter had increased significantly. At that time, the pulmonary arterial pressure had only increased 4 mmHg. By 2 weeks, the pressure had increased to 37 mmHg (from 18 mmHg), 75% of the alveolar duct arteries were muscular, and the ratio of wall thickness to diameter had doubled.

Meyrick and Reid[69] also examined rat pulmonary vessels following a similar period of hypoxia and observed an increase in both large and small vessel wall thickness. $^3$H-thymidine labelling studies indicate that the small vessels are labelled more, suggesting that the increase in small vessel wall thickness is due to hyperplasia, while that of large vessels is due to hypertrophy. To test this further, they examined the pulmonary vasculature of rats following 10 days of hypoxia using quantitative morphology.[70] They observed that in association with an increase in vessel wall thickness, the vascular smooth muscle cells increase in diameter, but the production of non-cellular material increases more, so that the proportion of the wall area composed of muscle cells, decreases. Within the muscle cell, all components increased in proportion to the increase in cell size, except for the rough and smooth endoplasmic reticulum, golgi, and dense bodies, which increased more. The greater increase in rough ER and golgi is consistent with the observed increase in extracellular material deposition.

Following 2 weeks of normoxia, Meyrick and Reid[70] observed that the wall thickness, the proportion of extracellular material, and the diameter of muscle cells decreases, but the cellular content of myofilaments does not return to control values until the third week of normoxia.

As indicated above, the pulmonary arterial vasculature decreases in wall thickness during maturation. In general if the neonate is exposed to hypoxia, this thinning of the vessel wall is retarded.[56,58,71] Meyrick and Reid[58] exposed 1-week-old rats to 3 weeks of hypoxia and observed that the increase in smooth muscle cell diameter seen in the adult following hypoxia[70] did not occur, suggesting that the maturing pulmonary artery smooth muscle cell may not be capable of hypertrophy. As an adult, there is an increase in the amount of extracellular connective material; however, in the absence of an increase in cell diameter, this results in a greater reduction in the percentage of the wall area made up of muscle cells in the neonate.

Table 3

## HYPOXIA INDUCED CHANGES IN THE PULMONARY VASCULATURE

|  | Adult pulmonary vasculature | Neonatal pulmonary vasculature |
|---|---|---|
| Morphological changes | ↑ Wall thickness. | The ↓ in wall thickness associated with normal development is retarded. |
|  | ↓ Proportion of wall area composed of SMCs. SMC hyperplasia in small diameter vessels; hypertrophy in large.[68-71] | No SMC hypertrophy; ↓ proportion of wall composed of SMCs.[69-71] |
| Chemical changes | — | — |
| Functional changes | ↓ Resting membrane potential in large pulmonary arteries; ↑ membrane potential in small.[72] | — |

*Abbreviations:* SMC, smooth muscle cell.

Few studies have been performed to assess the contractile properties of the pulmonary vasculature following hypoxia induced hypertension. Suzuki and Twarog[72] measured the resting membrane potential of pulmonary smooth muscle cells from adult rats after 8 weeks of hypoxia. They observed that the resting membrane potential of the main pulmonary artery decreased while it increased in small pulmonary arteries (100 to 200 $\mu$m). They attributed the decrease in the large artery to an increase in the permeability of the cell to chloride, and the increase in the small artery to an increase in the activity of the electrogenic Na-K pump. It is not known whether these different changes in membrane potential are related to the observation that large pulmonary vessels hypertrophy while small vessels undergo hyperplasia.[69]

## C. Summary (Table 3)

As with the previous model of hypertrophy (strain induced), hypoxia induced pulmonary hypertension is characterized by an increase in vessel wall thickness in both large and small arteries. In large pulmonary arteries, the increase in wall thickness appears to be due to an increase in the proportion of extracellular connective material as well as smooth muscle cell hypertrophy. The majority of muscle cell constituents increase in proportion to the increase in cell size, with those involving protein synthesis increasing more. Small pulmonary vessels appear to respond differently than large, for smooth muscle hyperplasia rather than hypertrophy appears to contribute to the increased wall thickness along with an increase in the production of extracellular material. Finally, the hypertrophic response of smooth muscle cells does not appear to occur in the neonate.

## V. SYSTEMIC ARTERIAL HYPERTENSION

Three models of systemic hypertension will be considered: genetic, renal, and DOCA-salt. They will be considered separately since each has a unique etiology and therefore each might produce different hypertrophic responses. Contrasting the vascular response in each model would thereby provide insight into the variability of the hypertrophic response. No attempt will be made to determine the particular cause of the hypertrophic response in each model. Such stimuli are considered in other chapters of this volume. In addition, the hypertrophic response of individual vessels will be compared within a given model of systemic hypertension. As indicated in the section on Normal Development, there appears to be intervessel heterogeneity at both the biochemical and functional levels and therefore, it would not be surprising if systemic arterial hypertension induces different hypertrophic responses in different vessels.

## A. Genetic Hypertension

Since hypertension in this model is the result of a genetic defect, the choice of control animals is very difficult and therefore, it is not clear whether differences in the characteristics of vessels from hypertensive and control animals represent a "hypertrophic response" or a basic difference in tissue characteristics between the genetic hypertensive rat and the selected control unrelated to the elevated pressure. As reviewed by Gray,[73] there may not be a "prehypertensive" stage in the SHR which could be used to differentiate between genetic and pressure induced differences.

### 1. Morphological and Protein Changes

Olivetti et al.[74] examined aorta from SHR and WKY rats between 3 and 7 weeks of age. Compared to normotensive controls, the smooth muscle cells exhibit both hypertrophy and hyperplasia. At 4 weeks of age, hypertrophy was evident, but no further increase in cell size occurred over the next week and the increase in media mass was due to hyperplasia. This was followed by another period of hypertrophy (from 5 to 7 weeks of age) which resulted in a final increase in mean cell volume over the period of study of 68% compared to a 21% increase in the number of smooth muscle cells. It was therefore concluded that cellular hypertrophy is the predominate cause of wall thickening during the first 7 weeks after birth. During this same period there is also an increase in the collagen and elastin contents.

The suggestion that smooth muscle cell hyperplasia does not contribute substantially to wall hypertrophy during early development of the SHR aorta is confirmed[7,12] by DNA determination at ages where polyploidy is not significant.[75] However, hyperplasia may contribute to aortic wall hypertrophy in older SHR rats,[7,12] and it has been determined by Yamori et al.[76] that the growth rate of cultured aortic smooth muscle cells isolated from SHR rats 10 to 44 weeks after birth was greater than for cells isolated from normotensive rats. This greater rate of cell division was maintained through several cell passages.

The exact time course of cellular hypertrophy is not clear. The morphological work of Olivetti et al.[74] suggests that it begins 4 weeks after birth; however direct volume measurement of isolated smooth muscle cells by Bucher et al.[7] suggests that it does not occur until at least 12 weeks after birth.

Anversa et al.[77] examined the morphological characteristics of the left anterior descending coronary artery of SHR and WKY rats 3 to 7 weeks old, and observed a greater smooth muscle cell volume and proportion of connective material in SHR rats, but not until 7 weeks after birth.

Direct chemical analysis of aortic proteins indicates that the absolute amount of total protein is not significantly elevated until 12 weeks after birth, the time at which media cross-sectional area becomes greater, and the absolute amounts of actin and myosin are not significantly elevated until 23 weeks after birth.[12] At no time did the weight ratio of actin to myosin change. Owens et al.[78] have shown that the amount of actin and total protein per cell is elevated in 20-week-old SHR aorta. Yamori et al.[79] observed that the incorporation of $^3$H-lysine into collagen is greater in aorta from 8-week-old SHR rats suggesting an increase in collagen turnover; however, in a subsequent publication from the same laboratory,[80] no change in lysine incorporation was observed. The incorporation into elastin or nonconnective tissue protein is not different. Iwatsuki et al.[81] observed an increased proline incorporation into collagen of aorta from 20-week-old SHR rats as well as an increase in the amount of collagen per wet weight. Cox,[82] however, determined that the percentage of aortic dry weight made up of collagen decreased in 20-week-old SHR rats.

The chemical changes occurring during the hypertrophic response of other large diameter vessels from the SHR rats suggests that changes are not always similar between vessels. Yamori et al.[79] and Nakada and Lovenberg[80] observed that the incorporation of [3]H-lysine was elevated only in the nonconnective tissue protein fraction of large mesenteric arteries from 8-week-old SHR rats, which is different than the aorta where no increase in incorporation of lysine into nonconnective tissue protein is observed.[79]

Since the chemical changes that various large vessels undergo during the development of hypertension in the SHR are not the same, they may not all represent the "hypertrophic response" of vascular smooth muscle or, because of intervessel heterogeneity, the hypertrophic response varies from vessel to vessel.

Examination of the literature describing changes in vessels less than 200 $\mu$m in diameter from various vascular beds also indicates intervessel heterogeneity. At 6 weeks of age, the thickness of the media of mesenteric arteries less than 200 $\mu$m in diameter is either unchanged[83] or increased[80] in the SHR compared to normotensive controls; however, after 10 weeks of age an increase in media thickness is observed[30,31,83-86] without an increase in smooth muscle cell cross-sectional area.[30,31,85,86] At 25 weeks of age, the absolute amount of actin and myosin as well as the amount per wet weight is higher in mesenteric arteries from SHR rats while collagen and elastin contents are unchanged.[87] The DNA content of the mesenteric artery is also higher and expressing the actin and myosin contents relative to the DNA content indicates no change in the SHR. This suggests that cellular hyperplasia rather than hypertrophy is occurring at this age, with each cell containing a normal amount of contractile protein. Recent morphometric studies[86] substantiate this conclusion.

These data suggest that after 6 weeks of age the 150 $\mu$m mesenteric artery wall hypertrophies in the SHR and that the response involves smooth muscle cell hyperplasia[84,86,87] with an increase in the wall content of contractile but not connective tissue protein.[87] These changes are different than those observed in larger vessels where smooth muscle cell hypertrophy predominate[7,74,75,78] and where the net production of connective tissue protein and contractile protein increase in parallel.[12]

In the skeletal muscle vascular bed, media hypertrophy may develop more slowly than in the mesenteric vascular bed and be a function of vessel diameter. Mulvaney et al.[88] observed that the media of 150 $\mu$m diameter skeletal muscle arteries from 14-week-old SHR rats is significantly thicker than in control studies. However, Bohlen and Lobach[89] observed no change in wall thickness in skeletal arteries less than 130 $\mu$m at either 5 to 6 or 16 to 18 weeks of age, while Prewitt et al.[90] observed a greater wall thickness in 25 $\mu$m arteries only if the SHR rat is at least 16 to 18 weeks of age. Unfortunately, no chemical data exist on vessels from this vascular bed.

Small (200 $\mu$m) cerebral vessels from 25-week-old SHR rats have been examined morphologically and chemically. Brayden et al.[87] observed that at this age, the media thickness is greater, while the percentage of the media made up of smooth muscle cells and the amount of DNA per wet weight are unchanged. The actin to myosin weight ratio is higher and the amount of actin and myosin relative to the DNA content is lower. The amount of collagen per wet weight is higher while the amount of elastin is unchanged. These observations are different than those reported for small mesenteric arteries from SHR rats of the same age.[87] Both vessel types undergo wall hypertrophy; however, in the mesenteric artery this is characterized by an increase in contractile protein content without a change in connective tissue protein content, while in the cerebral artery there is a reduction in contractile protein content and an increase in the connective tissue content. The lack of a change in DNA content or in the percentage of muscle cells in the media of cerebral artery suggests that the wall hypertrophy is due to smooth muscle cell hypertrophy and an increase in extracellular protein. In the mesenteric artery, the increase in DNA content without a change in extracellular protein suggests that the increase in wall thickness is due in part to smooth muscle hyperplasia.

### 2. Functional Changes

Just as there are indications of intervessel differences in chemical changes in association with genetic hypertension, differences in contractile properties are also evident. A decrease in the maximum contractile response per wall cross-sectional area has been observed in the femoral artery[91] and in the thoracic aorta.[12] Even though the actin and myosin content of thoracic aorta increases during development of genetic hypertension,[12] normalization of maximum force development to contractile protein content indicates that force per myosin is reduced.[92] In contrast, Arner and Uvelius[93] observed no change in maximum contractile force per cell area or in maximum velocity of shortening of abdominal aorta, while Cox has shown that the maximum contractile response per cell area of carotid artery is unchanged and for tail arteries is increased.[82] Metabolic studies of aorta indicate that oxygen consumption at rest and during force maintenance is higher in the SHR.[94,95] However, lactate production is not different at rest and either decreases[94] or increases[95] during tension maintenance. Normalization of ATP utilization to force developed indicates no change with genetic hypertension, suggesting no alteration in the ATP consumption of the contractile proteins.

The ability of vessels from genetically hypertensive rats to relax is impaired,[96,97] which has been attributed to an alteration in Ca sequestration.[98,99] However, data suggests that these alterations in relaxation are not a response to the hypertension and therefore are not part of the hypertrophic response, but rather a characteristic of the genetically hypertensive rat.[96]

Examination of microvessels (150 $\mu$m diameter) from the mesenteric bed of 19 to 25 week old SHR rats indicates that maximum contractile force is unchanged when normalized to wall or cell cross-sectional area,[30,83,85,86] or contractile protein content.[87] Since the wall hypertrophy of small mesenteric arteries at this age is due to hyperplasia, these observations suggest that the contractile properties of these newly formed cells are not changed.

The same does not seem to be true for microvessels from the cerebral bed. Winquist and Bohr[100] observed that 200 $\mu$m diameter cerebral vessels from 12- to 20-week-old SHR rats develop less maximum contractile response than similar vessels from normotensive animals. Brayden et al.[87] reported similar findings in 25-week-old SHR rats when force is expressed relative to cell cross-sectional area, and determined that the actin and myosin contents relative to DNA content is lower, suggesting that the reduction in force generation is due in part to a reduction in cellular contractile protein content. The explanation as to why, in the same hypertensive animal, the mesenteric artery and the cerebral artery change differently is unknown.

### 3. Summary (Table 4A and 4B)

In summary, this survey of the literature suggests that morphological, protein, and functional changes occur in both large and small arterial vessels during the development of genetic hypertension; however, the particular changes that occur are not the same in all vessels. Why such intervessel differences exist is unknown. Regardless of the nature of the stimulus responsible for the hypertrophic changes, the heterogeneity in vessel response suggests that the effect is not the same in all arterial vessels. Either the stimulus varies with the vascular bed, which is unlikely, or smooth muscle cells respond differently. Experiments need to be performed that determine the effect of a

## Table 4A
### GENETIC HYPERTENSION CHANGES IN LARGE ARTERIES

| | Rat thoracic aorta | Rat mesenteric | Rat carotid | Rat tail |
|---|---|---|---|---|
| Morphological changes | SMC hypertrophy predominantly with some hyperplasia.[7,12,74,75] | — | — | — |
| Chemical changes | ↑ SMC content of actin and TP.[78]<br>↑ A² incorporation into collagen, but not elastin or non-connective tissue.[79] | ↑ A² incorporation into non-connective tissue followed by an ↑ in connective tissue.[79,80] | — | — |
| Functional changes | ↓ Max force/wall area or myosin content.[92]<br>↑ O² consumption during relaxed and contracted states.[94,95]<br>↑ Lactate production only during contraction.[94,95] | — | No change in max force/cell area.[82] | ↑ Max force/cell area.[82] |

*Abbreviations:*  SMC, smooth muscle cell; A², amino acid; TP, total protein.

Table 4B

## GENETIC HYPERTENSION CHANGES IN SMALL ARTERIES

| | Rat mesenteric (< 200 $\mu$m o.d.) | Rat skeletal (150 $\mu$m o.d.) | Rat cerebral (200 $\mu$m o.d.) |
|---|---|---|---|
| Morphological changes | ↑ Media thickness.[30,31,83-86] SMC hyperplasia.[30,31,85,86] | ↑ Media thickness.[88] | ↑ Media thickness, SMC hypertrophy.[87] |
| Chemical changes | ↑ A and M per wet wt., but no change per cell.[87] ↑ DNA suggesting hyperplasia; C and E unchanged.[87] | | ↑ A/M.[87] ↓ A and M per cell.[87] ↑ C/ww; E unchanged.[87] |
| Functional changes | No change in max force/cell area.[30,83,85,86] | | ↓ Max force/cell area.[87,100] |

*Abbreviations:* SMC, smooth muscle cell; A, actin; M, myosin; C, collagen; E, elastin.

particular hypertrophic stimulus (e.g., increased strain) on vessels from different vascular beds.

## B. Renal Hypertension

In this and the DOCA-salt model of systemic hypertension, comparisons are complicated by the fact that hypertension can be induced at different stages of animal development as well as be of varying duration and intensity. Therefore, both the age at onset of hypertension as well as the duration of hypertension will be indicated so that comparisons can be made between vessels exposed to similar stimuli.

### 1. Morphological and Protein Changes

Using the two-kidney, one-clip model of renal hypertension, Wolinsky[16,19] studied the morphological and chemical changes that occurred 10, 20, and 64 weeks after producing renal artery stenosis in rats 6 to 7 weeks of age. The aortic dry weight, media thickness, and the absolute amounts of collagen, elastin, and alkali soluble protein are larger in aortas from hypertensive animals; however, the number of lamellar units does not change but the lamellae become thicker. During the duration of hypertension, the absolute amounts of collagen and elastin increase while the amount of alkali soluble protein does not change after the 10th week. Total dry weight and media thickness, elevated at 10 weeks, do not increase further until after 20 weeks. When the percentage increase above control is calculated for the alkali soluble protein, collagen, and elastin, it is evident that the net production of these three proteins changes differently during the maintenance of renal hypertension. The alkali soluble fraction (primarily contractile protein) is elevated to 175% of the control value at 10 weeks and remains fairly constant at over 100% at 64 weeks. Collagen also increases (164% over control) at 10 weeks and then remains elevated at a constant percentage (130%) of the control. Elastin, on the other hand, does not increase as much as the other proteins at 10 weeks (136%), but continues to increase over the duration of hypertension (152% at 64 weeks).

Using the same model of renal hypertension, Mangiarua et al.[101] observed a significant increase in the absolute amounts of elastin and collagen in the aorta within 4 weeks of artery clipping. The percentage of the dry weight composed of these proteins remained unchanged[101,102] after 4 weeks of hypertension, but at 12 weeks the percentage of elastin had decreased.[102] In the one-kidney, one-clip renal model, where blood pressure rises very rapidly (within 2 weeks), Rorive et al.[103] observed a significant increase in aortic collagen content within 3 weeks of clipping and an 11% increase in aortic weight within 1 week. These observations suggest that hypertrophic changes can occur soon after a reduction in renal blood flow.

Several studies have reported increases in aortic DNA content and the rate of thymidine incorporation within 1 week[103] or 2 weeks[101] after renal artery clipping. In addition, smooth muscle cells isolated from aorta of rats subjected to 2 weeks of renal hypertension have a greater proliferation rate[106] which is maintained through the 6th subculturing. If the hypertension is of a longer duration (4 to 6 weeks), the increased proliferation rate is not maintained beyond the 2nd subculturing. These studies suggest that smooth muscle cell hyperplasia may occur early in the hypertrophic response. However, measurement of the DNA content of nuclei from aortic smooth muscle cells of renal hypertensive rats[14] indicates a significantly greater frequency of tetraploid and octaploid cells 4 weeks after renal artery clipping. The frequency of tetraploid cells increases with the duration of hypertension (20 weeks). Measurement of the total protein content of individual cells indicates that it more than doubles in aorta 20 weeks after renal artery clipping. These data suggest that polyploidy is contributing to the elevated values for total DNA content and the rate of thymidine incorporation, rather than hyperplasia, and that smooth muscle cell hypertrophy is occurring. However, these results do not explain the increased proliferative ability of cells isolated from aorta of hypertensive animals.

Grunwald et al.[105] observed that smooth muscle cells cultured from aorta of rats subjected to 4 weeks of cellophane-induced perinephritic hypertension are larger and have a higher proliferative rate. Two cell populations can be identified based on size in aorta from normal and hypertensive rats. The large muscle cell population from hypertensive aorta is bigger than the similar populations from control aorta, while the population of small smooth muscle cells is unchanged from control.

Changes in the chemical composition of other arteries from renal hypertension animals have also been observed. As described for the thoracic aorta, no change in the percentage of collagen or elastin is observed in carotid, tail,[102] or mesenteric arteries[101] after 4 weeks of hypertension; however, in contrast to the aorta, at 12 weeks the percentage of collagen is decreased while the percentage of elastin is unchanged in the carotid.[102] In the tail artery, the relative amount of collagen is unchanged and the percentage of elastin decreased as in the aorta at 12 weeks.[102] Again these differences in the changes of chemical composition between vessels emphasizes the heterogeneity in response of the vessel wall to hypertension.

Apparently, no studies have been performed that characterize the protein composition of microvessels from renal hypertensive rats. Since in the genetic hypertensive model, differences in response are observed between conduit and microvessels, such observations are needed.


### 2. Functional Changes

Cox[102] observed that the maximum contractile force per cell cross-sectional area of the tail artery is greater than control 4 weeks after renal artery clipping, but equal to control after 12 weeks. The maximum force of the carotids, however, is elevated at both 4 and 12 weeks. This is different than is observed for similar vessels from SHR rats[82] where maximum force of the carotid is unchanged and maximum force of the tail artery is increased. In aorta from the one-kidney, one-clip renal model, no change in maximum contractile force per wall cross-sectional area is observed 2 to 5 weeks after clipping;[95,106] however, alterations in metabolic activity are detected.[95] Under resting conditions, basal oxygen consumption but not lactate production is elevated, while during force maintenance both oxygen consumption and lactate production are significantly higher. Normalization of ATP utilization to force developed indicates no change with hypertension. This suggests that the higher metabolic activity of the aorta from hypertensive animals is not due to an alteration in energy utilization by the contractile proteins and therefore may be the result of the greater rates of ion pumping reported for this tissue.[107]

### 3. Summary (Table 5)

As in the genetic model, arterial wall hypertrophy occurs in association with renal hypertension which is due to smooth muscle cell hypertrophy and an increase in extracellular connective material. Because of intervessel heterogeneity, a more detailed characterization of the response is not possible at this time.

### C. DOCA-Salt Hypertension
#### 1. Morphological and Protein Changes

Wolinsky[15] determined the collagen and elastin contents of thoracic aorta from rats that had been hypertensive for 6, 9, or 18 weeks beginning at 12 weeks of age and observed that there is no change in the percentage of those proteins even though the absolute amounts are elevated. Media thickness is significantly elevated after 6 weeks but does not increase further with the duration of hypertension. Berry and Greenwald[17] induced hypertension in rats 4 weeks of age and followed the changes in collagen and elastin over the next 16 weeks. It was observed that the percentage of the dry weight made up by collagen is less, while that made up by elastin is greater in aorta from hypertensive animals at all time periods. More recently Cox[108] made similar determinations in aorta from rats made hypertensive at 10 weeks of age and observed that by 4 weeks of hypertension, the percentage of the dry weight made up of either collagen or elastin decreased. The explanation for these divergent observations is not clear except that in the work of Berry and Greenwald, hypertension was induced in young rats, and therefore, their response may be different than that observed in older rats.

Moreland et al.[109] observed an increase in the amount of total protein per wet weight in aorta from rats hypertensive for 8 weeks (beginning at 8 weeks of age); however, there was no change in the amount of actomyosin per wet weight or per cell. Since total protein per wet weight increased but actomyosin does not, these observations suggest that noncontractile protein increased more than contractile protein. These observations are different than those made on aorta from SHR rats[12] where contractile proteins are synthesized in parallel with noncontractile proteins but similar to aortas from renal hypertensive rats.[16,19] Cox[108] also examined the connective tissue composition of carotid and tail arteries from the same animals used for the study of the thoracic aorta (see above). As in the aorta, the percentages of collagen and elastin decreased; however, the time course is different for the three vessels. The decrease in the percentage of collagen begins within 2 weeks after DOCA-salt treatment in the carotid, but not until 4 weeks in the aorta and 8 weeks in the tail artery. With regard to elastin changes, the carotid has a similar time course to that observed in the aorta; however, a decrease in elastin percentage is not seen in the tail artery until after 12 weeks of hypertension.

As with the renal hypertensive model, no information is available about changes in the chemical composition of microvessels from the DOCA-salt model.

#### 2. Functional Changes

A decrease in the maximum contractile response per wall cross-sectional area has

## Table 5
## RENAL HYPERTENSION, CHANGES IN THE VASCULATURE

| | Rat aorta | Rat mesenteric a. | Rat carotid a. | Rat tail a. |
|---|---|---|---|---|
| Morphological changes | ↑ Media thickness; SMC hypertrophy.[14,16,19,104] | — | — | — |
| Chemical changes | ↑ In absolute amounts of E and nonconnective tissue protein; however, each changes at a different rate. Amount per dry wt. unchanged.[16,19] | % C and E unchanged.[102] | % C and E unchanged at 4 wks; at 12 wks % C ↓.[102] | % C and E unchanged at 4 wks; at 12 wks % E ↓.[102] |
| Functional changes | ↑ $O_2$ consumption under relaxed and contractile states.[95] ↑ Lactate production only during contraction.[95] | — | ↑ Max force/cell area at 4 and 12 weeks.[102] | ↑ Max force/cell area at 4 weeks; unchanged at 12.[102] |

*Abbreviations:* C, collagen; E, elastin.

## Table 6
## DOCA-SALT HYPERTENSION, CHANGES IN THE VASCULATURE

| | Rat aorta | Rat carotid | Rat tail |
|---|---|---|---|
| Morphological changes | ↑ Media thickness.[15] | — | — |
| Chemical changes | ↑ Absolute amounts of C and E, but no change in % during 18 weeks,[15] or ↓ % C and ↑ % E during 16 wks,[17] or ↓ % C and % E at 4 wks.[108] No change in AM/cell.[109] | ↓ % C at 2 weeks; ↓ % E at 4 weeks.[103] | ↓ % C at 8 weeks; ↓ % E at 12 weeks.[108] |
| Functional changes | No change in max force/wall area.[112] ↑ O₂ consumption under relaxed and contracted states.[95,112] ↑ or no changes in lactate production during contraction.[95,112] | No change in max force/wall area.[108] | ↑ Max force/area at 4 weeks.[108] ↓ Max force/area at 12 weeks.[108] |

*Abbreviations:* AM, actomyosin; C, collagen; E, elastin.

been observed in femoral arteries,[91,110,111] while others[112] observed no change in aorta. Cox[108] observed no change in the force per cell area of the carotid, while that of the tail artery is greater after 4 weeks of hypertension, but becomes less than control after 12 weeks. Several investigators have observed an increase in sensitivity to contractile agents in a number of arteries from DOCA-salt hypertensive rats[91,113] and a decrease in ability to relax to isoproterenol and adenosine.[113] The decrease in relaxation response has been suggested to be due to impaired membrane regulation of intracellular Ca concentration.[99]

McMahon and Paul[112] and Seidel and Strong[95] observed that resting oxygen consumption but not lactate production is elevated in aorta from 4- to 5-week hypertensive rats. During force development in response to KCl, the increase in oxygen consumption is greater in the hypertensive aorta, while the increase in lactate production is the same[112] or elevated.[95] These observations suggest that under both resting and stimulated conditions, the metabolic activity of the thoracic aorta from the DOCA-salt hypertensive rats is elevated. As in the other forms of systemic hypertension, estimation of the metabolic cost of tension maintenance indicates no change with DOCA-salt hypertension.

### 3. Summary (Table 6)

The changes in the vasculature that occur in association with DOCA-salt hypertension are not as well documented as in the other models of systemic arterial hypertension. Changes in connectve tissue protein content occur; however, there is sufficient variability in the data that conclusions cannot be drawn. As in the other models, there is an increased metabolic rate both at rest and during tension maintenance. Changes in maximum force development may be present, but this depends upon the vessel examined and the duration of hypertension.

## VI. CONCLUSION

The objective of this chapter was to delineate a "basic" response of the vascular smooth muscle cell during wall hypertrophy upon which specific characteristics could be added that were related to the initiating stimulus of wall hypertrophy. Based on the current literature, this goal cannot be achieved and only a very superficial picture exists as to the nature of this "basic" response. The characteristics of this response include: (1) cell hypertrophy and/or hyperplasia; (2) an increased net production of various extracellular and intracellular proteins; and (3) minimal change in contractile function. Variations from this basic pattern occur because of differences between hypertrophic models in the type, magnitude, and duration of the initiating stimulus, and the developmental state of the smooth muscle cell when the stimulus is applied. Finally, variations occur because of vascular smooth muscle heterogeneity with regard to the response to a hypertrophic stimulus. It cannot be assumed that all hypertrophic models or smooth muscle cells are the same. Only by controlling for intervessel and intermodel differences can the current superficial picture of the hypertrophic response of the smooth muscle cell be clarified.

## ACKNOWLEDGMENTS

I would like to express my appreciation to Dr. Julius Allen for his assistance in reviewing this manuscript and to Ms. Rainy Lane-Greenleaf for its preparation.

## REFERENCES

1. Ross, R. and Klebanoff, S. J., The smooth cell, *J. Cell Biol.*, 50, 159, 1971.
2. Cliff, W. J., The aortic tunica media in growing rats studied with the electron microscope, *Lab. Invest.*, 17, 599, 1967.
3. Cliff, W. J., The aortic tunica media in aging rats, *Exp. Mol. Pathol.*, 13, 172, 1970.
4. Olivetti, G., Anversa, P., Melissari, M., and Loud, A. V., Morphometric study of early postnatal development of the thoracic aorta in the rat, *Circ. Res.*, 47, 417, 1980.
5. Gerrity, R. G. and Cliff, W. J., The aortic tunica media of the developing rat, *Lab. Invest.*, 32, 585, 1975.
6. Bucher, B., Travo, P., Laurent, P., and Stoclet, J. C., Vascular smooth muscle cell hypertrophy during maturation in rat thoracic aorta, *Cell. Biol. Int. Rep.*, 6, 883, 1982.
7. Bucher, B., Travo, P., and Stoclet, J. C., Smooth muscle cell hypertrophy and hyperplasia in the thoracic aorta of spontaneously hypertensive rats, *Cell. Biol. Int. Rep.*, 8, 567, 1984.
8. Travo, P., Weber, K., and Osborn, M., Coexistence of vimentin and desmin type intermediate filaments in a subpopulation of adult rat vascular smooth muscle cells growing in primary culture, *Exp. Cell. Res.*, 139, 87, 1982.
9. Gabbiani, G., Rungger-Brandle, E., De Chastonay, C., and Franke, W. W., Vimentin-containing smooth muscle cells in aortic intimal thickening after endothelial injury, *Lab. Invest.*, 47, 265, 1982.
10. Berry, C. L., Looker, T., and Germain, J., The growth and development of the rat aorta, *J. Anat.*, 113, 1, 1972.
11. Looker, T. and Berry, C. L., The growth and development of the rat aorta, *J. Anat.*, 113, 17, 1972.
12. Seidel, C. L., Aortic actomyosin content of maturing normal and spontaneously hypertensive rats, *Am. J. Physiol.*, 237, H34, 1979.
13. Seidel, C. L. and Murphy, R. A., Changes in rat aortic actomyosin content with maturation, *Blood Vessels*, 16, 98, 1979.
14. Owens, G. K. and Schwartz, S. M., Vascular smooth muscle cell hypertrophy and hyperploidy in the Goldblatt hypertensive rat, *Circ. Res.*, 53, 491, 1983.
15. Wolinsky, H., Response of the rat aortic wall to hypertension, *Circ. Res.*, 26, 507, 1970.

16. Wolinsky, H., Effects of hypertension and its reversal on the thoracic aorta of male and female rats, *Circ. Res.*, 28, 622, 1971.
17. Berry, C. L. and Greenwald, S. E., Effects of hypertension on the static mechanical properties and chemical composition of the rat aorta, *Cardiovasc. Res.*, 10, 437, 1976.
18. McCloskey, D. I. and Cleary, E. G., Chemical composition of the rabbit aorta during development, *Circ. Res.*, 34, 828, 1974.
19. Wolinsky, H., Long-term effects of hypertension on the rat aortic wall and their relation to concurrent aging changes, *Circ. Res.*, 30, 301, 1972.
20. Seidel, C. L. and Allen, J. C., Pharmacologic Characteristics and Actomyosin Content of Aorta from Neonatal Rats, *Am. J. Physiol.*, 237, C81, 1979.
21. Kocher, O., Skalli, O., Cerutti, D., Gabbiani, F., and Gabbiani, G., Cytoskeletal features of rat aortic cells during development, *Circ. Res.*, 46, 829, 1985.
22. Cohen, M. L. and Berkowitz, B. A., Vascular contraction: effect of age and extracellular calcium, *Blood Vessels*, 13, 139, 1976.
23. Cohen, M. L. and Berkowitz, B. A., Age-related changes in vascular responsiveness to cyclic nucleotides and contractile agonists. *Pharmacol. Exp. Ther.*, 191, 147, 1974.
24. Cohen, M. L., Blume, A. S., and Berkowitz, B. A., Vascular adenylate cyclase: role of age and guanine nucleotide activation, *Blood Vessels*, 14, 25, 1977.
25. Ericsson, E. and Lundholm, L., Adrenergic B-receptor activity and cyclic AMP metabolism in vascular smooth muscle; variations with age, *Mech. of Aging and Develop.* 4, 1, 1975.
26. Fleisch, J. H. and Hooker, C. S., The relationship between age and relaxation of vascular smooth muscle in the rabbit and rat, *Circ. Res.*, 38, 243, 1976.
27. Altura, B. M. and Altura, B. T., Some physiological factors in vascular reactivity — aging in vascular smooth muscle and its influence on reactivity, in *Factors Influencing Vascular Reactivity*. Carrier, O. and Shibata, S., Eds., Iga-Ku-Shoin, New York, 169, 1977.
28. Su, C., Bevan, J. A., Assali, N. S., and Brinkman, C. R., Development of neuroeffector mechanisms in the carotid artery of the fetal lamb, *Blood Vessels*, 14, 12, 1977.
29. Cox, R. H., Effects of age on the mechanical properties of rat carotid artery, *Am. J. Physiol.*, 233, H256, 1977.
30. Warshaw, D. M., Mulvany, M. J., and Halpern, W., Mechanical and morphological properties of arterial resistance vessels in young and old spontaneously hypertensive rats, *Circ. Res.*, 45, 250, 1979.
31. Warshaw, D. M., Root, D. T., and Halpern, W., Effects of antihypertensive drug therapy on the morphology and mechanics of resistance arteries from spontaneously hypertensive rats, *Blood Vessels*, 17, 257, 1980.
32. Miller, B. G., Overhage, J. M., Bohlen, H. G., and Evan, A. P., Hypertrophy of smooth muscle cells in the rat small intestine during maturation, *Microvasc. Res.*, 29, 56, 1985.
33. Duckles, S. P., Carter, B. J., and Williams, C. L., Vascular adrenergic neuroeffector function does not decline in aged rats, *Circ. Res.*, 56, 109, 1985.
34. Cox, R. H., Jones, A. W., and Swain, M. L., Mechanics and electrolyte composition of arterial smooth muscle in developing dogs, *Am. J. Physiol.*, 231, 77, 1976.
35. Gray, S. D., Reactivity of neonatal canine aortic strips, *Biol. Neonate*, 31, 10, 1977.
36. Lai, F. M., Tanikella, T., Thibault, L., Chan, P. S., and Cervoni, P., Effects of different stages of aortic coarctation hypertension on aortic contraction and relaxation in rats, *J. Pharmacol. Exp. Ther.*, 214, 388, 1980.
37. Bevan, J. A., Bevan, R. D., Chang, P. C., Pegram, B. L., Purdy, R. E., and Su, C., Analysis of changes in reactivity of rabbit arteries and veins two weeks after induction of hypertension by coarctation of the abdominal aorta, *Circ. Res.*, 37, 183, 1975.
38. Bevan, R. D., An autoadiographic and pathological study of cellular proliferation in rabbit arteries correlated with an increase in arterial pressure, *Blood Vessels*, 13, 100, 1976.
39. Hume, W. R. and Bevan, J. A., Amino acid uptake in rabbit blood vessels two weeks after induction of hypertension by coarctation of the abdominal aorta, *Cardiovasc. Res.*, 12, 106, 1978.
40. Bevan, R. D., Eggena, P., Hume, W. R., Van Marthens, E., and Bevan, J. A., Transient and persistent changes in rabbit blood vessels associated with maintained elevation in arterial pressure, *Hypertension*, 2, 63, 1980.
41. Olivetti, G., Anversa, P., Melissari, M., and Loud, A. V., Morphometry of medial hypertrophy in the rat thoracic aorta, *Lab. Invest.*, 42, 559, 1980.
42. Owens, G. K. and Reidy, M. A., Hyperplastic growth response of vascular smooth muscle cells following induction of acute hypertension in rats by aortic coarctation, *Circ. Res.*, 57, 695, 1985.
43. Johansson, B., Structural and functional changes in rat portal veins after experimental portal hypertension, *Acta Physiol. Scand.*, 98, 381, 1976.
44. Uvelius, B., Arner, A., and Johansson, B., Hypertrophic venous smooth muscle, *Acta Physiol. Scand.*, 112, 463, 1971.

45. Sigurdsson, S. B. and Uvelius, B., Membrane potential in smooth muscle cells from hypertrophic rat portal vein, *Experientia,* 39, 1288, 1983.
46. Arner, A. and Uvelius, B., Oxygen dependence and energy turnover in normal and hypertrophic rat portal vein, *Acta Physiol. Scand.,* 113, 341, 1971.
47. Berner, P. F., Somlyo, A. V., and Somlyo, A. P., Hypertrophy-induced increase of intermediate filaments in vascular smooth muscle, *J. Cell Biol.,* 88, 96, 1981.
48. Junker, J. L., Wasserman, A. J., Berner, P. F., and Somlyo, A. P., Electron probe analysis of sodium and other elements in hypertrophied and sodium-loaded smooth muscle, *Circ. Res.,* 54, 254, 1984.
49. Seidel, C. L., Lewis, R. M., Bowers, R., Bukoski, R. D., Kim, H. S., Allen, J. C., and Hartley, C., Adaptation of canine saphenous veins to grafting, *Circ. Res.,* 55, 102, 1984.
50. Hume, W. R., Proline and thymidine uptake in rabbit ear artery segments in vitro increased by chronic tangenital load, *Hypertension,* 2, 738, 1980.
51. Leung, D. Y. M., Glagov, S., and Matthews, M. B., Cyclic stretching stimulates synthesis of matrix components by arterial smooth muscle cells in vitro, *Science,* 191, 475, 1976.
52. Leung, D. Y. M., Glagov, S., and Mathews, M. B., A new in vitro system for studying cell response to mechanical stimulation, *Exp. Cell Res.,* 109, 285, 1977.
53. Dobrin, P. B., Mechanical properties of arteries, *Physiol. Rev.,* 58, 397, 1978.
54. Sottiurai, V. S., Kollros, P., Glagov, S., Zarins, C. K., and Mathews, M. B., Morphologic alteration of cultured arterial smooth muscle cells by cyclic stretching, *J. Surg. Res.,* 35, 490, 1983.
55. Seidel, C. L., Ives, B. V., and Eskin, S. L., The effect of strain on vascular muscle cells in culture, *Physiologist,* 27, 236, 1984.
56. Haworth, S. G. and Hislop, A. A., Adaptation of the pulmonary circulation to Extra-uterine life in the pig and its relevance to the human infant, *Cardiovasc. Res.,* 15, 108, 1981.
57. Haworth, S. G. and Hislop, A. A., Effect of hypoxia on adaptation of the pulmonary circulation to extra-uterine life in the pig, *Cardiovasc. Res.,* 16, 293, 1982.
58. Meyrick, B. and Reid, L., Normal postnatal development of the media of the rat hilar pulmonary artery and its remodeling by chronic hypoxia, *Lab. Invest.,* 46, 505, 1982.
59. Leung, D. Y. M., Glagov, S., and Mathews, M. B., Elastin and collagen accumulation in rabbit ascending aorta and pulmonary trunk during postnatal growth, *Circ. Res.,* 41, 316, 1977.
60. Nuwayhid, B., Brinkman, C. R., Su, C., Bevan, J. A., and Assali, N. S., Systemic and pulmonary hemodynamic responses to adrenergic and cholinergic agonists during fetal development, *Biol. Neonate,* 26, 301, 1975.
61. Joiner, P. D., Kadowitz, P. J., Hughes, J. P., and Hyman, A. L., NE and ACh responses of intra-pulmonary vessels from dog, swine, sheep, and man, *Am. J. Physiol.,* 228, 1821, 1975.
62. Su, C. and Bevan, J. A., Pharmacology of pulmonary blood vessels, *Pharmacol. Ther.,* 2, 275, 1976.
63. O'Donnell, S. R. and Wanstall, J. C., Beta-1 and beta-2 adrenoceptor-mediated responses in preparations of pulmonary artery and aorta from young and aged rats, *J. Pharmacol. Exp. Ther.,* 228, 733, 1984.
64. Fishman, A. P., Hypoxia on the pulmonary circulation, *Circ. Res.,* 38, 221, 1976.
65. Fishman, A. P., Vasomotor regulation of the pulmonary circulation, *Ann. Rev. Physiol.,* 42, 211, 1980.
66. Lloyd, T. C., Hypoxic pulmonary vasoconstriction: role of perivascular tissue, *J. Appl. Physiol.,* 25, 560, 1968.
67. Davis, M. J., Joyner, W. L., and Gilmore, J. P., Microvascular pressure distribution and responses of pulmonary allografts and cheek pouch arterioles in the hamster to oxygen, *Circ. Res.,* 49, 125, 1981.
68. Robinovitch, M., Gamble, W., Nadas, A. S., Miettinen, O. S., and Reid, L., Rat pulmonary circulation after chronic hypoxia: hemodynamic and structural features, *Am. J. Physiol.,* 236, H818, 1979.
69. Meyrick, B. and Reid, L., Hypoxia and incorporation of $^3$H-thymidine by cells of the rat pulmonary arteries and alveolar wall, *Am. J. Pathol.,* 96, 51, 1979.
70. Meyrick, B. an Reid, L., Hypoxia-induced structural changes in the media and adventitia of the rat hilar pulmonary artery and their regression, *Am. J. Pathol.,* 100, 151, 1980.
71. Robinovitch, M., Gamble, W. J., Miettinen, O. S., and Reid, L., Age and sex influence on pulmonary hypertension of chronic hypoxia and on recovery, *Am. J. Physiol.,* 240, H62, 1981.
72. Suzuki, H. and Twarog, B. M., Membrane properties of smooth muscle cells in pulmonary arteries of the rat, *Am. J. Physiol.,* 242, H900, 1982.
73. Gray, S. D., Spontaneous hypertension in the neonatal rat, *Clin. Exp. Hyper. — Theory Pract.,* A6, 755, 1984.
74. Olivetti, G., Massimo, M., Germano, M., and Anversa, P., Quantitative structural changes of the rat thoracic aorta in early spontaneous hypertension, *Circ. Res.,* 51, 19, 1982.

75. Owens, G. K. and Schwartz, S. M., Alterations in vascular smooth muscle mass in the spontaneously hypertensive rat, *Circ. Res.,* 51, 280, 1982.
76. Yamori, Y., Igawa, T., Kanbe, T., Kihara, M., Nara, Y., and Horie, R., Mechanisms of structural vascular changes in genetic hypertension: analyses on cultured vascular smooth muscle cells from spontaneously hypertensive rats, *Clin. Sci.,* 61, 121S, 1981.
77. Anversa, P., Melissari, M., Tardini, A., and Olivetti, G., Connective tissue accumulation in the left coronary artery of young SHR, *Hypertension,* 6, 526, 1984.
78. Owens, G. K., Rabinovitch, P. S., and Schwartz, S. M., Smooth muscle cell hypertrophy versus hyperplasia in hypertension, *Proc. Natl. Acad. Sci.,* 78, 7759, 1981.
79. Yamori, Y., Nakada, T., and Lovenberg, W., Effect of antihypertensive therapy on lysine incorporation into vascular protein of the spontaneously hypertensive rat, *Eur. J. Pharm.,* 38, 349, 1976.
80. Nakada, T. and Lovenberg, W., Lysine incorporation in vessels of spontaneously hypertensive rats: effects of adrenergic drugs, *Eur. J. Pharm.,* 48, 87, 1978.
81. Iwatsuki, K., Cardinale, G. J., Spector, S., and Udenfriend, S., Hypertension: increase of collagen biosynthesis in arteries but not in veins, *Science,* 198, 403, 1977.
82. Cox, R. H., Basis for the altered arterial wall mechanics in the spontaneously hypertensive rat, *Hypertension,* 3, 485, 1981.
83. Mulvany, M. J., Aalkjaer, C., and Christensen, J., Changes in noradrenaline sensitivity and morphology of arterial resistance vessels during development of high blood pressure in spontaneously hypertensive rats, *Hypertension,* 2, 664, 1980.
84. Lee, R. M. K. W., Garfield, R. E., Forrest, J. B., and Daniel, E. E., Morphometric study of structural changes in the mesenteric blood vessels of spontaneously hypertensive rats, *Blood Vessels,* 20, 57, 1983.
85. Mulvany, M. J., Hansen, P. K., and Aalkjaer, C., Direct evidence that the greater contractility of resistance vessels in spontaneously hypertensive rats is associated with a narrowed lumen, a thickened media, and an increased number of smooth muscle cell layers, *Circ. Res.,* 57, 854, 1978.
86. Mulvany, J. J., Baandrup, U., and Gundersen, H. J. G., Evidence for hyperplasia in mesenteric resistance vessels of spontaneously hypertensive rats using a three-dimensional disector, *Circ. Res.,* 57, 794, 1985.
87. Brayden, J. E., Halpern, W., and Brann, L. R., Biochemical and mechanical properties of resistance arteries from normotensive and hypertensive rats, *Hypertension,* 5, 17, 1983.
88. Mulvany, M. J., Nilsson, H., Nyborg, N., and Mikkelsen, E., Are isolated femoral resistance vessels or tail arteries good models for the hindquarter vasculature of spontaneously hypertensive rats?, *Acta Physiol. Scand.,* 116, 275, 1982.
89. Bohlen, H. G. and Lobach, D., In vivo study of microvascular wall characteristics and resting control in young and mature spontaneously hypertensive rats, *Blood Vessels,* 15, 322, 1978.
90. Prewitt, R. L., Chen, I. I. H., and Dowell, R., Development of microvascular rarefaction in the spontaneously hypertensive rat, *Am. J. Physiol.,* 243, H243, 1982.
91. Hansen, T. R. and Bohr, D. F., Hypertension, transmural pressure, and vascular smooth muscle response in rats, *Circ. Res.,* 36, 590, 1975.
92. Seidel, C. L., Allen, J. C., and Bowers, R. L., Mechanical and biochemical alterations of aorta induced by hydralazine hypotension, *J. Pharmacol. Exp. Ther.,* 213, 514, 1980.
93. Arner, A. and Uvelius, B., Force-velocity characteristics and active tension in relation to content and orientation of smooth muscle cells in aortas from normotensive and spontaneously hypertensive rats, *Circ. Res.,* 50, 812, 1982.
94. Arner, A. and Hellstrand, P., Energy turnover and mechanical properties of resting and contracting aortas and portal veins from normotensive and spontaneously hypertensive rats, *Circ. Res.,* 48, 539, 1981.
95. Seidel, C. L. and Strong, M. S., Metabolic characteristics of aorta from SHR, renal and DOCA-salt hypertensive rats, *Hypertension,* 8, 103, 1986.
96. Cheng, J. B. and Shibata, S., Vascular relaxation in the spontaneously hypertensive rat, *J. Cardiovasc. Pharm.,* 3, 1126, 1981.
97. Asano, M., Aoki, K., and Matsuda, T., Reduced beta adrenoceptor interactions of norepinephrine enhance contraction in the femoral artery from spontaneously hypertensive rats, *J. Pharmacol. Exp. Ther.,* 223, 207, 1982.
98. Bhalla, R. C., Webb, R. C., Singh, S., Ashley, T., and Brock, T., Calcium fluxes, calcium binding, and adenosine cyclic $3', 5'$-monophosphate-dependent protein kinase activity in the aorta of spontaneously hypertensive and kyoto wistar normotensive rats, *Mol. Pharm.,* 14, 468, 1978.
99. Kwan, C. Y., Belbeck, L., and Daniel, E. E., Abnormal biochemistry of vascular smooth muscle plasma membrane as an important factor in the initiation and maintenance of hypertension in rats, *Blood Vessels,* 16, 259, 1979.

100. Winquist, R. J. and Bohr, D. F., Structural and functional changes in cerebral arteries from spontaneously hypertensive rats, *Hypertension,* 5, 292, 1983.
101. Mangiarua, E., Basso, N., Dubner, D., Ruiz, P., and Taquini, A. C., Evidence of early structural change in the artery wall of two-kidney one-clip Goldblatt hypertensive rats., *Clin. Exp. Hyper. — Theory and Pract.,* A4, 1271, 1982.
102. Cox, R. H., Changes in arterial wall properties during development and maintenance of renal hypertension, *Am. J. Physiol.,* 242, H477, 1982.
103. Rorive, G. L., Carlier, P. J., and Foidart, J. M., Hyperplasia of rat arteries smooth muscle cells associated with development and reversal of renal hypertension, *Clin. Sci.,* 59, 335S, 1980.
104. Mey, J., Grunwald, J., and Hauss, W. H., Growth rate differences between arterial smooth muscle cells cultivated from rat impaired by short- or long-term hypertension respectively, *Artery,* 8, 348, 1980.
105. Grunwald, J., Robenek, H., Mey, J., and Hauss, W. H., In vivo and in vitro cellular changes in experimental hypertension: electronmicroscopic and morphometric studies of aortic smooth muscle cells, *Exp. Mol. Path.,* 36, 164, 1982.
106. Arner, A., Malmqvist, U., and Uvelius, B., Structural and mechanical adaptations in rat aorta in response to sustained changes in arterial pressure, *Acta Physiol. Scand.,* 122, 119, 1984.
107. Overbeck, H. W. and Grissette, D. E., Sodium pump activity in arteries of rats with Goldblatt hypertension, *Hypertension,* 4, 132, 1982.
108. Cox, R. H., Time course of arterial wall changes with DOCA plus salt hypertension in the rat, *Hypertension,* 4, 27, 1982.
109. Moreland, R. S., Webb, R. C., and Bohr, D. F., Vascular changes in DOCA hypertension, *Hypertension,* 4(supp. III), III-99, 1982.
110. Pang, C. C. Y. and Sutter, M. C., Contractile responses of aortic and portal vein strips during the development of DOCA/salt hypertension, *Blood Vessels,* 17, 281, 1980.
111. Hansen, T. R., Dineen, D. X., and Pullen, G. L., Orientation of arterial smooth muscle and strength of contraction of aortic strips from DOCA-hypertensive rats, *Blood Vessels,* 17, 302, 1980.
112. McMahon, E. G. and Paul, R. J., Metabolic and mechanical properties of aortas from aldosterone-salt hypertensive rats, *Circ. Res.,* 55, 349, 1984.
113. Hagen, E. C. and Webb, R. C., Coronary artery reactivity in deoxycorticosterone acetate hypertensive rats, *Am. J. Physiol.,* 247, H409, 1984.
114. Bosher, L. P., Autogenous greater saphenous vein as a vascular conduit, in *Vascular Grafting,* Wright, C. B., Hobson, R. W., Hiratzka, L. F., and Lynch, T. G., Eds., John Wright — PSG Inc., Boston, 1983, 117.

Chapter 3

# MUSCLE HYPERTROPHY IN THE PARTIALLY OBSTRUCTED INTESTINE

Giorgio Gabella

## TABLE OF CONTENTS

## I. INTRODUCTION

The physiological growth of the organism is harmonious, if not uniform, and it unfolds according to a flexible plan that regulates the relative size of each organ at each stage up to maturity. Growth in excess to this plan is hypertrophic growth, and the process whereby an organ grows above the size it has at maturity or beyond the relative size it averages at any stage of development is hypertrophy. It would be impossible to give an exact measure of the normal size of an organ or of its normal level of functional capacity. However, it is possible to estimate some average values, to analyze factors that limit growth (physical constraints, economy of materials, competition between parts), and to assess the normal functional load, i.e., that which is evolutionarily anticipated by the organism.

Hypertrophy generally occurs when increased functional demands are imposed upon an organ, or when there is an imbalance (even if transient) between functional capacity and demand (or load). Thus, organs hypertrophy to compensate for the failure of a synergistic organ, or, in general, when the functional load is markedly increased, either by accident or by experimental design. Hypertrophy is not an acute response, and a stimulus effective to produce it must be intense and persistent, but not such as to cause acute organ failure.

In hollow organs hypertrophic growth is not uncommon, a fact related to the great variability in the volume and the density of their contents and in the resistance of their outflow. The contents can drop to zero, or, on the contrary, the resistance of the outflow can become insurmountable. The obvious ability of segments of the gut, of the biliary pathways, and of the urinary tract to hypertrophy is probably to some extent a property acquired through evolution, as one could say for most forms of adaptability. The occurrence of the "risk" (e.g., possibility of an increase in volume or in density of the contents) is detected evolutionarily, and the adaptability (e.g., by recurrence to hypertrophy or to atrophy) is an attempt to optimize the fitness of the organ under the conditions imposed by the factors that limit growth (see above).

The medical interest in the hypertrophy of hollow viscera is due to its common occurrence in connection with congenital malformations (such as Hirsphrung's disease,[1] infantile hypertrophic pyloric stenosis[2]), with chronic obstructions (such as with stones in the ureter or the bile duct, or with prostate enlargment, or with tumors of the gut), or after surgery (e.g., resections, anastomosis, adherences, denervation). A case of muscular hypertrophy of the small intestine without a known initiating factor has been published.[3]

Hypertrophy of the gut was already being studied experimentally during the second half of last century. Herczel,[4] for example, produced an experimental stenosis in the small intestine of rabbits and observed an increase in diameter of the intestine on the oral side and an increase in the thickness of the muscle coat. In many of the previous studies, the main question has been whether the increase of the muscle coat is due to hypertrophy of the muscle cells or to hyperplasia (increase in cell number); however, the structural features of a smooth muscle cannot be studied properly by light microscopy and it has been only recently that the morphology of the hypertrophic intestinal musculature has been characterized.

A similar process of hypertrophy in partially obstructed ducts has been studied in the ureter,[5] the urinary bladder,[6,7,8,9] the vas deferens,[10] and in blood vessels of hypertensive animals.[11] A special case of smooth muscle hypertrophy is the myometrium in the pregnant uterus. Surprisingly, the ultrastructure of this tissue has not been studied in great detail;[12,13] however, it will not be discussed in this chapter since it is examined elsewhere in this book (see Chapter 5).

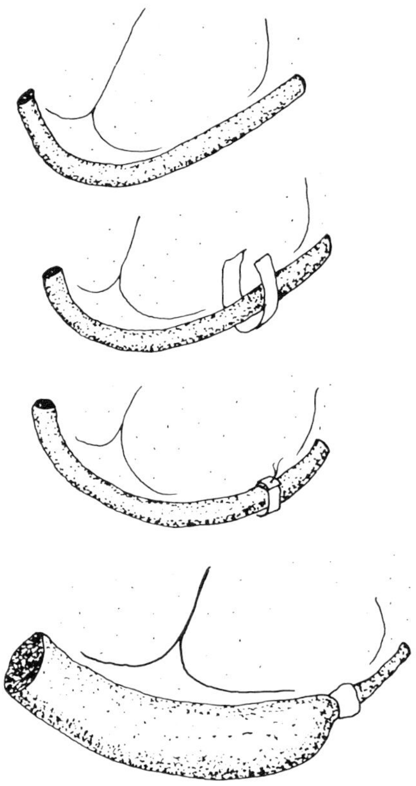

FIGURE 1.  Drawing illustrating the surgical procedure to produce stenosis of the small intestine with a ring of cellophane (middle panels). A few weeks after the operation the intestine orad to the stenosis is distended and hypertrophied (lowest panel).

It should be stressed that the reaction of the intestinal wall to a partial obstruction is a complex one: in addition to growth of tissue there is a substantial enlargement of the lumen which strictly speaking is not hypertrophy, although it may be a stimulus for it. Other forms of intestinal hypertrophy occur during lactation,[14] after intestinal resections,[15-17] and in hyperphagia.[18,19] These changes involve mainly the mucosa, the ultrastructural features of which have been extensively studied and reviewed in Chapter 4, although more recent studies show that the musculature responds too (Chapter 4).

## II. EXPERIMENTAL PROCEDURE

The experiments reviewed in this article were carried out on adult guinea pigs, weighing 400 to 600 grams. Other experiments were carried out on rats. (All the illustrations in this chapter are from the guinea pig intestine, except for Figure 7 which is from a rat.) The surgical procedure aims at producing a partial obstruction of the terminal part of the ileum. Under general anaesthesia the terminal ileum is exposed and a small window is cut in the mesentery close to the gut. A 5 to 8 mm wide, 60 mm long strip of acetate film (cellophane) is passed through this window and around the gut, and it is rolled on so as to form a ring around the gut. The ring is then fixed by two silk stitches and the excess length of the strip is cut off (Figure 1). The ring has an inner diameter about 2 mm wider than the diameter of the intestine when distended, and is therefore free to move, being restrained only by the peritoneum and its blood vessels.

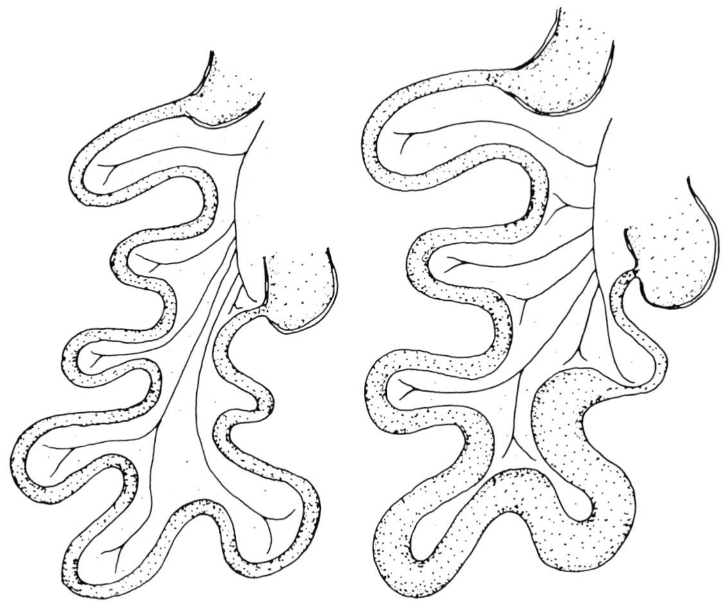

FIGURE 2.   Schematic drawing illustrating the small intestine with the mesenteric
arteries and part of the stomach and caecum, in the control animal (left), and a few
weeks after the experimental stenosis.

For a few days after the operation there is no impairment to the transit of ingesta.
Subsequently, the serosa of the gut under the acetate film reacts and proliferates, first
anchoring the acetate film to the gut and then slowly and progressively reducing the
lumen of the intestine inside the ring. Impaired transit of ingesta and stasis within the
loops orad to the stenosis ensue (Figure 2). A slow development of the partial obstruction is essential for the growth of the musculature to keep pace with the distension, so
that orad to the stenosis, the gut is expanded and its wall is thickened. The process
develops over a period of 3 to 5 weeks (in rats the process is faster and it is usually
maximal within 3 weeks).

Previous investigators, who have worked chiefly on dogs, have produced a stenosis
by applying a sleeve of cellophane,[20] silk tape,[21] or teflon mesh[22] around a loop of
ileum; the sleeve is 20 to 50 mm long and is tightened so as to reduce the diameter of
the lumen by about half from the beginning of the experiment[22] and to enhance this,
stitches are placed on the intestinal wall under the sleeve so as to form a conspicuous
longitudinal fold.[20] This approach is unsuitable for small animals such as rats and
guinea pigs. Brent[23] obtained hypertrophy of the large intestine of rabbits by constrict-
ing the anal opening with surgical stitches.

## III. HYPERTROPHY OF THE INTESTINE

Within a few weeks of the operation, the intestine orad to the stenosis is markedly
enlarged (Figure 3), and is hereafter referred to as the hypertrophic intestine. The loops
are distended and smooth-surfaced, and are full with a semi-fluid material. The total
length of the small intestine, measured at the level of the attachment of the mesentery,
is the same in control and in operated animals, i.e., the hypertrophy affects the diameter but not the length of the intestine. The mesenteric blood vessels supplying the
hypertrophic intestine are enlarged and more prominent than in controls. The loop of

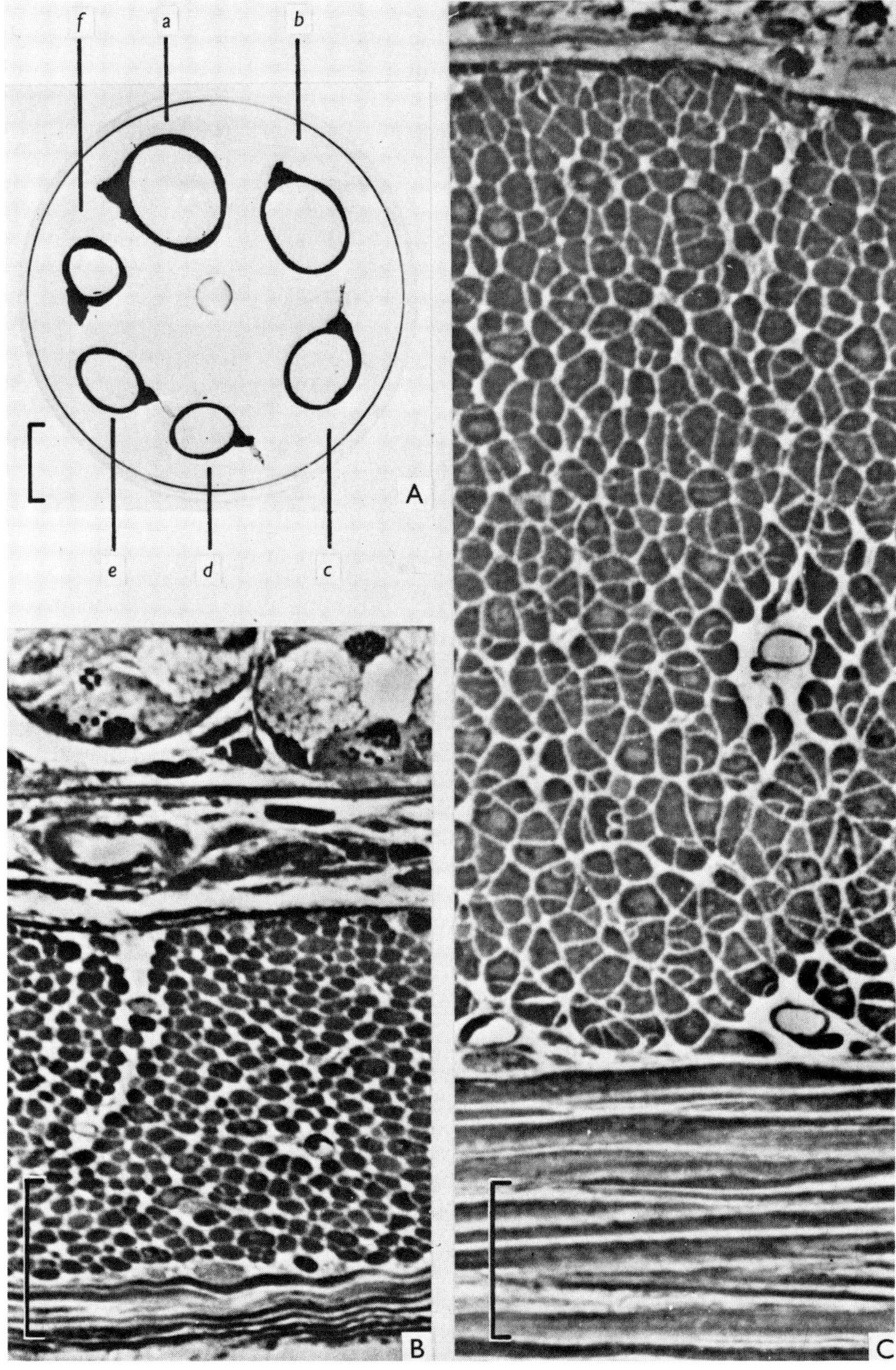

FIGURE 3. (A) A block of Araldite with embedded rings of ileum taken at different distances from the site of stenosis. Guinea pig 4 weeks after the operation. a, immediately orad to the stenosis; b, about 10 cm more orad; c, d, e, more orad segments, about 10 cm apart from each other; f, segment aborad to the stenosis, (B and C) Phase contrast micrographs of transverse sections of intestinal wall of the control intestine and of the maximally hypertrophic intestine (C). Of the latter only the muscle coat is shown. Calibration bars: (A), 10 mm; (B) and (C), 25 $\mu$m.

ileum aboral to the stenosis, when present, is empty and of smaller diameter than in controls. No changes are observed in the large intestine, except that the caecum is usually empty.

The hypertrophy is maximal 5 to 10 cm orad to the stenosis, where the ileum can reach an outer diameter of 15 mm (i.e., about three times the original diameter), and is progressively less extensive in the more oral segments; the process, however, can spread, and be noticeable under the naked eye, almost over the entire length of ileum and jejunum.

While a segment of small intestine does not change in length after the operation of stenosis, the total weight of its wall (without content) increases up to 10-fold, the outcome of a very intense process of hypertrophy.[24,25] The increase in the mass of the intestinal wall is due to a thickening of the wall and to a distension of the wall across the long axis of the gut (increase in the circumference of the gut). In the maximally hypertrophied intestine, the wall is more than three times as thick as in controls, a difference which can be clearly appreciated during the dissection. All the layers of the wall are affected by the process. The villi of the mucosa are taller (although not more numerous) and the epithelium is higher than in controls. The submucosa maintains the same thickness as in the control intestine or is slightly thickened; its volume, of course, is always substantially increased on account of the larger diameter of the hypertrophic intestine. In some preparations the thickness of the serosa is markedly increased, with the appearance of new blood vessels and of isolated bundles of circular musculature, while in other preparations it is hardly changed at all.

The most dramatic structural changes occur in the muscle coat (muscularis externa) (Figure 3). The extent of the hypertrophy of the muscle coat can be estimated by multiplying its increase in thickness by the increase in the circumference of the intestine. The musculature can increase up to 15-fold its volume in the maximally hypertrophied intestine. Both the circular and the longitudinal muscle layer hypertrophy, although in some experiments the longitudinal muscle did so to a lesser extent than the circular muscle, and in most cases the hypertrophy attains the same intensity in both muscle layers. It is known that in the normal guinea pig, the ratio in thickness between the longitudinal and the circular muscle layers is characteristic of each portion of the small intestine:[26] the ratio changes gradually from 1:4.5 in the duodenum to 1:2 in the terminal ileum (this change amounts to an increase in the proportion of longitudinal musculature in the muscle coat along an orad-aboral gradient). In spite of the marked changes in total thickness of its muscle coat, each segment of hypertrophic intestine maintains the ratio between longitudinal and circular muscles that is characteristic of that segment in the control intestine. All the structural changes described are found to affect the entire circumference of the intestinal wall to the same extent. However, for sake of uniformity, especially for the quantitative work, the observations were focussed on tissue blocks located at three o'clock and at 9 o'clock around the circumference of the intestine.

The fact that the increase in thickness of the muscle coat always accompanies an increase in the diameter of the intestine underlines the connection between distension of the lumen and growth of the musculature. This contrasts with the observations of Brent,[23] who reports that in the colon of rabbits with anal stenosis dilatation preceeds the thickening of the wall.

## IV. SMOOTH MUSCLE CELLS

Very substantial structural changes are apparent in the muscle cells of both muscle layers of the hypertrophic intestine. The following description refers to the muscula-

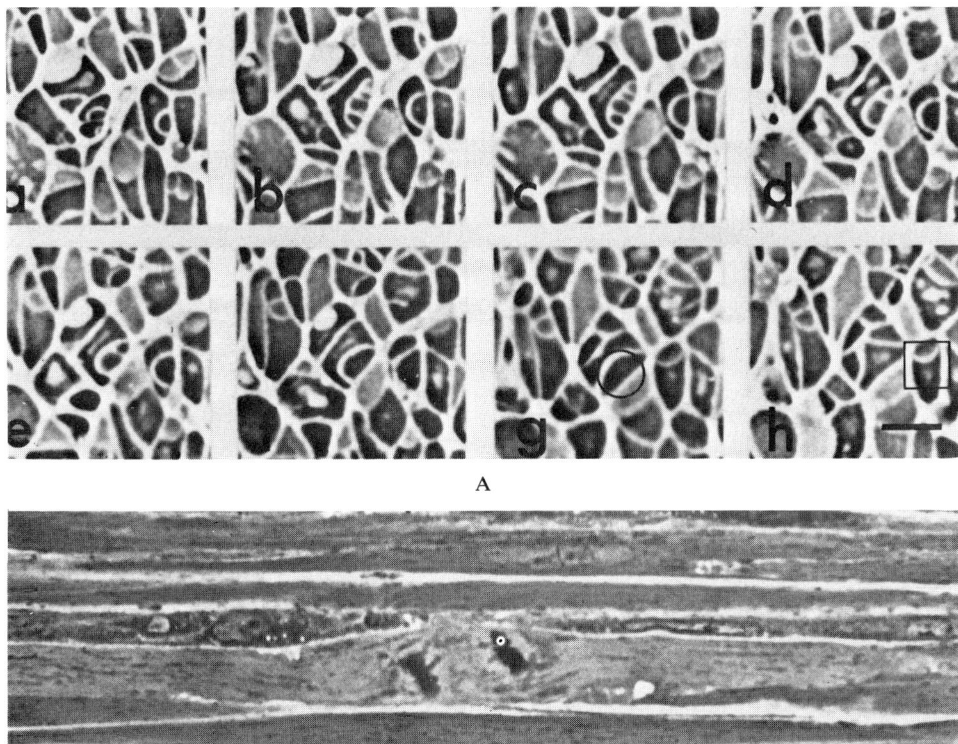

A

B

FIGURE 4.   (A) Transverse section of the circular muscle layer of the hypertrophic intestine of a guinea pig. An example of multiple branching of a muscle cell: the profile indicated with a circle in g splits into two in e and into four in c. As an example of the changes in shape and relations along the muscle length, note how the pair of cells indicated by a square in h change their appearance through the series. Calibration bar: 10 µm (B, C, D). Muscle cells in mitosis in the circular muscle layer (in longitudinal section). In D, a muscle cell in mitosis is also seen in the longitudinal muscle layer (bottom right). Calibration bar: 30 µm. (From Gabella, G., Hypertrophic smooth muscle. Size and shape of cells, occurrence of mitoses. Sarcoplasmic reticulum, caveolae and mitochondria, *Cell Tissue Res.*, 201, 63, 1979. With permission.)

ture in the portion of maximal hypertrophy; progressively less extensive changes are found in the more oral portions of the intestine, farther away from the stenosis.

## A. Cell Volume

The marked increase in muscle cell volume is apparent in light micrographs of transversely sectioned muscle layers (Figure 3). Since both the control and the experimental materials were fixed in a condition of rest, the difference in size of the muscle cell profiles is not attributable to an isotonic contraction of one of the two tissues. The volume of a muscle cell in the control ileum, measured by stereology, is about 3,500 $\mu m^3$. In the hypertrophic intestine the muscle cells grow in volume up to an average of 13,000 $\mu m^3$, a 3- to 4-fold increase. There is a greater variability in size among the cells of the hypertrophic musculature than among the cells of the control tissue, reflecting a certain variation in the response of individual cells (and to some extent of individual animals) to the experimental condition.

## B. Mitoses

Smooth muscle cells in mitosis are present at all stages of hypertrophy and in both muscle layers, while they are absent from the control muscle (Figure 4, B to D). In the

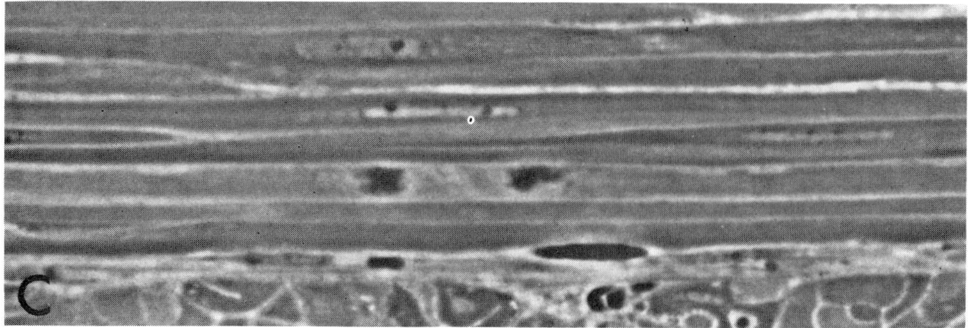

FIGURE 4C.

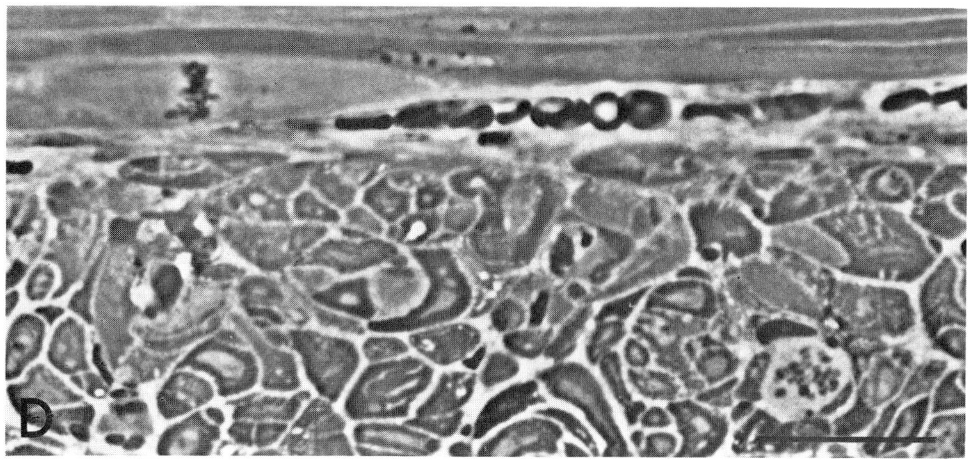

FIGURE 4D.

longitudinal muscle, dividing muscle cells are found only in the innermost half of the layer; in the circular muscle, dividing muscle cells are predominantly located in the middle portion of the layer, approximately at the same distance from the longitudinal muscle and the submucosa.

The muscle cells in prophase are somewhat longer than adjacent cells and their equatorial plate is invariably oriented transversely to the length of the muscle. The dividing muscle cell displays a central portion, ovoid, occupied by the chromosomes and at a later time by the mitotic spindle. On either side of this portion the cell appears as a fully differentiated muscle cell: it is packed with myofilaments, contains organelles such as mitochondria and sacroplasmic reticulum, and forms gap junctions and intermediate junctions with neighboring cells. In the hypertrophic gut, the dividing muscle cells are thus not undifferentiated cells or muscle cells that have dedifferentiated (at least not to the extent of losing their myofilaments or their specialized junctions), but they are mature cells that have retained their complement of myofilaments and their intercellular junctions.

The occurrence of mitosis in mature smooth muscle cells has already been reported in the literature. They were observed in the taenia coli of the guinea pig[27,28] and in the gastric musculature of the dog,[29] in the vicinity of a surgical cut; they were found in the hypertrophic musculature of the gallbladder of the guinea pig,[30] of the ureter of the dog,[31] and of the portal vein of the rat.[32]

Moreover, mitoses are found in the musculature of the chicken gizzard, around the time of hatching, when muscle cells, although still growing, are already structurally and functionally well differentiated,[33] and they are also observed in the aorta of young pigs.[34] It seems, therefore, that both during normal development and in adult animals in experimental conditions, the presence of myofilaments (and of cell junctions) is compatible with the occurrence of mitoses. This property is shared also by the developing cardiac muscle cells.[35,36,37,38] In hypertrophy, however, cardiac muscle cells do not undergo hyperplasia.[39,40,41] In contrast, in skeletal myoblasts no mitoses occur after the appearance of myofilaments; new nuclei appear in the growing muscle fiber by addition fom satellite cells.[42,43]

Since the growth in muscle cell size (cell hypertrophy) accounts for only part of the total increase in volume of the muscle, it is reasonable to assume that the remaining part of the increase is due to muscle cell divisions (hyperplasia); however, there have not yet been studies on the rate of cell division in the hypertrophic gut. Among the questions that remain open as regards mitosis of mature muscle cells, is that of the growth of the daughter cells. Soon after mitosis, the daughter cells are highly asymmetric, since one of their ends is "blunt" or truncated, and is close to the nucleus. It is not known how the subsequent growth and remodelling of the cell takes place.

It also remains to be established whether the number of muscle cells can increase only by occurrence of mitosis. Some hypertrophic muscle cells branch or are split along part of their length — as described in the next section — the split sometimes reaching the middle portion of the cell and its nucleus. Muscle cell profiles suggestive of a complete division of the cell along its length are not rare, so that the possibility of such an unusual form of cell division cannot be ruled out. The occurrence of amitotic divisions in smooth muscle cells has been reported and discussed in an early paper by Stohr.[44] For skeletal muscle it has been suggested that overloading (as by weight-lifting exercise) produces complete splitting of some fibers, whose number is therefore increased.[45]

It has been noted in previous investigations[23] and in the present work[25] that mitoses appear more commonly in the early stages of hypertrophy than in the more advanced stages. If this observation is confirmed by quantitative studies, it will help to interpret one of the most puzzling aspects of intestinal muscle hypertrophy, namely the occurrence of both cell hypertrophy and hyperplasia. In preliminary studies carried out on the urinary bladder of the rabbit,[46] it has been claimed that with a moderate obstruction of bladder voiding, there is hyperplasia followed by hypertrophy of muscle cells in adult subjects, whereas in growing subjects there is early hyperplasia followed by hypertrophy. These results are not yet confirmed by quantitation and the role played by age in the process of intestinal hypertrophy remains to be established. In the hypertrophic myometrium, the question of hypertrophy vs. hyperplasia has not yet been fully clarified. Here hypertrophy is the predominant process,[47] and according to some authors there is no hyperplasia at all.[48] There are, however, reports of mitoses among uterine muscle cells.[49,50,51]


## C. Cell Shape

Hypertrophic muscle cells have a much larger transverse sectional area than control muscle cells, an indication of their greater volume. They measure 500 to 750 $\mu$m in length (while in control they measure about 525 $\mu$m) and lie parallel to one another. In transverse sectioned muscles, the profiles of the hypertrophic muscle cells are also more irregular in shape than those of control cells (Figure 5); many are flattened, crescent-shaped, dumbell-shaped, or scalloped (Figures 4A, 4D, 6A). A number of muscle cells

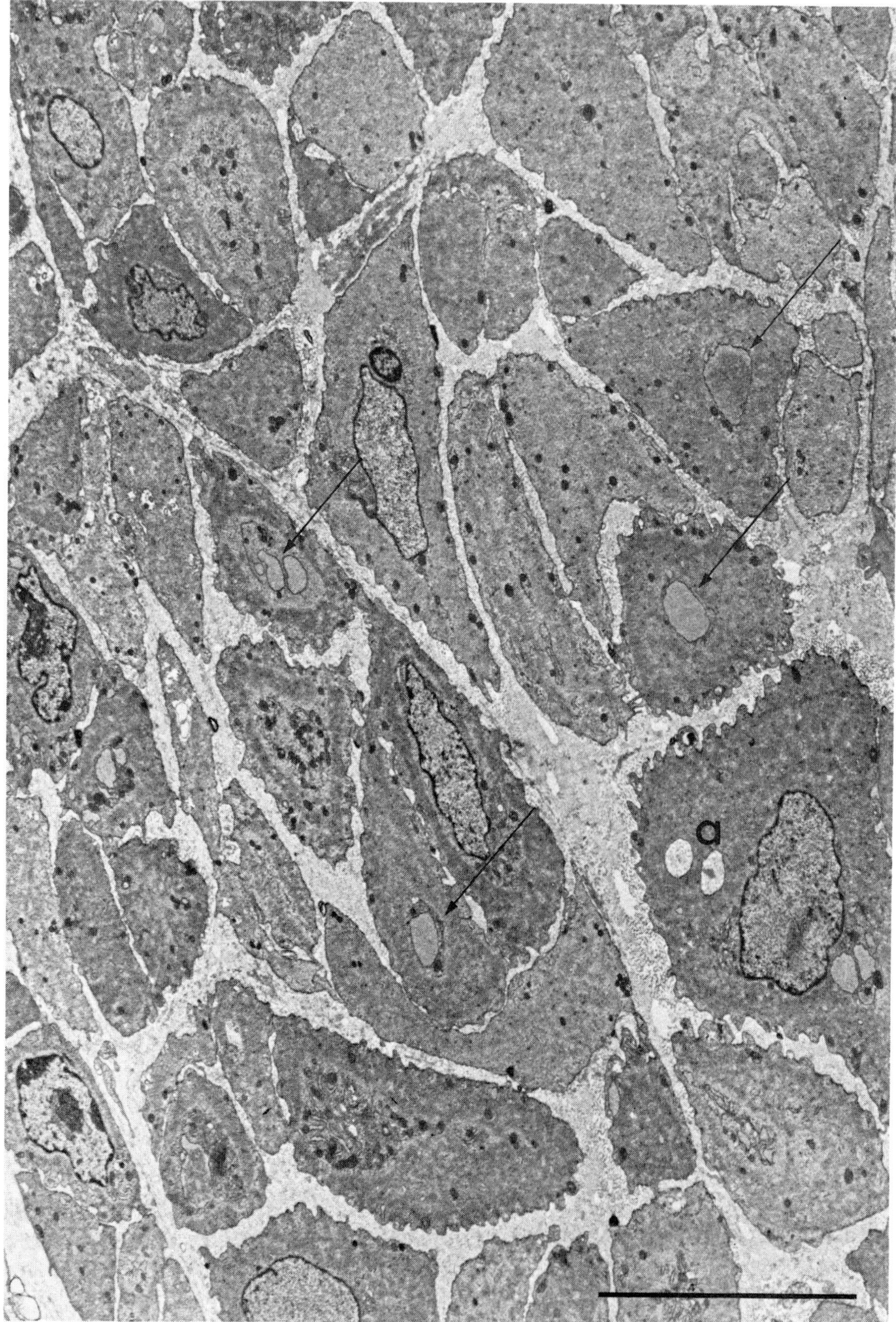

FIGURE 5.    Transverse section of the longitudinal muscle layer of a hypertrophic intestine. The micrograph illustrates the large size and the irregular shape of the hypertrophic muscle cells. Several cells have large vacuoles (arrows) originating from rough endoplasmic reticulum. a are two invaginations of the cell membrane tunnelling along the length of a muscle cell. Note the unusual appearance of the material in the intercellular space. Calibration bar: 10 μm. (From Gabella, G., Hypertrophic smooth muscle. Size and shape of cells, occurrence of mitoses. Sarcoplasmic reticulum, caveolae and mitochondria, *Cell Tissue Res.*, 201, 63, 1979. With permission.)

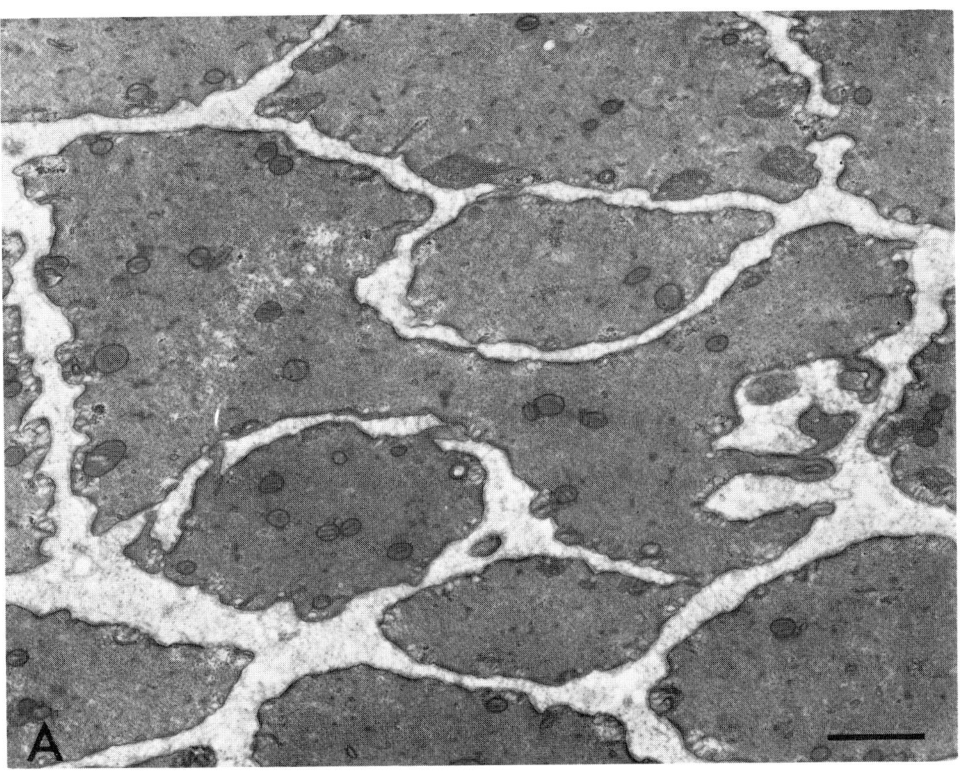

FIGURE 6. Transverse sections of the hypertrophic circular muscle. (A) Muscle cell profile deeply indented by adjacent cells. Calibration bar: 1 μm. (B) The cell at bottom, sectioned at the level of its nucleus, shows deep invaginations of the cell membrane, lined by a basal lamina and bearing caveolae and dense bands. Calibration bar: 1 μm.

are split longitudinally along part of their length into two or more "branches". These are usually of similar size and remain parallel to the long axis, but can become separated by interposition of another muscle cell, as can best be appreciated by looking at serial sections (Figure 4A). In the most advanced states of hypertrophy more than half of the muscle cells have undergone this process, and various patterns of longitudinal interdigitation arise between the branches of several cells. A similar phenomenon is known to occur in skeletal muscle fibers exposed to overload. Normal skeletal fibers are uniformly cylindrical; only about 1% of them (in the extensor digitorum of the rat[52]) show some form of branching. However, when a skeletal muscle is denervated or exposed to chronic overload, several of its fibers become split along the length.[53,54] It is possible that a similar mechanism is at work in the musculature of the hypertrophic intestine, i.e., an overload of the musculature (or indeed of the entire wall) which manifests itself in the increase of the circumference of the gut. Several observations suggest that in this condition the wall is exposed to intense mechanical stresses, including compression from the lumen by the accumulated ingesta, stretch along the length of the musculature, and reflex contractures arising from the abnormal stimulation of the wall. This aspect of the gut hypertrophy is probably best demonstrated in Figure 7, where a muscle cell profile within the circular layer continues into a transverse process passing between neighboring cells and, once having reached the longitudinal layer and turned at a right angle, runs longitudinally. A possible interpretation of this appearance is that this cell was originally located at the border between longitudinal and

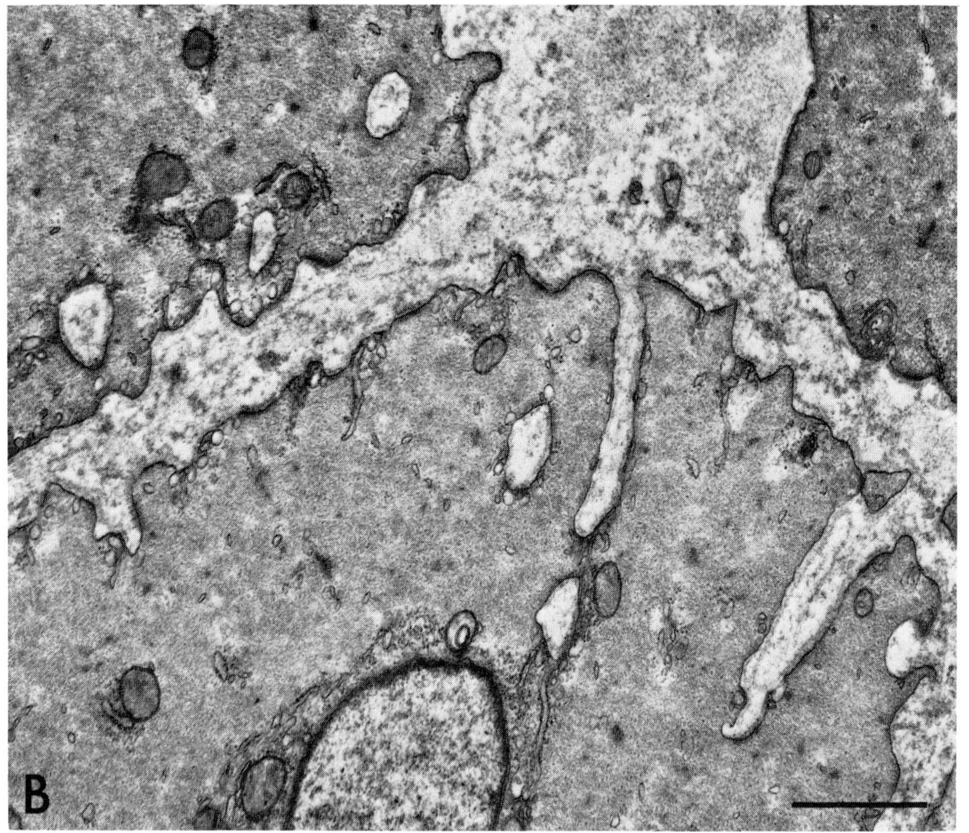

FIGURE 6B.

circular layers, running in either of them: growth of the tissue and mechanical distortion, such as compression and stretch, pulled a part of the cell with the longitudinal muscle and another part with the circular muscle.

Similar changes in cell shape and an increase in cell volume are found in the hypertrophic urinary bladder of the rat, examined 10 days after excision of the pelvic ganglion (and consequent urine retention). There is a 5-fold increase in the weight of the bladder, and muscle cells grow from 2200 $\mu$m³ to 3800 $\mu$m³, while cell length decreases from 325- to 226 $\mu$m.[55] However, one could argue that, since both control and hypertrophic bladders were distended prior to fixation with the same amount of fluid,[55] the hypertrophic muscle cells were not stretched by the same amount as the control cells; this procedure may explain the apparent shorter cell length and the higher value (0.50 vs. 0.30) of the ratio between the length of an unloaded and stimulated strip in vitro and its length *in situ*.[56]

## D. Cell Surface

The surface of hypertrophic muscle cells is more irregular and more corrugated than in control muscle cells. Deep invaginations of the cell membrane, running parallel to the cell length, are common in hypertrophic muscle cells. Some membrane infoldings form a long and narrow groove at the cell surface, others are long, cylindrical invaginations which may extend for tens of microns and reach the nuclear portion of the cell (Figure 6). The invaginations are usually related to a thickened basal lamina and to

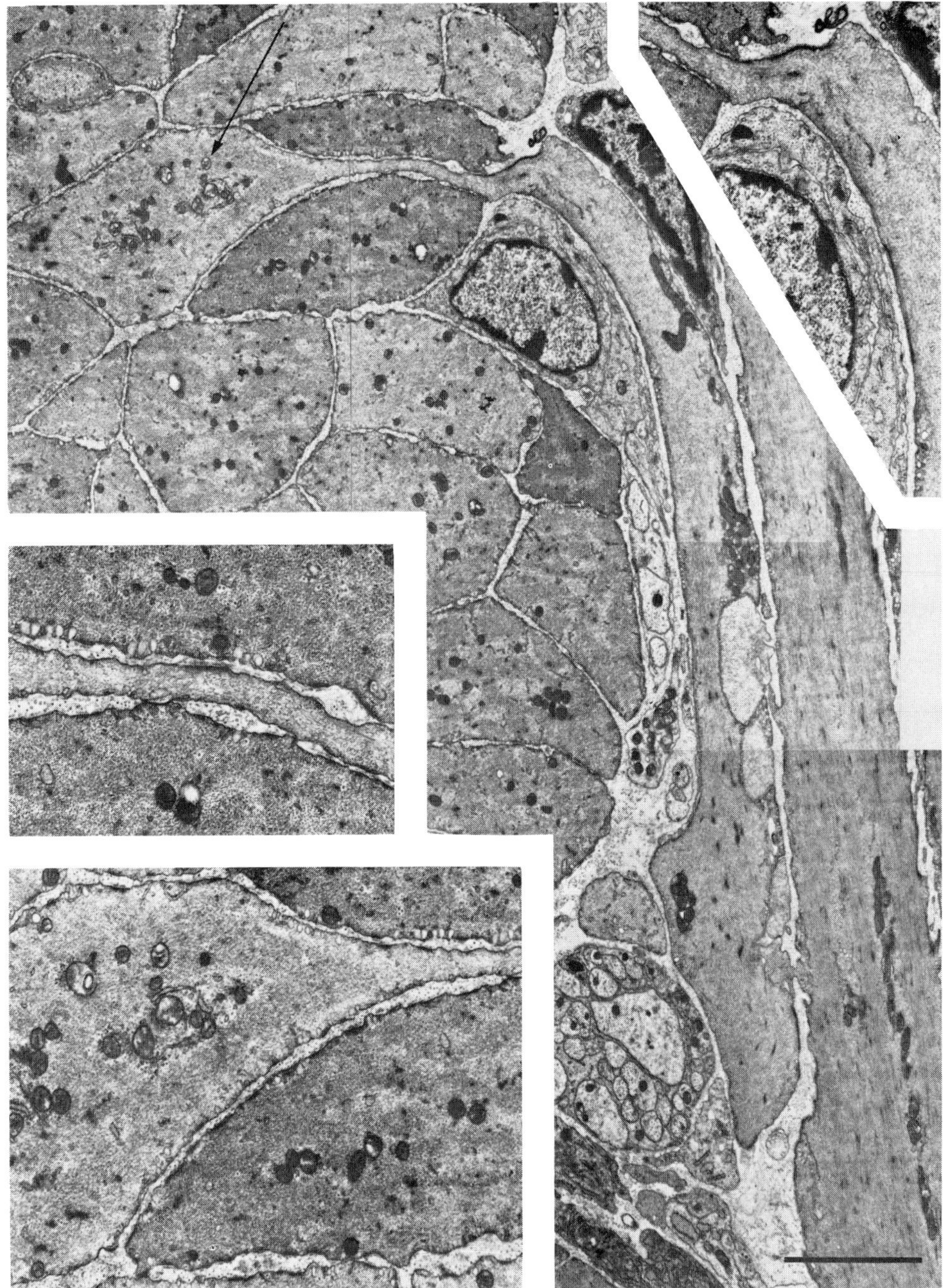

FIGURE 7.   Photographic montage of the circular muscle layer, in transverse section, and the longitudinal muscle layer, in longitudinal section, from the hypertrophic intestine of a rat. A muscle cell (arrow) is located in the circular layer and then, after acquiring an orthogonal orientation, extends into the longitudinal layer. The insets show details of the same cell at higher magnification. Calibration bar: 5 μm.

special stromal components (see Section V). Their membrane bears dense bodies which provide additional attachment sites for bundles of actin filaments. They also provide additional surface for metabolic exchanges.

Cell hypertrophy is accompanied by a fall in the surface to volume ratio, as, geometrically, the surface grows by the power of two-thirds of the volume. In normal muscle cells of the guinea pig ileum, the surface to volume ratio (square microns of cell surface per cubic micron of cell volume) is about 1.4. In hypertrophic muscle cells the ratio falls to 0.8,[25] a value that is not as low as expected. It is the presence of membrane infoldings and the splitting of muscle cells that prevent a fall of the ratio to the value of 0.5 that would be expected on the basis of the increase in cell volume. Obviously, there are limits to the permissible fall of the surface to volume ratio that accompanies a cell growth in size. It seems that a value of 0.8 is still sufficient to provide enough surface for the attachment of myofilaments and enough surface for metabolic exchanges. A further reduction would probably be incompatible with the survival of the cell.

## E. Cell Membrane and Caveolae

The cell membrane of the hypertrophic muscle cells, visualized in freeze-fracture preparations, contains a larger number of intramembrane particles both on the P-face and the E-face, when compared with control muscle cells (Figure 8A).[57] In many preparations their basal lamina appears thicker and more conspicuous than in controls.

Caveolae are flask-shaped invaginations of the cell membrane, mostly distributed in bands parallel to the cell length. They are similar in size and shape in control and hypertrophic muscle cells, and their packing density, measured in freeze-fracture preparations, shows a small decrease from $19.2/\mu m^2$ of cell surface in control muscle cells to $16.5/\mu m^2$ in hypertrophic muscle cells. However, on account of the increase of cell size, the total number of caveolae is substantially increased above the control value of about 170,000 per cell.

## F. Gap Junctions

Gap junctions (nexuses) are commonly found between muscle cells of the circular layer of the control intestine, while they are absent in the longitudinal layer. The same is true of the hypertrophic intestine (Figure 8B). In freeze-fracture preparations, gap junctions are recognized as clusters of intramembrane particles or connexons (on the P-face) and of intramembrane pits (on the E-face) (Figure 8C). The clusters have smooth and sharp outlines and are round or oval in shape: when elongated, their long axis is usually parallel to the length of the cell. Their size (expressed as surface area or as number of connexons) in hypertrophic muscle cells ranges from less than 0.01 $\mu m^2$ (equivalent to less than 60 particles) to over 0.3 $\mu m^2$ (equivalent to over 2000 particles), while in the control muscle they do not exceed 0.15 $\mu m^2$ (or about 1000 particles).[57] The spatial frequency of gap junctions over the cell surface is low and is similar in control and hypertrophic muscle cells (46 and 47 nexuses, respectively, per 1000 $\mu m^2$ of cell membrane). The total number of nexuses per cell is therefore 3 to 4 times greater in the experimental condition. The percentage of the cell surface occupied by nexuses also increases (from 0.22% to 0.49%), as accounted for by the increase of the average size of the nexuses. These values were obtained with morphometric study of freeze-fracture preparations and represent averages from large expanses of cell membrane; they may, however, hide variations from cell to cell. In the hypertrophic musculature there are also a few gap junctions between processes from the same cell[57] (reflexive gap junctions[58] (Figure 9B)). Reflexive gap junctions can hardly be formed in the control ileum where the muscle cells tend to have round or oval profiles. In the hypertrophic ileum, however, the muscle cells have a more irregular surface with numerous processes, a condition that must favor the formation of reflexive junctions. Junctions of this type have been observed in muscle cells of the human aorta in a case of congenital coarctation.[59]

These observations indicate that many new gap junctions are formed in a muscle cell

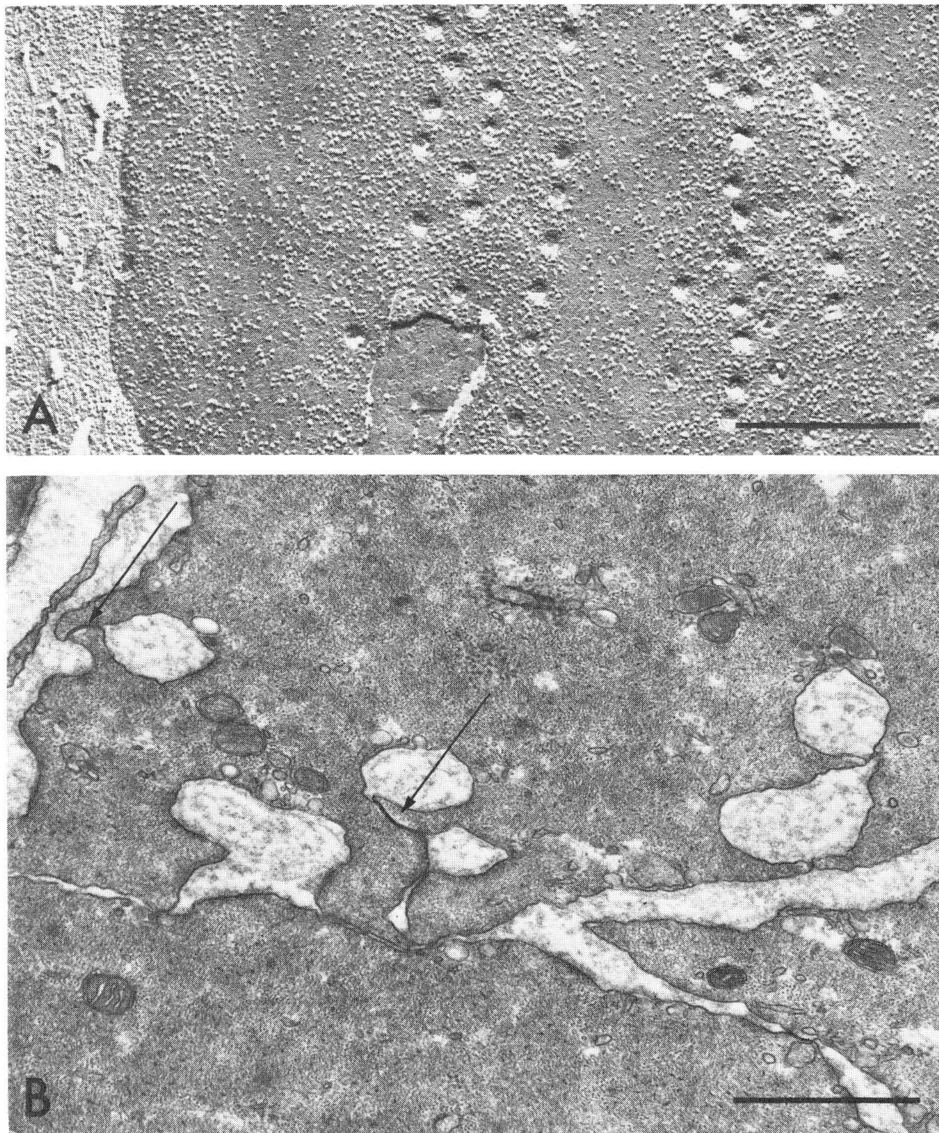

FIGURE 8. (A) Freeze-fracture preparation of the circular muscle layer of the hypertrophic intestine. The plasma membrane, exposed on its P-face, shows a greater number of intramembrane particles than muscle cells of controls. The small craters on the membrane are necks of caveolae. A discontinuity in the fracture face (bottom of center) exposes the membrane of an underlying cisterna of sarcoplasmic reticulum. To the left is the extracellular space with a few fractured collagen fibrils. Calibration bar: 0.5 $\mu$m. (B) Transverse section of the circular muscle layer of a hypertrophic intestine. The cell membrane is invaginated into deep grooves and tunnels. A gap junctions occurs between two muscle cells (bottom right). Two more gap junctions occur between processes from the same muscle cell (arrows). Calibration bar: 1 $\mu$m. (C) Freeze-fracture preparation of the circular muscle layer of a hypertrophic intestine, exposing the P-face of a muscle cell. The cell displays three large clusters of intramembrane particles, identified as gap junctions. Calibration bar: 0.5 $\mu$m.

during hypertrophy, and both the newly formed junctions and the preexisting ones grow to a size greater than in the control. By contrast, the longitudinal muscle, devoid of gap junctions in the control, remains so in hypertrophy. There are no data on the mechanism by which these junctions are formed and are enlarged. Other examples of

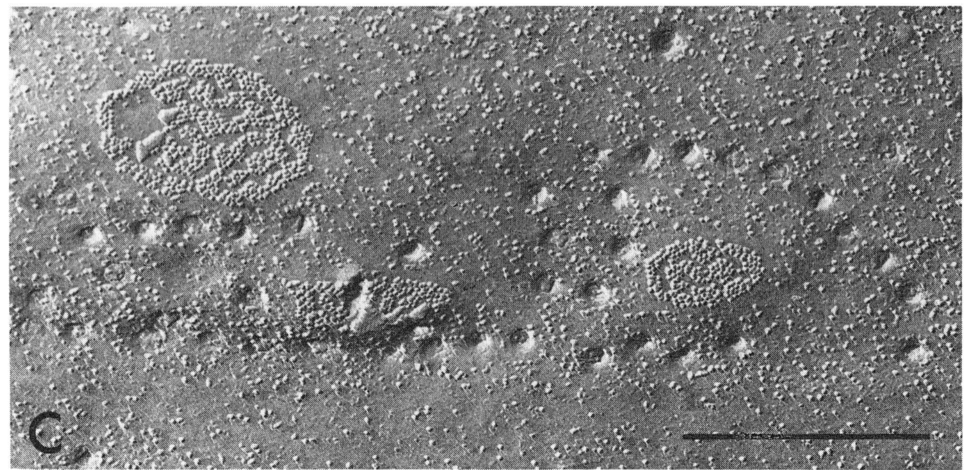

FIGURE 8C.

increase in size of the gap junctions are known — e.g., the granulosa cells of the ovary during hormonal stimulation[60] or embryonic epithelial cells stimulated in vitro with vitamin A.[61] The formation of new gap junctions (and their disappearance) can be a relatively rapid process. For example, new junctions are formed in the rat myometrium a few days before parturition and they disappear after delivery.[62] Yee and Revel[63] have shown that 28 hr after partial hepatectomy in the rat, gap junctions are absent in the surviving hepatic cells, but they are fully reformed by the 48th hour. An even more rapid increase in the number of gap junctions has been described in the canine tracheal muscle stimulated in vitro with potassium conductance blockers.[64] More relevant to the present experiments is the increase in the number of gap junctions in the vas deferens of the rat after post-ganglionic denervation.[65] There may be a similarity between the two conditions in that the hypertrophic intestinal musculature, although not denervated, has a reduced density of innervation (see Section VII).

## G. Mitochondria

The percentage volume occupied by mitochondria is about 6% in control muscle cells and it decreases to an average of 2.9% (with a range of 1.7 to 4.1% in five experiments) in the hypertrophic muscle cells. In spite of the halving of the mitochondrial percentage volume, a hypertrophic muscle cell contains on average almost twice as much total mitochondrial volume as a control muscle cell.

## H. Sarcoplasmic Reticulum

Hypertrophic muscle cells show a conspicuous increase in the amount of sarcoplasmic reticulum.[25] This organelle appears in different forms, which probably represent different functional specializations and which seem to predominate in different cells. The small cisternae and tubules of sarcoplasmic reticulum in control muscle cells are localized mainly underneath the cell membrane and the caveolae. In all hypertrophic muscle cells, but in some more than in others, there are large numbers of small tubules of reticulum scattered through the sarcoplasm (Figure 9A). This reticulum has been implicated with calcium storage and release;[66] however, the significance of its increase in hypertrophic muscle cells has not yet been explored.

Cisternae of rough endoplasmic reticulum are usually found in each hypertrophic cell profile, and are thus far more abundant than in control muscle cells (Figure 9B).

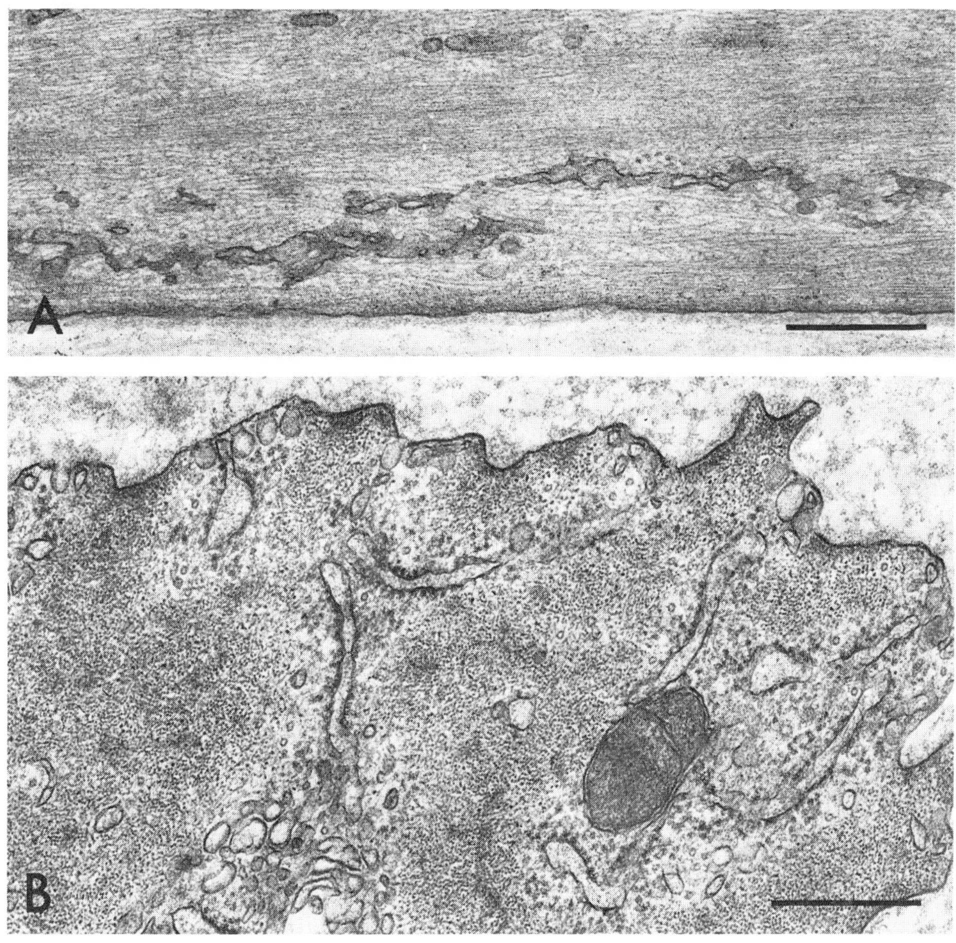

FIGURE 9.  (A) Longitudinal section of a hypertrophic muscle cell of the circular layer, showing fenestrated cisternae of sarcoplasmic reticulum of great extension and complex shape. Calibration bar: 0.5 $\mu$m. (B) Transverse section of a hypertrophic muscle cell of the longitudinal layer, showing numerous cisternae of rough sarcoplasmic reticulum. Calibration bar: 0.5 $\mu$m.

Many cisternae are radially arranged and reach with one edge the cell membrane. Some cell profiles display numerous cisternae of rough reticulum as if they were in an intense phase of synthesis and secretion. In other muscle cells (especially in the longitudinal layer) there are grossly swollen cisternae of reticulum with a fine granular content of medium electron density.

## J. Other Organelles

These include ribsomes which are probably increased over the control. Lyosomes found in hypertrophic muscle cells do not appear more common than in the normal tissue. It is common in the hypertrophic musculature to find large aggregates of glycogen granules.

## K. Intermediate Filaments

Intermediate filaments measure 10 nm in diameter and in visceral muscle cells are made of a 50,000 to 55,000 dalton protein called desmin.[67] In muscle cells of the control ileum, intermediate filaments are found around dense bodies, close to some dense bands and in small bundles between the myofilaments. In hypertrophic muscle cells intermediate filaments become more conspicuous; their number per unit sectional area is markedly increased at the expense of the relative number of myofilaments (Figure 10A). Intermediate filaments abound in the proximity of dense bodies and close to dense bands, and large bundles of them are found over the entire cell profile, and especially in the central portion of the cell. The amount of intermediate filaments varies between different cells in the same hypertrophic muscle and probably along the length of individual cells as well; their relative frequency, however, is always markedly greater than in control muscle cells. Similar changes have been observed in a hypertrophic vascular smooth muscle in the portal anterior mesenteric vein of the rabbit.[68] In this muscle, the ratio of actin to intermediate filament is 15:1.1 in control cells and 15:3.5 in the hypertrophic cells. In absolute terms, the figures show that the number of intermediate filaments grows by more than three times as much as that of actin filaments.[68]

It is known from the literature that intermediate filaments increase in muscle cells explanted and grown in vitro[69] and are generally more abundant in developing muscles.[70,71] Therefore, the increase in the number of these filaments in hypertrophic muscles may not represent a specific aspect of the process of hypertrophy. It is, however, a most prominent structural change, and one that must affect the mechanical properties of the muscle. There have not yet been studies on the chemical composition of the newly formed intermediate filaments. The assumption that they are made of desmin as the preexisting filaments is not yet warranted.

## L. Myofilaments

Thick (myosin) and thin (actin) filaments are readily recognized in hypertrophic muscle cells and they are not appreciably different in appearance from those in control muscle cells. Myosin filaments measure about 15 nm in diameter and have a rather irregular profile. They are sensitive to preparative procedure and easily disrupted, displaced, or destroyed. Actin filaments measure 7 to 8 nm in diameter and are usually aggregated in bundles or cables of a few tens of filaments: a hexagonal pattern is often apparent in the packing of filaments within a bundle. By contrast with the intermediate filaments, the relative frequency of actin and myosin filaments decreases with hypertrophy, although their total number per cell is greatly increased. Berner et al.[68] in their study of the hypertrophic portal vein of the rabbit, report that the number of actin filaments per cell profile increases in proportion to the increase in cell volume, whereas the number of myosin filaments per cell profile remains approximately the same as in the control. The actin to myosin filament ratio increases from 1:15 to 1:30,[68] a result implying that the hypertrophic vascular muscle cells synthesize new actin but not new myosin. Counts of myofilaments have not yet been carried out in the hypertrophic musculature of the gut, and there are no data on possible biochemical changes in the myofilament proteins.

The measurements of maximal force generated by the muscle are in accordance with these observations. As one would expect from the decrease of the number of myofilaments per unit sectional area, the force generated by the hypertrophic muscle, when expressed per unit sectional area of the tissue, is lower than in the control muscle.[57] The reduction affects the longitudinal and the circular musculature to a similar extent.[57] Although the total force generated by the hypertrophic muscle is much greater than in the control, on account of its greater mass, the "efficiency" of the hypertrophic muscle is somewhat reduced. This fall in efficiency is in agreement with the studies of Johansson[72] who finds that the longitudinal force developed by the rat portal vein increases from 10 to 25 mN during hypertrophy; the force developed by 1 cm² of

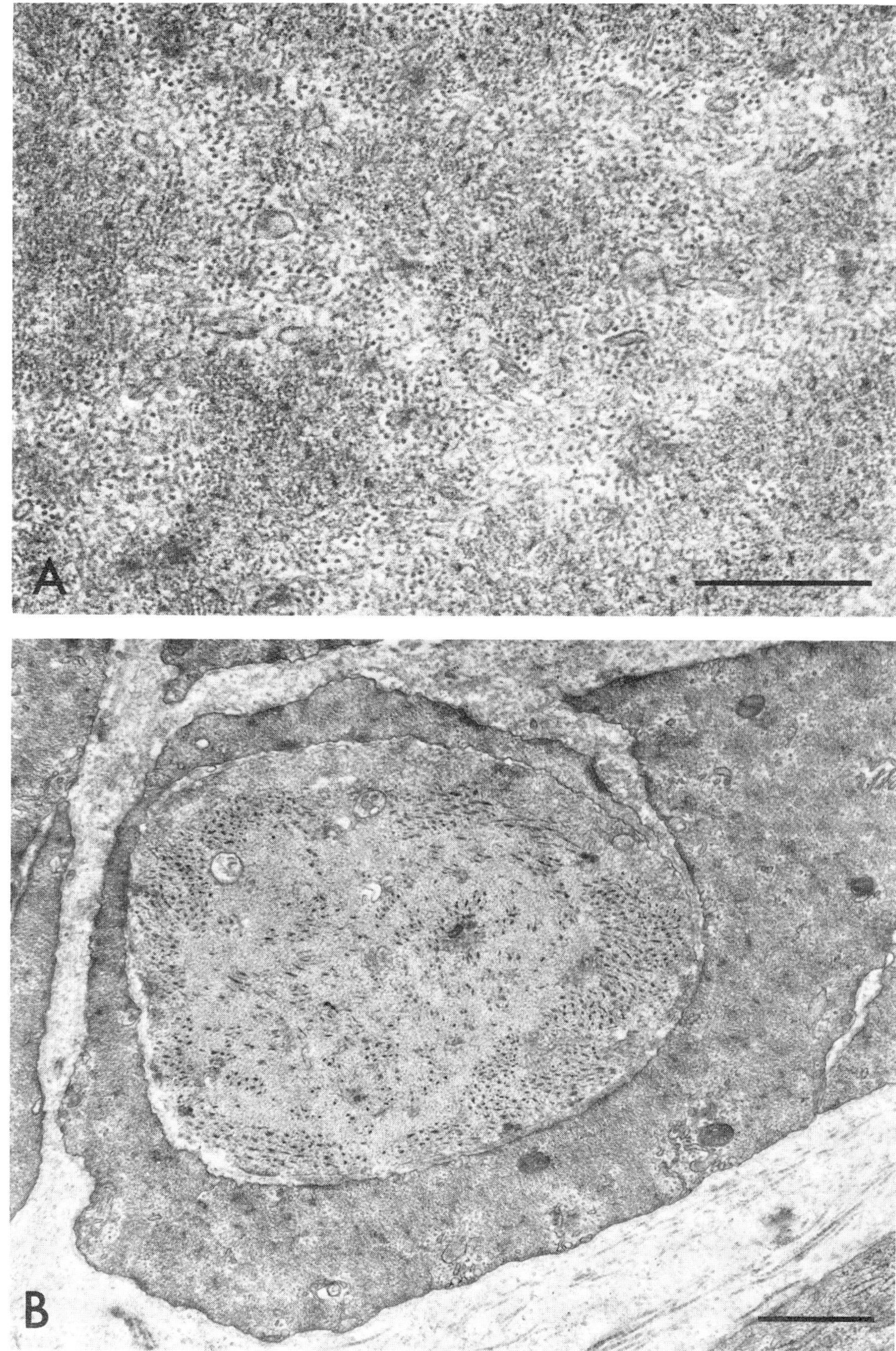

FIGURE 10.   (A) Transverse section of a hypertrophic muscle cell of the circular layer. In addition to thin and thick filaments, there are conspicuous areas occupied by bundles of intermediate filaments. These areas are of lighter electron density than those occupied by myofilaments. Calibration bar: 1 $\mu$m. (B) Transverse section of the longitudinal muscle layer of a hypertrophic intestine. A muscle cell profile, with an incomplete cell membrane and filaments of abnormal appearance, and partly engulfed by an adjacent muscle cell, is interpreted as a degenerating muscle cell. Calibration bar: 1 $\mu$m.

muscle layer, however, falls from 9 to 5 N. Johansson[72] suggests that the decrease in active force may be due to (1) impairment of the contractile elements due to distension, (2) immaturity of the proliferating muscle cells, and (3) a possible reorientation of the longitudinal cells in a more circular direction. Some of these factors may be at work also in the hypertrophic intestine, especially the extreme distension of the circular musculature. In addition, the relative reduction of myofilaments must also play a major role. Another important factor bearing on the mechanical properties of the hypertrophic musculature is the substantial change in the composition of the stroma (see Section V).

### M. Degenerating Muscle Cells

A small number of degenerating muscle cells are found in both layers of the hypertrophic musculature.[57] In these cells most of the organelles have disappeared, whereas the filaments remain clearly visible. Characteristically, the cell membrane is discontinuous or missing altogether (Figure 10B).

## V. INTERCELLULAR MATERIAL

In the control intestine a small number of collagen fibrils are found around the muscle cells. These fibrils measure about 35 nm in diameter and have a cross-banding of about 60 nm period. They approach the basal lamina and, although they do not penetrate it, they are linked to it by a fine web of microfibrils. There are only very few elastic fibers.

In the hypertrophic musculature the collagen fibrils are similar to those in controls in size and banding; however, they are more dispersed and are surrounded by an amorphous material of medium electron density which pervades the spaces between muscle cells[73] (Figure 6B). Bundles of 10 to 20 microfibrils of 10 nm diameter are also common. In some preparations, there is an increase in the relative number of elastic fibers; these are often increased over the controls in the gap between longitudinal and circular muscle layers, and especially around the myenteric ganglia. In addition, there is a large amount of electron-dense extracellular material in the form of streaks parallel to the cell axis and lamellae that are parallel to or radiate from the basal lamina. The chemical nature of this material, which is linked to the basal laminae and resembles them in texture, is unknown.

The concentration of collagen (measured as hydroxyproline concentration and expressed in mg/g of wet weight) undergoes a significant increase during hypertrophy. The increase affects the whole wall, but it is particularly noticeable in the muscle layer.[74]

The increase in the collagen concentration within the musculature, added to the large increase in the mass of the muscle, indicates that synthesis of collagen on a large scale occurs during hypertrophy. There are no data on the turnover of collagen in the intestinal musculature of the control or the experimental animals. An increase in the concentration of collagen may be due to an increased rate of synthesis as well as to a decrease in the rate of degradation. It has recently been shown that in a skeletal muscle (the anterior latissimus dorsi of the chicken), stretch-induced hypertrophy is accompanied by increase in collagen concentration — this is due to a 5-fold increase in the rate of collagen synthesis and to a 60% decrease in the rate of degradation of newly synthesized collagen.[75] Increased synthesis and deposition of collagen is also found in arteries of hypertensive rats,[76-79] and in the pregnant myometrium.[80,81]

As to the source of the newly-synthesized collagen, the structural evidence suggests that it is the muscle cells themselves. The intramuscular fibroblasts may be involved in the process but they are very few in number and show little sign of activation in the hypertrophic musculature, and they are not likely to make an important contribution. The consistent presence of rough endoplasmic reticulum in hypertrophic muscle cells and the apparent continuity of their basal lamina and fibrillar elements among the cells, supports the view that the muscle cells are involved in the synthesis of extracellular material.[73,74] There is good evidence that vascular smooth muscle cells in culture synthesize collagen.[82-84] This collagen is mainly of type I and III (the latter being predominant[85,86]) and it is therefore similar to that present in arteries *in situ*.[87-89] The evidence that a similar process occurs *in situ* or in the nonvascular smooth muscles is less compelling; however, Ross and Klebanoff[90] have shown that muscle cells of the developing aorta and of the estrogen-stimulated myometrium synthesize and secrete connective tissue proteins in vivo. Increased synthesis and deposition of collagen occurs in arteries of hypertensive rats[76-79] and the muscle cells of the media are probably involved in this process. The studies on the myometrium are particularly interesting because it has been shown that mechanical factors (notably stretch) are the primary stimulus for muscle hypertrophy[48,91,92] and for collagen synthesis.[93] Both processes can be stimulated by introducing an inert bolus into the lumen of the uterus. Moreover, Leung et al.[94] have grown vascular muscle cells in vitro over an elastic membrane and have found that cyclic stretching of the latter (repeated elongation and relaxation) results in increased rates of collagen synthesis over those of quiescent cultures or of cultures that are agitated but not stretched.

## VI. VASCULARIZATION

The circular muscle layer of the control ileum is well supplied with blood vessels, mainly blood capillaries. Up to 1,000 blood vessels are found in a square mm of cross-sectioned musculature, depending on the degree of contraction. In contrast, blood vessels are usually not present in the longitudinal muscle layer.

In the hypertrophic gut the density of intramuscular blood vessels is somewhat reduced (to 250 to 500/mm² in the maximally hypertrophic musculature). However, on account of the increase in muscle volume, the number of blood vessels is greatly increased.[73] No data are available on the length of these intramuscular vessels, and their absolute number is therefore unknown. Some of the intramuscular capillaries in the hypertrophic gut are fenestrated. In spite of its increase in thickness, the hypertrophic longitudinal muscle remains devoid of blood vessels. The serosa is more richly vascularized than in the controls.

## VII. INTRAMUSCULAR NERVES

In the unoperated control intestine the intramuscular nerve fibers are distributed in a characteristic pattern.[95,96] Nerve bundles run almost parallel to the muscle bundles and are present throughout the circular muscle layer; they are, however, more abundant in its innermost part in the proximity of a layer of small and dark muscle cells at the border with the submucosa than in other portions. All the axons are unmyelinated. Individual axons, as opposed to bundles, are found only very rarely.

In the hypertrophic intestine, intramuscular nerve bundles within the circular muscle layer are more sparsely distributed than in the control tissue — the more advanced the condition of hypertrophy, the rarer their occurrence (Figure 11). This results in a marked decrease in the density of innervation of the muscle; however, the nerve bundles have a distribution similar to that in controls, i.e. they are mainly located near the innermost cells of the circular layer (Figure 11). Those located in the bulk of the cir-

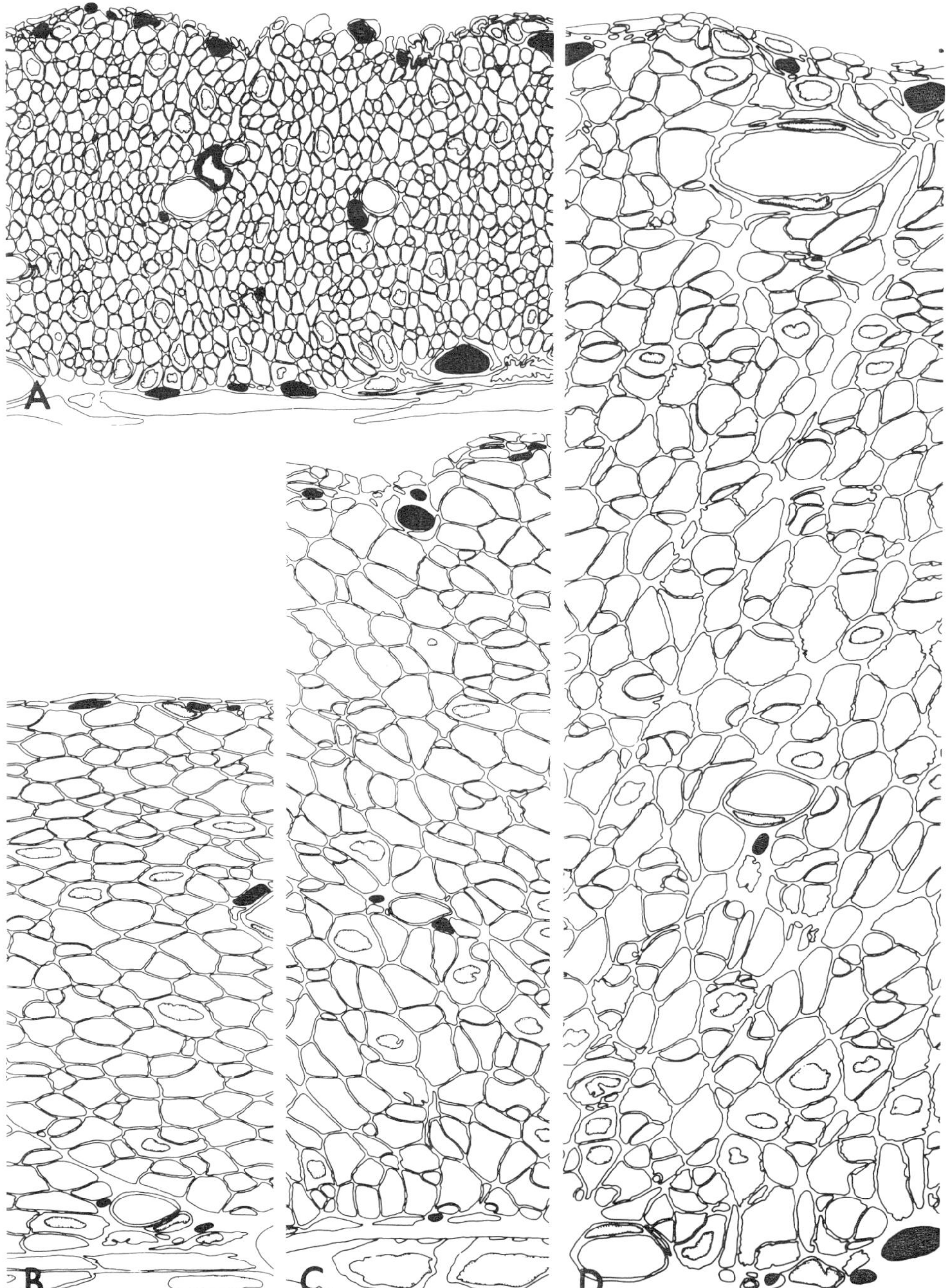

FIGURE 11.   Distribution of nerve bundles in control and hypertrophic circular muscle of the guinea pig. Transverse sections of the circular muscle layer of a control ileum (A) and of three segments of the ileum from a stenosis experiment, taken 25 cm (B), 15 cm (C), and 5 cm (D) orad to the obstruction. (D) is the segment in which the hypertrophy was maximal. Drawings traced from photographic montages originally at 9,000 × then reduced xerographically to 1,700 ×. All at the same magnification. They are selected areas of the montages used for Table 1. Each drawing displays the full thickness of the circular muscle layer with the boundary to the submucosa at the top and the boundary to the longitudinal muscle at the bottom. Muscle cell (some nucleated), interstitial cells, blood vessels, and nerve bundles are outlined. For sake of clarity the nerve bundles have been blocked in. A ganglion of the myenteric plexus with two neurons is partially visible in (C) (bottom). Note the differences in sizes and shapes of the muscle cell profiles. A single array of muscle cells smaller than those of the bulk of the layer is visible at the submucosal surface of the circular muscle (top in all the drawings).

cular musculature are often perivascular; however, while intramuscular blood vessels (mainly capillaries) are plentiful in the hypertrophic circular musculature, (see Section VI), many of them are not accompanied by a nerve bundle — a rather unusual occurrence in the control intestine. All the axons are unmyelinated. Individual axons are very rare, and by comparison with the control intestine, large bundles appear to predominate over small ones. These observations are confirmed by morphometric studies of electron microscopic montages of the circular muscle layer, such as the one reported in Table 1. The increase in the diameter of the intestine orad to the stenosis is associated with an increase in the thickness of the wall and with a reduction in the relative number of nerve bundles and axons (number of nerves per unit sectional area of circular musculature). There is also an increase in the average number of axons per nerve bundle, from 21 in the control to about 50 in the maximally hypertrophic muscle, with intermediate values in the more moderately hypertrophic muscles. With the increase in the average number of axons per nerve bundle, there is also an increase in the average transverse sectional area of the bundles. The counts show that in the hypertrophic circular musculature, there is a dramatic decrease in the number of nerve bundles per unit transverse sectional area by comparison with the control muscle (notice, however, how the decrease in the relative number of axons is less dramatic than the decrease in the relative number of nerve bundles).

A different approach consists in showing the density of innervation per unit "width" of circular musculature (e.g., per 100 $\mu$m). A unit width of circular muscle corresponds to a unit length of the intestine and is comparable in the control and the hypertrophic intestine, because the length of the intestine is not altered during hypertrophy; of course, that amount of musculature present in such a sectional unit is greater in the hypertrophic intestine because the wall is thicker. The data are shown in the last two lines of Table 1. With this approach, the number of nerve bundles is reduced in the hypertrophic intestine, whereas there is only a marginal reduction in the number of axons.

The longitudinal musculature is usually devoid of nerves both in the control and in the hypertrophic intestine. In spite of the increase in thickness of this layer, the nearest axons likely to influence its muscle cells are located in the myenteric plexus or in nerve bundles running in the interstice between circular and longitudinal muscle.

The main conclusions suggested by this study are as follows:

1. There is no loss of nerve bundles when a segment of musculature hypertrophies.
2. Most of the axons succeed in growing in length by the same extent as the musculature, so that in a projection as in Figure 11 there appears to be no significant loss of axons per unit width of muscle.
3. There is a rearrangement of the branching pattern of the intramuscular nerve bundles and/or a different growth of nerve bundles of different size. As a result of this in the hypertrophic musculature, large nerve bundles become relatively more frequent and small bundles less frequent by comparison with the control.
4. The absolute density of innervation (as expressed, e.g., by the number of varicosities per number of muscle cells) is unknown for either the control or the hypertrophic intestine. It is clear, however, that the hypertrophic growth of the musculature far exceeds that of the nerves, and there is as a result a reduction in the density of innervation: fewer muscle cells are in contact with axons in the hypertrophic intestine than in the control.
5. The pattern of innervation is unchanged: most of the nerve bundles are perivascular, they are most abundant in the innermost portion of the circular layer, and they are absent in the longitudinal muscle.

Table 1
QUANTITATIVE DATA ON THE INNERVATION OF A HYPERTROPHIC AND A CONTROL INTESTINE

| Hypertrophy[a] | A (max) | B | C | D (min) | Control |
|---|---|---|---|---|---|
| Diameter of intestine[b] | 12 mm | 11 mm | 10 mm | 8.5 mm | 5.5 mm |
| Thickness of circular muscle[c] | 180 $\mu$m | 164 $\mu$m | 113 $\mu$m | 86 $\mu$m | 50 $\mu$m |
| Montage | | | | | |
| Area of circular muscle studied[d] | 13,000 $\mu$m$^2$ | 15,300 $\mu$m$^2$ | 8,100 $\mu$m$^2$ | 7,200 $\mu$m$^2$ | 9,000 $\mu$m$^2$ |
| Number of nerves[e] | 6 | 11 | 11 | 9 | 40 |
| Number of axons[e] | 297 | 382 | 288 | 243 | 858 |
| Average area of nerves[f] | 11.2 $\mu$m$^2$ | 9.5 $\mu$m$^2$ | 7.3 $\mu$m$^2$ | 5.3 $\mu$m$^2$ | 4.5 $\mu$m$^2$ |
| Average number of axons per nerve[f] | 49.5 | 34.7 | 24.5 | 26.9 | 21.5 |
| Number of nerves in a 100-$\mu$m width of circular muscle | 7.7 | 11.8 | 14.8 | 13.0 | 21.6 |
| Number of axons in a 100-$\mu$m width of circular muscle[g] | 381 | 411 | 365 | 352 | 464 |

[a]   All the specimens (A to D) from the same animal, at increasing distance oral to the obstruction (approximately 5, 10, 15, and 25 cm).

[b]   Measured on the embedded specimens.

[c]   Measured on the histological sections.

[d]   Selected parts of four montages (A, C, D, and control) are shown in Figure 11.

[e]   Total number of nerve bundles and axons in the area surveyed.

[f]   Additional nerve bundles lying outside the area of the montage were also used to calculate these figures.

Table 2

## RELATIVE VOLUME OF MYENTERIC PLEXUS IN A HYPERTROPHIC AND TWO CONTROL INTESTINES

| | Diameter of intestine (mm)[b] | Thickness of muscle coat ($\mu$m)[c] | Volume ratio of myenteric plexus and muscle coat | Sectional area examined ($\mu$m$^2$) |
|---|---|---|---|---|
| | | Hypertrophy[a] | | |
| A | 15.5 | 170 | 0.61:100 | $2.7 \times 10^6$ |
| B | 13.5 | 147 | 0.97:100 | $4.0 \times 10^6$ |
| C | 11.5 | 120 | 1.22:100 | $1.5 \times 10^6$ |
| D | 11.0 | 110 | 1.26:100 | $1.1 \times 10^6$ |
| E | 10.0 | 68 | 2.08:100 | $0.5 \times 10^6$ |
| F | 9.5 | 59 | 1.68:100 | $0.7 \times 10^6$ |
| G | 7.5 | 46 | 2.11:100 | $0.2 \times 10^6$ |
| | | Controls | | |
| 1 | 5.5 | 50 | 3.10:100 | $1.0 \times 10^6$ |
| 2 | 7.0 | 40 | 3.86:100 | $1.6 \times 10^6$ |

[a] All the hypertrophic specimens (A to G) are from the same animal, at increasing distance oral to the stenosis (approximately 6 cm apart from one another). Different animal from that used for Table 1.

[b] Measured on the embedded specimens.

[c] Thickness of both muscle layers, measured on the histological sections.

Other studies have shown that the concentration of choline acetyltransferase, an enzyme localized in neuronal elements, remains unchanged in the hypertrophic musculature of the rabbit ileum.[97] This result does not contradict the present observations, since we observe both a decrease in the density of intramuscular innervation and a hypertrophy of the enteric ganglia (see next section). Other authors, however, have reported a 50% decrease in catecholamine concentration and in the spatial density of adrenergic fibers in the hypertrophic rat ileum (20 days after a 50% enterectomy).[98]

## VIII. NERVE GANGLIA

The intramural ganglia of the intestine are assembled in two ganglionated plexuses, the myenteric and the submucosal plexuses, which lie parallel and close to the inner and the outer surface of the circular muscle layer. The intramural ganglia undergo major structural changes in the hypertrophic intestine.[99] In the ileum of control guinea pigs, the volume of the myenteric plexus as a whole, measured on histological sections, is 3 to 4% of the volume of the muscle coat (Table 2). In the maximally hypertrophic intestine, the volume of the plexus is only about 0.6% that of the much enlarged musculature; in absolute terms, however, this value still represents a substantial increase in volume of the plexus in an intestinal segment, even if the neuronal growth is not as intense as that of the musculature (10-fold or more). At all degrees of hypertrophy — which are available in each experiment along the length of the small intestine — the myenteric plexus does increase in total volume; however, it clearly does not keep pace with the growth of the musculature (Table 2).

The ganglia of the myenteric plexus are firmly anchored to the muscle layers by the surrounding connective tissue; during muscle contraction, the ganglia are altered in shape and dimensions.[99] With the long-term change in the musculature due to hypertrophy, the myenteric ganglia remain anchored to the musculature, and therefore become extensively stretched along the long axis of the circular muscle (increase in circumference of the intestine), while the separation of ganglia along the length of the gut is not appreciably changed. The general pattern of the plexus remains similar, except for being considerably stretched along the transverse axis of the intestine (this axis is also the long axis of most of the individual ganglia). A similar deformation is noted in the submucosal plexus.

Previous studies have shown that in the hypertrophic intestine orad to a partial obstruction in rats,[100] dogs,[20,22,23] and rabbits,[97] the myenteric neurons undergo hypertrophy, i.e., there is an increase in the size of nerve cell bodies (in all these studies the size measurements were actually done on nerve cell nuclei). Filogamo[101] described an increase in size also in the neurons of the submucosal plexus. As to the more contentious question of a possible increase in number of neurons, Benninghoff[100] was convinced it did occur, and he reports, in agreement with an early study of Miura,[102] that there is both hypertrophy and hyperplasia of myenteric neurons. Filogamo and Vigliani[20] counted an increased number of neurons in the hypertrophic ileum, but they attributed the change to the fact that some very small neurons were not adequately stained in the control intestine but became visible during hypertrophy, a suggestion that had first been put forward by Ambrosi.[103] Other authors have since confirmed the absence of neuronal hyperplasia,[21,22] and Okada and Okamoto[22] noted an increase in the affinity of neurons of the hypertrophic intestine for silver.

Recent studies on myenteric and submucosal ganglia of the guinea pig hypertrophic intestine have confirmed the occurrence of neuronal hypertrophy.[99] By measuring neuronal cell body sizes (rather than nucleus sizes) in standard conditions of wall distension and orientation, it becomes clear that the neuronal hypertrophy is extremely intense (Figure 12). The average area of nerve cell outlines is twice as large as in controls. The entire population of neurons, which is very heterogeneous in terms of cell sizes, grows and the histogram of cell sizes is progressively shifted to the left. Nearly half of the neuronal profiles measure over 500 $\mu m^2$ (compared with less the 3% in the controls), and less than 1% measure less than 150 $\mu m^2$ (compared with 21% in the controls).

The glial cells of the enteric ganglia are also affected in the hypertrophic intestine. Their relative number is unchanged or slightly decreased in the myenteric ganglia. In contrast, in the submucosal ganglia the relative number of glial cells is nearly doubled over the control value. Mitoses of glial cells were not observed within the ganglia in the present study and have not been reported by previous investigators. The origin of the extra number of glial cells in the submucosa ganglia is, therefore, unknown.

In the hypertrophic ganglia there is also a small, but consistent increase in the size of the glial cell nuclei. Since the size of a glial nucleus is considered to reflect the size of the whole cell, the observations point to an increase in glial cell volume concomitant with the neuronal hypertrophy.

## IX. SUMMARY

A synopsis of experiments on the hypertrophic response of the small intestine orad to a partial obstruction of the lumen has been presented. The experimental condition studied is a particular case of hypertrophy and it does not represent a general model for the many known examples of hypertrophic smooth muscles. The stenosis imposes an overload on the musculature of a part of the intestine, and the hypertrophy partially overcomes it. In addition, another important factor at work is the distension of the wall due to accumulation of ingesta: the circular muscle, especially, is stretched to the

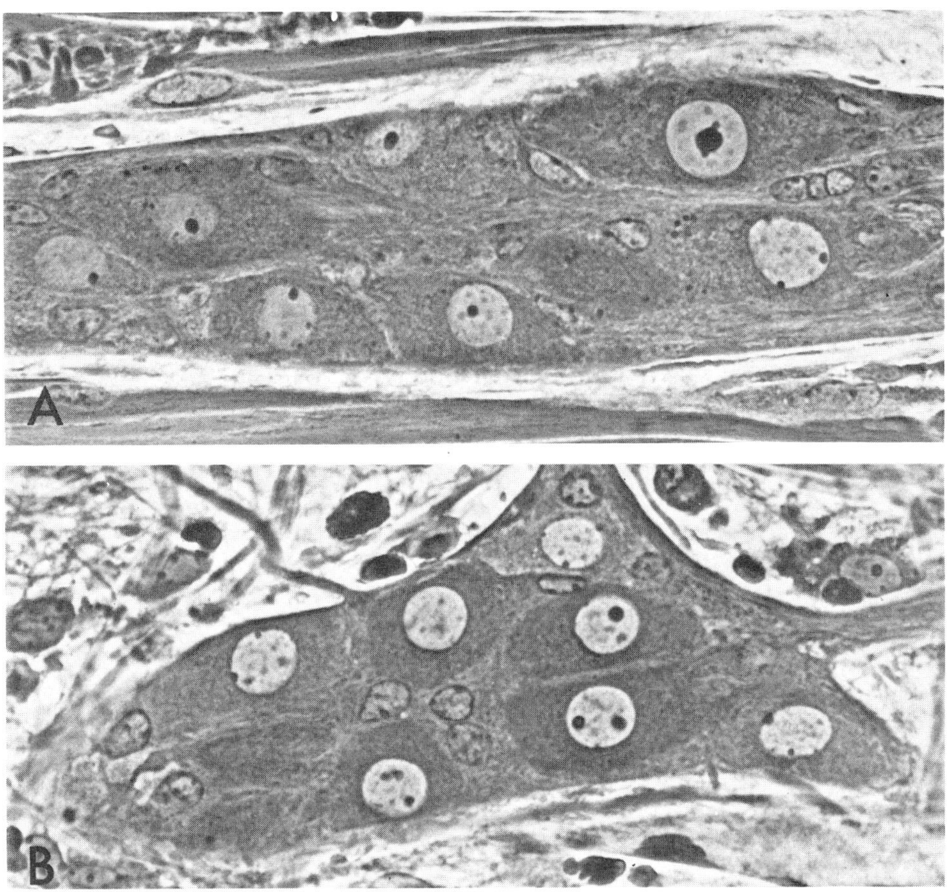

FIGURE 12.   (A) Tangential section of a ganglion of the myenteric plexus of a hypertrophic intestine. Note the large size of neurons and their dense packing. Circular musculature is visible on both sides of the ganglion. (B) A submucosal ganglion of the hypertrophic ileum, packed with neurons and glial cells, and surrounded by connective tissue of the submucosa. (C) A myenteric ganglion from a control. The longitudinal muscle is at the bottom left, the circular muscle to the right. (D) A submucosal ganglion from a control. All ganglia at the same magnification. Calibration bar: 30 μm. (From Gabella, G., Size of neurons and glial cells in the intramural ganglia of the hypertrophic intestine of the guinea pig, *J. Neurocytol.*, 13, 73, 1984. With permission.)

point that its length (i.e., the circumference of the intestine) is three times that in controls.

The extent of the increase in mass of the intestinal musculature in the hypertrophic intestine is truly remarkable. Provided the obstruction sets in slowly and does not lead to a full occlusion, a 15-fold increase in muscle volume can be achieved in a period of a few weeks. This growth, which involves both muscle layers, is entirely in excess of the full physiological growth of the organ; without the stenosis (or some pathological occurrence), it would not have taken place during the life span of the animal. It remains unknown which factors limit the growth of the intestine during normal development and which mechanisms release the potential for further growth after the stenosis. It is possible that there is more than one causative mechanism at work. The distension of the wall and the stretch imposed on the muscle cells (and on the intramural nerve ganglia) are probably major factors inducing the reaction of the organ.

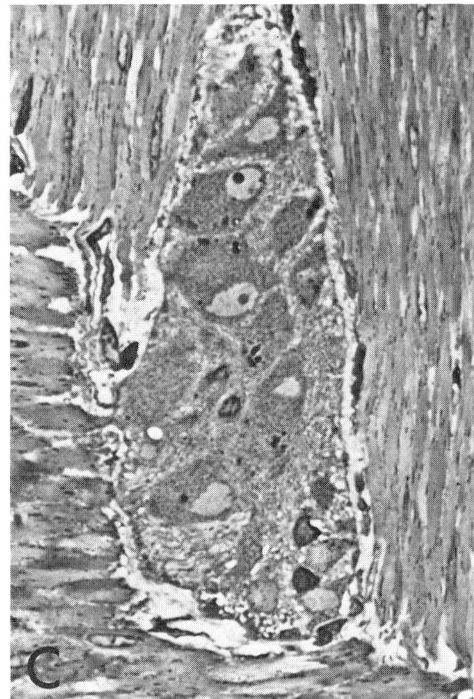
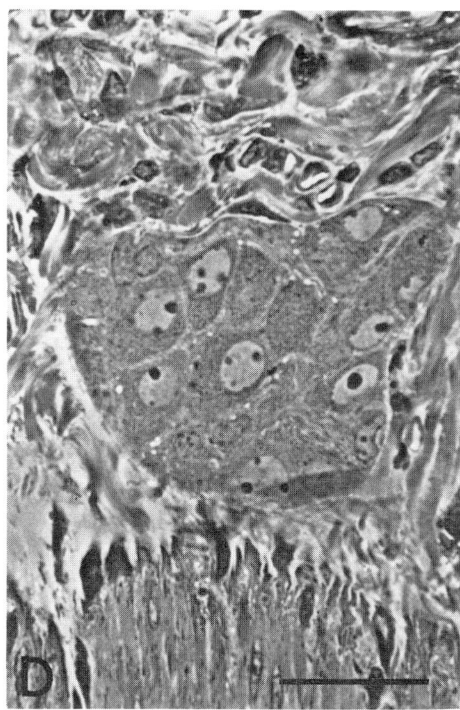

FIGURE 12C.                                        FIGURE 12D.

There are, however, also changes in the composition of the lumenal contents, hence changes in the chemical and mechanical stimulation of the mucosa and consequent alteration of intramural reflexes. It is possible that chemical substances are released from stimulated muscle cells which may affect adjacent structures (e.g., the hypertrophic musculature may stimulate the neuronal hypertrophy).

The hypertrophic response is not a generalized response of the gut to a partial obstruction; it is, rather, a localized response, whose intensity is maximal a short distance orad to the stenosis and progressively less intense further away from the stenosis. The causative mechanisms are local ones and operate only over a short distance, an observation that again stresses the important role of the distension of the wall. A response that in some respects is opposite to that of hypertrophy is observed in the segment of small intestine aborad to the stenosis.

The hypertrophic muscle cells studied in this model undergo cell division and a vast increase in cell size. Their surface to volume ratio decreases in spite of the development of many convolutions and invaginations of the cell membrane. The relative volume of mitochondria is decreased, while sarcoplasmic reticulum (in its many forms) is increased. There are a greater number of gap junctions per cell than in controls, and the average size of junctions is increased; the percentage surface occupied by gap junctions is also increased. Intramuscular nerve bundles do not follow the growth of the musculature, and they are dispersed in the hypertrophic tissue; the density of innervation is greatly reduced. Myenteric and submucosal ganglia grow much larger than in controls, mainly because of the increase in ganglion nerve cell size. The muscle coat remains well vascularized even when maximally hypertrophic.

Our understanding of the hypertrophy of the intestinal muscle coat is still very incomplete. In addition to the lack of more detailed data on the structure of hypertrophic muscle cells and nerve cells, we especially need basic data on the functional properties of the obstructed and the hypertrophic intestine[104,105] and on their biochemical properties.

## ACKNOWLEDGMENTS

The experimental work presented in this chapter has been supported by the Medical Reseach Council (U.K.).

## REFERENCES

1. Ehrenpreis, T., *Hirschprung's Disease,* Year Book Medical Publishing, Chicago, 1970.
2. Belding, H. H., III and Kernohan, J. W., A morphologic study of the myenteric plexus and musculature of the pylorus with special reference to the changes in hypertrophic pyloric stenosis, *Surg. Gynecol. Obstet,* 97, 322, 1953.
3. Schneider, J. E., Kennedy, G. A., and Leipold, H. W., Muscular hypertrophy of the small intestine in a horse, *J. Equine Med. Surg.,* 3, 226, 1979.
4. Herczel, E., Experimentelle histologische Untersuchungen über compensatorische Muskelhypertrophie bei Darmstenosen, *Z. Klin. Med.,* 11, 321, 1886.
5. Gee, W. F. and Kiviat, M. D., Ureteral response to partial obstruction. Smooth muscle hyperplasia and connective tissue proliferation, *Invest. Urol.,* 12, 309, 1975.
6. Peterson, C. M., Goss, R. J., and Atryzek, V., Hypertrophy of the rat urinary bladder following reduction of its functional volume, *J. Exp. Zool.,* 187, 121, 1973.
7. Goss, R. J., Liang, M. D., Weisholtz, S. J., and Peltzer, T. J., The physiological basis of urinary bladder hypertrophy, *Proc. Soc. Exp. Biol. Med.,* 142, 1332, 1973.
8. Carpenter, F. G., Histological changes in parasympathetically denervated feline bladder, *Am. J. Physiol.,* 166, 692, 1951.
9. Brent, L. and Stephens, F. D., The response of smooth muscle cells in the rabbit urinary bladder to outflow obstruction, *Invest. Urol.,* 12, 494, 1975.
10. Gillenwater, J. Y., Morris, S. A., Bryan, P. R., Attinger, F., Smith, H. C., and McManamy, J. L., In vivo smooth muscle function in normal and obstructed rabbit vas deferens, *Invest. Urol.,* 16, 207, 1978.
11. Wiener, J., Loud, A. V., Giacomelli, F., and Anversa, P., Morphometric analysis of hypertension-induced hypertrophy of rat thoracic aorta, *Am. J. Pathol.,* 88, 619, 1977.
12. Dessouky, D. A., Electron microscopic studies of the myometrium of the guinea pig, *Am. J. Obstet. Gynecol.,* 100, 30, 1968.
13. Dessouky, D. A., Ultrastructural observations of the human uterine smooth muscle cells during gestation, *Am. J. Obstet. Gynecol.,* 125, 1099, 1976.
14. Fell, B. F., Smith, K. A., and Campbell, R. M., Hypertrophic and hyperplastic changes in the alimentary canal of the lactating rat, *J. Pathol. Bacteriol.,* 85, 179, 1963.
15. Laplace, J. P., Compensatory hypertrophy of the residual small intestine after partial enterectomy, *Ann. Rech. Vet.,* 11, 165, 1980.
16. Williamson, R. C. N. and Buchholtz, T. W. M. R. A., Humoral stimulation of cell proliferation in small bowel after transection and resection in rats, *Gastroenterology,* 75, 249, 1978.
17. Booth, C. C., Evans, K. T., Menzies, T., and Street, D. F., Intestinal hypertrophy following partial resection of the small bowel in the rat, *Br. J. Surg.,* 46, 403, 1958-1959.
18. Brobeck, J. R., Tepperman, J., and Long, C. N. H., Experimental hypothalamic hyperphagia in the albino rat, *Yale J. Biol. Med.,* 15, 831, 1943.
19. Jervis, E. L. and Levin, R. J., Anatomic adaptation of the alimentary tract of the rat to hyperphagia of chronic alloxan-diabetes, *Nature,* 210, 391, 1966.
20. Filogamo, G. and Vigliani, F., Ricerche sperimentali sulla correlazione tra estensione del territorio di innervazione e grandezza e numero delle cellule gangliari del plesso mienterico (di Auerbach), nel cane, *Riv. Patol. Nerv. Ment.,* 75, 441, 1954.
21. Okada, A. and Okamoto, E., Myenteric plexus in hypertrophied intestine, *J. Neuro-Visc. Relat.,* 32, 75, 1971.
22. Earlam, R. J., Ganglion cell changes in experimental stenosis of the gut, *Gut,* 12, 393, 1971.

23. Brent, L., The response of smooth muscle cells in the rabbit colon to anal stenosis: a preliminary report, *Pathology,* 5, 209, 1973.
24. Gabella, G., Hypertrophy of intestinal smooth muscle, *Cell Tissue Res.,* 163, 199, 1975.
25. Gabella, G., Hypertrophic smooth muscle. Size and shape of cells, occurrence of mitoses. Sarcoplasmic reticulum, caveolae and mitochondria, *Cell Tissue Res.,* 201, 63, 1979.
26. Gabella, G., On the musculature of the gastro-intestinal tract of the guinea pig, *Anat. Embryol.,* 163, 135, 1981.
27. McGeachie, J., Ultra-structural specificity in regenerating smooth muscle, *Experientia,* 27, 436, 1971.
28. McGeachie, J., Smooth muscle regeneration: a review and experimental study, *Monogr. Dev. Biol.,* 9, 1, 1975.
29. Jurukova, Z. and Atanassova, E., Smooth muscle cell regeneration in repair of gastric anastomosis in the dog, *Res. Exp. Med.,* 162, 299, 1974.
30. Jacoby, F., *In Mitogenesis,* Ducoft, H. S. and Ehret, C. F., Eds., University of Chicago Press, Chicago, 1959, 55.
31. Cussen, L. J. and Tymms, A., Hyperplasia of ureteral muscle in response to acute obstruction of the ureter, *Invest. Urol.,* 9, 504, 1972.
32. Kaufman, O. Y., Mitosis in the hypertrophied muscle tissue of the posterior vena cava of rats, *Biull. Eksp. Biol. Med.,* (in Russian), 4, 485, 1976.
33. Cobb, J. L. S. and Bennett, T., An ultrastructural study of mitotic division in differentiated gastric smooth muscle cells, *Z. Zellforsch.,* 108, 177, 1970.
34. Grohmann, D., Mitotische Wachstumsintensität des embryonalen and fetalen Hühnchenherzens und ihre Bedeutung fur die Entstehung von Herzmissbildungen, *Z. Zellforsch.,* 55, 104, 1961.
35. Rumyantsev, P. P. and Snigirevskaya, E., Ultrastructure of differentiating cells of the heart muscle in the state of mitotic division, *Acta Morphol. Acad. Sci. Hung.,* 16, 271, 1968.
36. Kamio, A., Huang, W. Y., Imai, H., and Kummerow, F. A., Mitotic structures of aortic smooth muscle cells in swine and in culture: paired cisternae, *J. Electron Microsc.,* 26, 29, 1977.
37. Manasek, F. G., Mitosis in developing cardiac muscle, *J. Cell Biol.,* 37, 191, 1968.
38. Zak, R., Cell proliferation during cardiac growth, *Am. J. Cardiol.,* 3, 211, 1973.
39. Ljungqvist, A. and Unge, G., The proliferative activity of the myocardial tissue in various forms of experimental cardiac hypertrophy, *Acta Pathol. Microbiol. Scand. Sect. A,* 81, 233, 1973.
40. Korecky, B. and Rakusan, K., Normal and hypertrophic growth of the rat heart: changes in cell dimensions and number, *Am. J. Physiol.,* 234, H123, 1978.
41. Dammrich, J. and Pfeiffer, U., Cardiac hypertrophy in rats after supravalvular aortic constriction, *Virchows Arch., Cell Pathol.,* 43, 265, 1983.
42. Moss, F. and Leblond, C., Nature of dividing nuclei in skeletal muscle of growing rats, *J. Cell. Biol.,* 44, 459, 1970.
43. Moss, F. and Leblond, C., Satellite cells as the sources of nuclei in muscles of growing rats, *Anat. Rec.,* 190, 421, 1971.
44. Stohr, P. Jr., Bemerkungen über Amitose und über "Reihenstellung" bei den Kernen der glatten Muskelfasern, *Z. Mikrosk.-Anat. Forsch.,* 36, 525, 1834.
45. Goneya, W., Ericson, G. C., and Bonde-Petersen, F., Skeletal muscle fiber splitting by weight-lifting exercise in cats, *Acta Physiol. Scand.,* 99, 105, 1977.
46. Brent, L. and Stephens, F. D., The response of smooth muscle cells in the rabbit urinary bladder to outflow obstruction, *Invest Urol.,* 12, 494, 1975.
47. Stieve, H., Muskulatur und Bindegewebe in der Wand der menschlichen Gerbärmutter ausserhalb und während des Schwangerschaft, während der Geburt und des Wochenbettes, *Z. Mikrosk.-Anat. Forsch.,* 17, 371, 1929.
48. De Mattos, C. E. R., Kempson, R. L., Erdos, T., and Csapo, A., Stretch-induced myometrial hypertrophy, *Fertility and Sterility,* 18, 545, 1967.
49. Salvatore, C. A., The significance of the myometrial cell hypertrophy during pregnancy, *Ergeb. Allg. Pathol. Pathol. Anat.,* 42, 148, 1962.
50. De Feo, V. J. and Kleinfeld, R. G., Mitotic activity of uterine smooth muscle during decidualization in the rat: a hypothesis for embryo spacing, *Anat. Rec.,* 169, 305, 1971.
51. Krueger, W. A. and Maibenco, H. C., DNA replication and cell division in the hamster uterus, *Anat. Rec.,* 173, 229, 1972.
52. Bekoff, A. and Betz, W. J., Physiological properties of dissociated muscle fibres obtained from innervated and denervated adult rat muscle, *J. Physiol. (London),* 271, 25, 1977.
53. Van Linge, B., The response to strenuous exercise, *J. Bone Joint Surg.,* 44B, 711, 1962.
54. Miledi, R. and Slater, C. R., Electron-microscopic structure of denervated skeletal muscle, *Proc. Roy. Soc. London Ser. B,* 173, 253, 1969.

55. Ekstrom, J. and Uvelius, B., Changes in length and volume of smooth muscle cells in the hypertrophied rat urinary bladder, *Acta Physiol. Scand.*, 118, 305, 1983.
56. Ekstrom, J., Mattiason, A., and Uvelius, B., Force development in smooth muscle strips of the hypertrophied urinary bladder of the rat after autonomic decentralization, *Experientia*, 38, 948, 1982.
57. Gabella, G., Hypertrophic smooth muscle. Increase in number and size of gap junctions. Myofilaments, intermediate filaments and some mechanical properties, *Cell Tissue Res.*, 201, 263, 1979.
58. Herr, J. C., Reflexive gap junctions. Gap junctions between processes arising from the same ovarian decidual cell, *J. Cell Biol.*, 69, 485, 1976.
59. Iwayama, T., Nexus between areas of the cell surface membrane of the same arterial smooth muscle cell, *J. Cell Biol.*, 49, 521, 1971.
60. Albertini, D. F. and Anderson, E., Structural modifications of lutein gap junctions during pregnancy in the rat and the mouse, *Anat. Rec.*, 181, 171, 1975.
61. Elias, P. M. and Friend, D. S., Vitamin A-induced mucous metaplasia, *J. Cell Biol.*, 68, 173, 1976.
62. Garfield, R. E., Sims, S., and Daniel, E. E., Gap junctions: their presence and necessity in myometrium during parturition, *Science*, 198, 958, 1977.
63. Yee, A. G. and Revel, J. P., Loss and reappearance of gap junctions in regenerating liver, *J. Cell Biol.*, 78, 554, 1978.
64. Kannan, M. S. and Daniel, E. E., Formation of gap junctions by treatment in vitro with potassium conductance blockers, *J. Cell Biol.*, 78, 338, 1978.
65. Westfall, D. P., Lee, T. J. F., and Stitzel, R. E., Morphological and biochemical changes in supersensitive smooth muscle, *Fed. Proc.*, 34, 1985, 1975.
66. Johansson, B. and Somlyo, A. P., Electrophysiology and excitation-contraction coupling, in *Handbook of Physiology: Vascular Smooth Muscle*, Bohr, D. F., Somlyo, A. P., and Sparks, H. V., Eds., American Physiological Society, Baltimore, 1980, 301.
67. Bennett, G. S., Fellini, S. A., Croop, J. M., Otto, J. J., Bryan, J., and Holtzer, H., Differences among 100A filament subunits from different cell types, *Proc. Nat. Acad. Sci. USA*, 75, 4364, 1978.
68. Berner, P. F., Somlyo, A. V., and Somlyo, A. P., Hypertrophy-induced increase of intermediate filaments in vascular smooth muscle, *J. Cell Biol.*, 88, 96, 1981.
69. Campbell, G. R., Uehara, Y., Mark, G., and Burnstock, G., Fine structure of smooth muscle cells grown in tissue culture, *J. Cell Biol.*, 49, 21, 1971.
70. Ishikawa, H., Bischoff, R., and Holtzer, H., Mitosis and intermediate-sized filaments in developing skeletal muscle, *J. Cell Biol.*, 38, 538, 1968.
71. Uehara, Y., Campbell, G. R., and Burnstock, G., Cytoplasmic filaments in developing and adult vertebrate smooth muscle, *J. Cell Biol.*, 50, 488, 1971.
72. Johansson, B., Structural changes in rat portal veins after experimental hypertension, *Acta Physiol. Scand.*, 98, 381, 1976.
73. Gabella, G., Hypertrophic smooth muscle. V. Collagen and other extracellular materials. Vascularization, *Cell Tissue Res.*, 235, 275, 1984.
74. Gabella, G. and Yamey, A., Synthesis of collagen by smooth muscle in the hypertrophic intestine, *Q. J. Exp. Physiol.*, 62, 257, 1977.
75. Gibson, J., Laurent, G. J., and McAnulty, R. J., Collagen metabolism during muscle growth in the chicken: evidence for increased collagen deposition associated with a decreased degradation of newly synthesized collagen as well as an increased synthesis rate, *J. Physiol. (London)*, 364, 80p, 1985.
76. Wolinsky, H., Response of the rat aortic media to hypertension, *Circ. Res.*, 26, 507, 1970.
77. Ooshima, A., Fuller, G. C., Cardinale, G. S., Spector, S., and Udenfriend, S., Increased collagen synthesis in blood vessels of hypertensive rats and its reversal by antihypertensive agents, *Proc. Nat. Acad. Sci. USA*, 71, 3091, 1974.
78. Iwatsuki, K., Cardinale, G. J., Spector, S., and Udenfriend, S., Hypertension: increase of collagen biosynthesis in arteries but not in veins, *Science*, 198, 403, 1977.
79. Nissen, R., Cardinale, G. J., and Udenfriend, S., Increased turnover of arterial collagen in hypertensive rats, *Proc. Nat. Acad. Sci. USA*, 75, 451, 1978.
80. Harkness, M. L. R. and Harkness, R. D., The collagen content of the reproductive tract of the rat during pregnancy and lactation, *J. Physiol. (London)*, 123, 492, 1954.
81. Harkness, M. L. R. and Harkness, R. D., The distribution of the growth of collagen in the uterus of the pregnant rat, *J. Physiol. (London)*, 132, 492, 1956.
82. Layman, D. L. and Titus, J. L., Synthesis of type I collagen by human smooth muscle cells in vitro, *Lab. Invest.*, 33, 103, 1975.
83. Barnes, M. J., Morton, L. F., and Leven, C. I., Synthesis of collagen type I and III by pig medial smooth muscle cells in culture, *Biochem. Biophys. Acta*, 418, 93, 1976.
84. Burke, J. M. and Ross, R., Collagen synthesis by monkey arterial smooth muscle cells during proliferation and quiescence in culture, *Exp. Cell Res.*, 107, 387, 1977.

85. Scott, D. M., Harwood, R., Grant, M. E., and Jackson, D. S., Characterization of the major collagen species present in porcine aortae and the synthesis of their precursors by smooth muscle cells in culture, *Connect. Tissue Res.,* 5, 7, 1977.
86. Mayne, R., Vail, M. S., and Miller, E. J., Characterization of collagen chains synthesized by cultured smooth muscle cells derived from rhesus monkey thoracic aorta, *Biochemistry,* 17, 446, 1978.
87. Chung, E., Keele, E. M., and Miller, E. J., Isolation and characterization of the cyanogen bromide peptides from the $\alpha$1 (III) chain of human collagen, *Biochemistry,* 13, 3459, 1974.
88. Trelstad, R. L., Human aorta collagens: evidence for three distinct species, *Biochem. Biophys. Res. Commun.,* 57, 717, 1974.
89. McCullagh, K. A. and Balian, G., Collagen characterization and cell transformation in human atherosclerosis, *Nature (London),* 258, 73, 1975.
90. Ross, R. and Klebanoff, S. J., The smooth muscle cell. I. In vivo synthesis of connective tissue protein, *J. Cell Biol.,* 50, 159, 1971.
91. Reynolds, S. R. M. and Kaminester, S., The rate of uterine growth resulting from chronic distension, *Anat. Rec.,* 69, 281, 1937.
92. Csapo, A., Erdos, T., De Mattos, C. R., Gramss, E., and Moscowitz, C., Stretch-induced uterine growth, protein synthesis and function, *Nature (London),* 207, 1378, 1965.
93. Cullen, B. M. and Harkness, R. D., Collagen formation and changes in cell population in the rat's uterus after distension with wax, *Q. J. Exp. Physiol.,* 53, 33, 1968.
94. Leung, D. Y. M., Glagov, S., and Mathews, M. B., Cyclic stretching stimulates synthesis of matrix components by arterial smooth muscle cell in vitro, *Science,* 191, 475, 1976.
95. Richardson, K. C., Electron microscope observations on Auerbach's plexus in the rabbit with special reference to the problem of smooth muscle innervation, *Am. J. Anat.,* 103, 99, 1958.
96. Taxi, J., Contribution à l'étude des connexions des neurones moteurs du systéme nerveux autonome, *Ann. Sci. Nat. Zool. Biol. Anim.,* 7, 413, 1965.
97. Filogamo, G. and Marchisio, P. C., Choline acetyltransferase activity or rabbit ileum wall. The effects of extrinsic and intrinsic denervation and of combined experimental hypertrophy, *Arch. Int. Physiol. Biochem.,* 78, 141, 1970.
98. Toulokian, R. J., Aghajanian, G. K., and Roth, R. H., Adrenergic denervation of the hypertrophied gut remnants, *Ann. Surg.,* 176, 633, 1972.
99. Gabella, G., Size of neurons and glial cells in the intramural ganglia of the hypertrophic intestine of the guinea pig, *J. Neurocytol.,* 13, 73, 1984.
100. Benninghoff, V. A., Vermehrung und Vergrosserung von Nervenzellen bei Hypertrophie des Innervationsgebietes, *Z. Naturforsch.,* 6b, 38, 1951.
101. Filogamo, G., Iperplasia ed ipertrofia delle cellule gangliari del plesso sottomucoso (di Meissner) nel cane, in condizioni sperimentali, *Riv. Biol.,* 47, 331, 1935.
102. Miura, G., (quoted by Filogamo and Vigliani [see Reference 20].
103. Ambrosi, F., Sulle alterazioni istologiche dei gangli e dei nervi del tubo digerente nell'occlusione intestinale. Ricerche nell'uomo e sperimentali, *Arch. De Vecchi Anat. Patol.,* 5, 215, 1942.
104. Takita, S., Action current of alimentary canal. Observations in the normal and obstructed bowels, *Jpn. J. Physiol.,* 3, 176, 1953.
105. Summers, R. W., Anuras, S., and Green, J., Jejunal manometry patterns in health, partial intestinal obstruction, and pseudoobstruction, *Gastroenterology,* 85, 1290, 1983.

Chapter 4

# RESPONSE OF INTESTINAL MUSCLE TO RESECTION AND BYPASS

Norman W. Weisbrodt

## TABLE OF CONTENTS

## I. INTESTINAL OPERATIONS — EXPERIMENTAL MODELS

The small intestine is a long tubular structure whose normal function is necessary for the digestion and absorption of nutrients, electrolytes, and water. The average person has 400 to 600 cm of intestine through which material must be propelled as it is processed. This propulsion is accomplished by smooth muscle cells that form a major component of the intestinal wall. The propulsive forces generated by the smooth muscle cells must be coordinated with the digestive and absorptive functions of the mucosa or else maldigestion and malabsorption will occur. These propulsive forces depend not only on those factors which regulate the force and timing of intestinal contractions, but also on those factors that regulate the growth, organization, and number of smooth muscle cells. This chapter deals primarily with these latter factors.

There are several pathological states which are treated, in part, by resection or bypass of a large portion of the small intestine. During resection, the diseased intestine is removed and normal continuity of the bowel is reestablished by an anastomosis. Such conditions as intestinal volvulus, mesenteric thrombosis, inflammatory bowel disease, and abdominal trauma are treated in this manner. Intestinal bypass, on the other hand, is a procedure in which a large portion of the small intestine is removed from continuity with the proximal bowel, but yet is left *in situ* (Figure 1). Such an operation has been used in the treatment of pathological obesity. However, its popularity as a therapeutic intervention has waned recently, due to the severe complications which frequently accompany the operation.

Since removal of a large segment of intestine is not uncommon, there have been many studies on the clinical course of individuals subjected to such operations. Clinically, Pullan[1] has identified three phases of recovery in patients. The first phase is characterized by a postoperative crisis due to a profuse diarrhea which can result in the loss of large amounts of water and electrolytes. The second phase is a period of adaptation when the problems of water and electrolyte balance are replaced by problems of nutritional balance. The third phase is a period of balance when nutritional compensation becomes adequate.

Intestinal resection and bypass have been used experimentally in order to study small bowel adaptation. Both resection and bypass are associated with structural and functional changes in the functioning remnant (in-continuity segment) of small intestine as well as in the bypassed segment. As reviewed by Laplace,[2] Dowling,[3] and Johnson,[4] the in-continuity segment hypertrophies and undergoes an increase in its ability to absorb fluid, electrolytes, and nutrients. Although there is little doubt that adaptation takes place postoperatively, the mechanisms for this adaptation are not entirely clear. Most investigations have focused upon changes in mucosal structure and function. These studies have demonstrated increases in villus height, in brush border enzymes, and in absorptive capacity. So far, little attention has been paid to the possible influence of alterations in gastrointestinal smooth muscle. Structural and functional changes in intestinal smooth muscle could greatly influence absorption, since digestion and transport of nutrients across epithelia require time. If smooth muscle function remained constant after the operation, then the contents would be propelled rapidly through the shortened bowel and time wuld be too short for these processes to take place. On the other hand, if transit through the functioning small bowel were to slow down as an adaptive response, nutrient and electrolyte absorption would improve.

## II. CHANGES IN MUSCLE FUNCTION IN VIVO

Intestinal motility has been monitored in both patients and experimental animals by

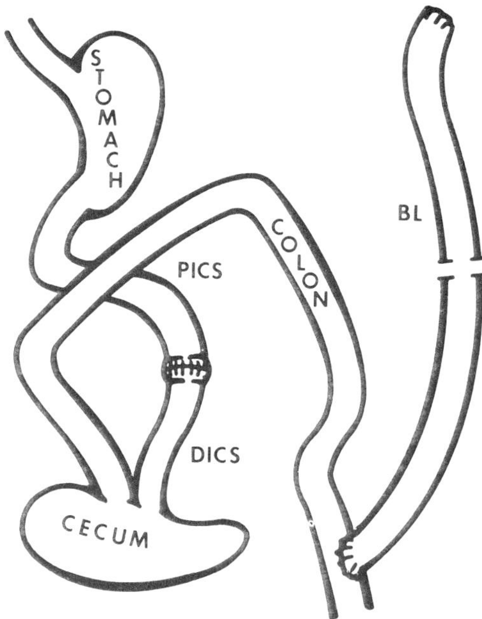


FIGURE 1. Schematic of intestinal bypass operation. In our studies, 70% of the mid small intestine is bypassed to form the bypassed loop (BL). It is closed at its oral end and anastomosed to the colon at its aboral end. Its vascular supply and extrinsic innervation remain intact. The duodenum (PICS) and terminal ileum (DICS) are anastomosed to form the in-continuity segment.

following the transit of a nonabsorbable marker through the gastrointestinal tract. In patients, Althansen et al.[5] found that the mouth to anus transit time increased from 3.5 hr at 9 days postenterectomy to 18 hr at 3 weeks postenterectomy. In rats, both Nygaard[6] and Nylander[7] saw similar changes in gastrointestinal transit. That is, transit was rapid at first after the operation, but then slowed down as adaptation to the operation occurred. Although these early studies indicated that there were adaptations in gastrointestinal motility after the operation, they did not specifically implicate adaptation of intestinal motility. In all cases, the marker was placed into the stomach so that the variable of gastric emptying could have had a major influence on the results. Also, the observations were made prior to the unequivocal demonstration of differences in patterns of intestinal motility between the fasted and fed state.[8]

In a more recent study, catheters were implanted so that the nonabsorbable marker could be placed directly into specific areas of the small bowel in unanesthetized animals. Nemeth et al.[9] found that if an isotope was placed in the proximal in-continuity segment in animals 3 days after bypass operation, then the isotope moved very rapidly from the proximal portion of the in-continuity segment to the cecum. This rapid transit occurred in both fasted as well as fed rats. By 35 days after the operation, however, transit through the in-continuity segment of fed animals had slowed markedly. Indeed, it took as long for the isotope to go through the shortened in-continuity segment as it took for the isotope to transverse the entire length of the small bowel of control animals. This slowing of intestinal transit with time after the operation was seen only in fed animals. In fasted animals, transit was still rapid at 35 days after bypass operation.

This difference in results between fasted and fed animals probably reflects the differing patterns of motility in the two states.[8]

There are very few studies on transit functions of the bypassed segment in patients and experimental animals. Abnormal decreases in propulsive activity (although never directly measured) have been implicated in some of the pathology seen after bypass operations in patients. Bacterial overgrowth has been demonstrated[10] and this has been blamed in part on an inability of the intestinal loop to clear itself of microorganisms. In experimental animals, Nemeth et al.[11] found that indeed intestinal transit is slowed immediately after the operation. By 35 days after the operation, some adaptation had taken place. Transit in the bypassed loop was similar to that seen in comparable areas of bowel in control animals when studied in the fasted state. However, when investigated in the fed condition, transit in the bypassed loop was markedly slowed compared to controls.

## III. CHANGES IN MUSCLE FUNCTION IN VITRO

There are few studies concerned with mechanical functions of intestinal smooth muscle from control animals, let alone animals that have undergone small bowel bypass. Such studies are needed in order to determine whether or not the changes seen in vivo are related to any changes in the basic properties of the smooth muscle cells. Also, structural changes seen in the musculature from bypassed animals (see below) indicate that there may be changes in the composition of the muscular wall. Mechanical studies on isolated muscle tissue could prove useful in characterizing these changes. For example, studies with hypertrophic smooth muscle from blood vessels,[12,13] urinary bladder,[14] and guinea pig intestine proximal to a stenosis[15] indicate that the ability of hypertrophic muscle to develop stress is reduced.

Although complete length-tension relationships were not characterized, Weisbrodt et al.[16] determined maximum active force development at $L_o$ for circularly oriented strips of rat small intestine. In strips taken from both the in-continuity and the bypassed segments of animals 3 days after the operation, no differences in active stress were noted when compared to values for tissue taken from unoperated animals. By 35 days after the operation, however, active stress was greater for strips taken from both the in-continuity and the bypassed segments. Bortoff et al.[17] recently studied strips of muscle from cat intestine. They found that strips of circular muscle from the in-continuity segment developed greater stress than did strips from comparable areas of control animals; however, muscle strips from the bypassed intestine demonstrated a decrease in the ability to generate stress. The reason for the differences in results for the bypassed segment between the cat and the rat model are not clear.

No studies have been conducted to characterize either membrane properties or calcium metabolism in intestinal smooth muscle from resected or bypass-operated animals. In vivo studies indicate that both slow waves and spike potentials can be recorded from the muscle;[18] however, these surface recordings give little indication as to the events at the cellular level.

## IV. CHANGES IN STRUCTURE OF INTESTINAL MUSCLE

Probably the most striking observation after intestinal resection or bypass is the change in diameter and mass of the in-continuity segment. There is a marked hypertrophy of the intestinal wall, especially evident in the ileum. Most investigations of this increase in mass have concentrated on the mucosa where increases in villus height, in brush border enzymes, and in absorptive capacity have been demonstrated.[2-4] Although these early studies did not specifically concentrate on the intestinal musculature, several investigators commented on a thickening of the muscular coat.[19,20] Nemeth et al.[9] removed the mucosa from segments of in-continuity intestine at various times after bypass operation. They found that the wet weight of the combined serosa and muscle per unit length of bowel increased significantly, especially in the distal regions. This increase in wet weight was also reflected in an increase in total protein content. Thus, there is a clear increase in the mass of the muscular tissue of the functioning small bowel that is left after resection or bypass.

The situation with the bypassed segment is less clear. Gleeson et al.[21] reported that histological sections from bypassed segments often showed increased thickness of the circular and longitudinal muscle layers, presumably a result of muscular contractions and narrowing of the intestinal lumen. Studies by Nemeth et al.[11] support the notion that there is no decrease in muscular tissue. In fact, they observed an increase in muscle layer wet weight and protein content per unit length of bypassed intestine. Thus, there appears to be an increase in mass. On the other hand, Bortoff et al.[17] reported that the bypassed segment in cats atrophied and that this atrophy involved the muscular layer. The reasons for the discrepancies are not clear but may involve a difference in species and/or a difference in the length of intestine being bypassed. Nemeth et al.[11] bypassed 70% of the small intestine whereas Bortoff et al.[17] bypassed approximately one third.

The increase in mass of the intestinal muscle layer seen in all areas of the bowel in rats with intestinal bypass appears to be due to an increase in functioning muscle rather than connective tissue. Intestinal muscle strips from bypassed animals generate as much if not more stress than strips from control animals (see above). Furthermore, we recently have determined the concentrations of actomyosin and collagen in segments of intestinal muscle from control animals and animals 35 days after bypass.[22] The actomyosin concentration averaged 14 mg/g of tissue wet weight with an actin-to-myosin ratio of 1.8. There were no differences in concentrations or ratios among strips taken from bypassed or control animals. Collagen concentrations averaged 48 mg/g tissue wet weight. There were no increases seen in bypassed animals; in fact, collagen concentrations were less in the distal in-continuity segments from bypassed animals. Thus, these biochemical measurements support the in vitro stress measurements and both indicate that the increased mass is due to functioning smooth muscle.

Thus far, the few studies that have been reported have concentrated on the changes in the overall muscular wall after intestinal resection and bypass. Little is known about the response of the individual smooth muscle cells to the operation. Bowers et al.[22] recently found that there was hypertrophy of the individual smooth muscle cells which comprise the circular muscle layer. Strips of muscle were prepared and $L_o$ was determined in vitro. The strips were then fixed at $L_o$ and prepared for microscopic observation. The number of cross-sectioned smooth muscle cells per mm² of tissue taken from bypassed animals was approximately one half that taken from control animals. Thus an element of hypertrophy is seen, even in the bypassed segment. Whether or not hyperplasia occurs is not known. Also, neither the time course of the changes nor the biochemical makeup of the smooth muscle cells has been studied.

## V. STIMULI RESPONSIBLE FOR THE CHANGES IN INTESTINAL MUSCLE

Intestinal resection and bypass induce many local and systemic alterations which could influence the intestinal muscle. One of the primary local changes is an alteration in intraluminal contents. The in-continuity segment, especially the distal portions, experiences a large increase in intraluminal contents. Normally, the processes of digestion and absorption occur so that minimal contents are delivered to the distal small bowel. After removal or bypass of a large portion of the bowel, digestion and absorption are impaired so that more contents are propelled through the distal gut. Also, once intestinal transit begins to slow, the contents are retained in the proximal bowel for a longer period of time, thus causing a relative increase in contents. This increased volume load may cause a work-induced hypertrophy as is seen for other types of muscle (see other chapters in this book).

Intraluminal contents are also altered in the bypassed segment. In this case, the volume of contents is decreased markedly since neither food nor gastric, pancreatic, or biliary secretions pass through this segment. The loops are not totally devoid of contents, however, since intestinal secretions such as mucus and some electrolytes, as well as desquamated cells are still being produced. In fact, Weisbrodt et al.[23] have reported that the bypassed segment in rats fills with a thick viscous material. The early slowing of intestinal transit in the bypassed segment seen after operation may be due to the presence of this material. If a mixture of pancreatic juice and bile salts are infused into the bypassed segment, transit through the loop at 3 days after the operation is nearly normal. Whether or not this is due to an action of the pancreatic juice on the mucus or to other effects of the pancreatic juice-bile salt mixture is not known. Also, it is not known whether the presence of this altered content is responsible for the muscular hypertrophy seen in the bypassed segment.

Intestinal resection and bypass also lead to altered neurohumoral factors. Transecting the bowel interrupts the enteric nervous system which is known to play an important role in controlling absorptive, secretory, and motility functions of the gut. Indeed, Smith[24] has reported that procedures which disrupt the enteric nervous system can cause a hypertrophy of the smooth muscle cells. Removal of a large portion of the small intestine can also directly affect the hormonal response to a meal. Many of the gastrointestinal hormones such as secretin, cholecystokinin, and GIP, as well as many biologically active peptides are found in the intestine. When these are removed, the hormonal response to a meal is altered.

In addition to these direct effects of the surgery, many indirect effects also occur. During the early postoperative period, the animal is in negative nutrient, electrolyte, and water balance. This poses a stress on the animal which elicits numerous neural and hormonal responses. Once the initial adaptation has taken place, several alterations still persist, e.g., hypergastrinemia is seen after intestinal resection.[25] The exact reasons for this are not known but it may result from a loss of inhibitory signals from the small bowel, a loss of the small intestines ability to clear the hormone, or an increase in the mass of cells that secrete gastrin. Furthermore, certain peptides such as enteroglucagon and neurotensin are released in greater quantities since food stuffs are now reaching the distal small bowel where these substances are located.[26]

If the hormonal response is altered, this can influence smooth muscle activity. Patterns of intestinal motility, in vivo, are altered by hormones.[8] Thus, alterations in transit may be caused by alterations in circulating or locally released hormones. It is not known whether any of these neural and/or humoral factors alter the smooth muscle. To date, it is generally believed that those hormones which are trophic to the gastrointestinal tract have no effects on intestinal smooth muscle. Thus it is difficult to implicate such hormones as gastrin and cholecystokinin. On the other hand, hormones are known to alter the growth of other smooth muscles (see chapter on uterine hypertrophy), therefore this possibility should not be ruled out just yet.

## VI. SUMMARY

Intestinal resection and bypass cause gross changes in the structure and function of the remaining intestine. Although many investigators have commented on these qualitative changes, very few studies have attempted to characterize in a quantitative manner those changes which involve the intestinal smooth muscle. Thus, we have little idea as to whether the increased muscle mass is due to hypertrophy, hyperplasia, or to a combination. Furthermore, little is known about the time course of the muscular changes. Indeed, next to nothing is known about the normal growth and turnover of intestinal smooth muscle tissue.

Since little is known about the actual changes that take place, it is no wonder that next to nothing is known about the factors that cause the gross changes in muscle that are seen. Intestinal bypass and resection offer reasonable models with which to study such changes and their stimuli. The operations are relatively easy and reproducibly executed. Thus, a constant animal model is available. Furthermore, by altering the animals' diet and by feeding intravenously it will be possible to vary the workload of the intestine. This especially is true for the bypassed segment. This segment can be catheterized and various materials can be introduced into the lumen. Also, many of the hormones and peptides that could be implicated in the process of adaptation are currently available and their effects on the response to operation can be investigated.

Results of such studies should further establish the nature of the changes in smooth muscle function and structure which may be involved in the adaptive changes that occur in intestinal segments which have been bypassed, or in segments of bowel which remain in continuity with the rest of the gastrointestinal tract. Also such studies should shed light on the mechanisms that are responsible for the changes in smooth muscle function and structure. Finally, such data should add to our very limited knowledge on the growth and development of intestinal smooth muscle.

## REFERENCES


1. Pullan, J. M., Massive intestinal resection, *Proc. Roy. Soc. Med.,* 52, 31, 1959.
2. Laplace, J.-P., Small bowel resections — exhaustive approach to a theory of adaptation, *World Rev. Nutr. Diet,* 23, 1, 1975.
3. Dowling, R. H., The influence of luminal nutrition on intestinal adaptation after small bowel resection and bypass, in *Intestinal Adaptation,* Dowling, R. H. and Dieken, E. O., Eds., Springer-Verlag, Stuttgart, 1974, 35.
4. Johnson, L. R., Regulation of gastrointestinal growth, in *Physiology of the Gastrointestinal Tract,* Johnson, L.R., Ed., Raven Press, New York, 1981, 169.
5. Althansen, T. L., Doiz, R. K., Uyeyamu, K., and Weiden, S., Digestion and absorption after massive resection of the small intestine, *Gastroenterology,* 16, 126, 1950.
6. Nygaard, K., Resection of the small intestine in rats, *Acta Chir. Scand.,* 133, 407, 1967.
7. Nylander, G., Gastric evacuation and propulsive intestinal motility following resection of the small intestine in the rat, *Acta Chir. Scand.,* 133, 131, 1967.
8. Weisbrodt, N. W., Motility of the small intestine, in *Physiology of the Gastrointestinal Tract,* Johnson, L. R., Ed., Raven Press, New York, 1981, 411.
9. Nemeth, P. R., Kwee, D. J., and Weisbrodt, N. W., Adaptation of intestinal muscle in continuity after jejunoileal bypass in the rat, *Am. J. Physiol.,* 241, G259, 1981.
10. Maxwell, J. D. and McGouran, R. C., Jejunoileal bypass: clinical and experimental aspects, *Scand. J. Gastroenterol.,* 17, 129, 1982.
11. Nemeth, P. R., Kwee, D. J., and Weisbrodt, N. W., Adaptation of intestinal muscle in bypassed loops after jejunoileal bypass in rat, *Am. J. Physiol.,* 244, G599, 1983.
12. Uvelius, B., Arner, A., and Johansson, B., Structural and mechanical alterations in hypertrophic venous smooth muscle, *Acta Physiol. Scand.,* 112, 463, 1981.
13. Seidel, C. L., Allen, J. C., and Bowers, R. L., Mechanical and biochemical alterations of aorta induced by hydralazine hypotension, *J. Pharmacol. Exp. Ther.,* 213, 514, 1980.

14. Uvelius, B., Relation between mechanical and morphological characteristics in urinary bladder smooth muscles, *Acta Physiol. Scand.*, 483, 5, 1980.
15. Gabella, G., Hypertrophic smooth muscle, *Cell Tissue Res.*, 201, 277, 1979.
16. Weisbrodt, N. W., Nemeth, P. R., Bowers, R. L., and Weems, W. A., Functional and structural changes in intestinal smooth muscle after jejunoileal bypass, *Gastroenterology*, 88, 958, 1985.
17. Bortoff, A., Arazozaa, E., and Sillin, L. F., Mechanical properties of functional and defunctionalized intestinal smooth muscle, *Dig. Dis. Sci.*, 29, 564, 1984.
18. Christopher, N., Nemeth, P. R., Eeckhout, C., and Weisbrodt, N. W., Intestinal myoelectric activity in rats with jejunoileal bypass, *Fed. Proc.*, 42, 758, 1983.
19. Nygaard, K., Resection of the small intestine in rats. III. Morphological changes in the intestinal tract, *Acta Chir. Scand.*, 133, 233, 1976.
20. Hanson, W. R., Osborne, J. W., and Sharp, J. G., Compensation by residual intestine after resection in the rat, *Gastroenterology*, 72, 701, 1977.
21. Gleeson, M. H., Cullen, J., and Dowling, R. H., Intestinal structure and function after small bowel by-pass in the rat, *Clin. Sci.*, 43, 731, 1972.
22. Bowers, R. L., Eeckhout, C., and Weisbrodt, N. W., Actomyosin, collagen, and cell hypertrophy in intestinal muscle after jejunoileal bypass, *Am. J. Physiol.*, 250, G70, 1986.
23. Weisbrodt, N. W., Green, G. M., Levan, V. H., and Dudrick, S. J., Effect of pancreatic secretions on transit in bypassed loops of intestine in rats, *Dig. Dis. Sci.*, 30, 78, 1985.
24. Smith, B., *The Neurophysiology of the Alimentary Tract*, Williams & Wilkins, Baltimore, 1972.
25. Atkinson, R. L., Dahms, W. T., Bray, G. A., Lemmi, C., and Schwartz, A. A., Gastrin secretion after weight loss by dieting and intestinal bypass surgery, *Gastroenterology*, 77, 696, 1979.
26. Bloom, S. R. and Polak, J. M., Symposium on hormones and food utilization: gut hormones, *Proc. Nutr. Soc.*, 27, 259, 1978.

Chapter 5

# HYPERTROPHY OF UTERINE SMOOTH MUSCLE

Barbara M. Sanborn

## TABLE OF CONTENTS

## I. CONDITIONS FOR UTERINE HYPERTROPHY

### A. General Comments

Uterine hypertrophy most commonly occurs as a consequence of pregnancy, although it is associated with other conditions as well,[1-4] as described below. Consequently, much of what is known about myometrial hypertrophy is deduced indirectly from comparison of the properties of uterine muscle in the pregnant and nonpregnant states. There are obvious drawbacks to this approach: the heterogeneity of the cell types involved and of their responses to a given stimulus, and the occurrence of both hypertrophy and hyperplasia in one or more of these cell types, at least during early pregnancy. Nonetheless, because of the eventual predominance of myometrial hypertrophy during pregnancy, the comparison of the uterus in the nonpregnant and pregnant states enables some general conclusions to be drawn concerning the hypertrophic state of uterine muscle.

Experimentally, uterine hypertrophy has been produced in response to distention.[1-4] The modulating effects of hormones on this response have been studied. Results of studies involving hormonal treatment can be interpreted as pertaining to hypertrophy only with some caution, however. These treatments are known to induce edema and hyperplasia as well as hypertrophy, especially in the short term. Nonetheless, such studies have given insight into hormonal influences on the uterine response to chronic distention.

The biology and physiology of the uterus has been amply reviewed,[1-4] and the reader is referred to these sources for original references. This chapter will focus on information (or lack thereof) on the properties of hypertrophic myometrium.

### B. Experimentally Induced Hypertrophy

The uterus is markedly affected by hormones. Both estrogens and progestins have pronounced effects on hyperplasia of different cellular elements,[1-4] but this is not the subject of the current review. Estrogen and progesterone treatments in ovariectomized animals can also bring about hypertrophy, judged not only by morphological appearance, but also by an increase in dry weight and RNA/DNA ratio.[1-10]

As summarized by Reynolds,[1] the phenomenon of hypertrophy in response to distention has been known clinically and experimentally for a long time. In the absence of ovarian hormones, significant mitoses as well as hypertrophy occur in myometrial cells in response to chronic distention.[1,11] Csapo and colleagues[12,13] demonstrated an increase in RNA/DNA ratio as well as morphological hypertrophy of myometrium in distended uteri from parturient and nonpregnant ovariectomized rats.

Hormones markedly alter the response of the myometrium to stretch. Estrogens limit the growth of the distended uterus,[14] while progesterone alters the stretch/response relationship, making the uterus less sensitive to the stretching stimulus.[15,16] Reynolds[1] suggested that this action of progesterone is important in enabling the uterus to accommodate an increased load (such as occurs during pregnancy) without undergoing continuing structural changes. Estradiol and progesterone also have a number of indirect effects on the uterus which may affect the response of the tissue to hypertrophic stimuli such as distention. For example, they alter the nature of uterine connective tissue[16,17] and have complex effects on filament assembly[18] and gap junction formation between myometrial cells.[19-21]

### C. Hypertrophy During Pregnancy

The accommodation of the uterus to the condition of pregnancy consists of an early period of proliferation of uterine cells with increased edema and vascularity. Under the influence of progesterone, the products of conception can initially increase in size with a minimum of stretch-induced hypertrophy of the myometrium. However, as pregnancy progresses, the increased distention triggers uterine enlargement in the form of myometrial hypertrophy. Finally, toward the end of pregnancy, the influence of estrogen limits further growth; this period is characterized by increased stretch with decreased growth.[1,3] Marshall[4] points out that this situation assures that the capacity for maximal tension development, which is achieved when the muscle attains resting length (close to term in the rat and human), is available for the task of parturition.

In the rabbit, enlargement of the uterus is in the longitudinal direction.[1] There is maximal tension per radius of conception when the conceptus is spherical, but less tension when the uterus converts to a cylindrical form. In contrast, changes in the human uterus during the first two trimesters are characterized by increases in the dorso-ventral and dorso-lateral diameters. During the third trimester, the dorso-ventral diameter remains the same but the dorso-lateral diameter continues to increase. As pregnancy progresses, the uterine wall becomes increasingly thinner and the conceptus more easily palpable.

In contrast to the corpus, the isthmus of the human uterus fails to undergo hypertrophy. This segment becomes quite distensible as the size of the uterus increases and is eventually taken up into the lower segment of the uterus.

## D. Other Instances of Hypertrophy

A number of other conditions have been associated with myometrial hypertrophy. Reports linking myometrial hypertrophy with the presence of an intrauterine device in rats[22,23] and humans[24] are difficult to evaluate in the absence of strong quantitative or morphological evidence. However, the suggestion of a role for prostaglandins in this process, as evidenced by the ability of aspirin and indomethacin to inhibit it,[23] warrants further study.

Uterine hypertrophy has also been associated with hyperandrogenism. Treatment of immature, ovariectomized rats with testosterone derivatives induced uterine hypertrophy which differed qualitatively and quantitatively from that induced by estradiol.[25,26] Diffusely thickened myometrium has been reported in a patient with an ovarian hilar cell tumor.[27] The fact that no hypertrophy or hyperplasia of the endometrium was present suggests that the effects may be due to excess androgens rather than peripheral conversion to estrogens or interaction of androgens with the uterine estrogen receptor.

Tetrahydrocannabinol has been reported to cause uterine hypertrophy in rats.[28] Finally, conditions termed "idiopathic hypertrophy", described as increased uterine wall thickness resulting from an increase in the number of muscle fibers with a corresponding increase in connective tissue, have been described.[29,30]

## II. CHANGES ACCOMPANYING HYPERTROPHY

### A. Morphological Changes

The most obvious change accompanying uterine hypertrophy is an increase in tissue mass. As mentioned earlier, this increase can in fact reflect edema, hypertrophy, hyperplasia, or all three, depending on the hormonal state of the tissue and the stimulus. Nonetheless, clear morphological evidence of hypertrophy of myometrial cells during pregnancy and in response to experimentally-induced stretch and/or hormone treatment has been noted.

Myometrial hypertrophy is pronounced after estrogen treatment.[1,10,31] In ovariectomized rats, estrogens do not increase the labeling index or cell number in the myometrium[9] but increase the thickness of the muscle layer.[10] Progesterone has little effect alone. In the rabbit, the effect of estrogen is less marked and progesterone produces a greater hypertrophy than in the rat.[5]

In the absence of ovarian hormones, stretch-induced hypertrophy in the rabbit uterus is accompanied by edema and enlargement of small blood vessels.[1] This is thought to reflect impaired blood flow triggered by increased resistance at the distention site. There is an increase in the thickness of the muscle layer, which is due to both hypertrophy and hyperplasia.[1,5,13] However, in contrast to the hypertrophy associated with hormone treatment or pregnancy, distention-induced hypertrophy is reported to be associated with a loss of well-defined longitudinal and circular components.[32] Vascular changes are less pronounced in distention-induced hypertrophy than in hormone-induced hypertrophy, suggesting that they may result primarily from the hormonal stimulus rather than the hypertrophic stimulus per se.

In pregnancy, uterine hypertrophy has been documented by measurement of increases in muscle cell length and thickness.[1,10,33] Term uteri weigh approximately 3.5, 4.5, and 3.4 times as much as nongravid uteri in mice, rabbits, and humans, respectively. Smooth muscle cell length increases 2 to 4, 4 to 8, and 17 to 40-fold, respectively, with 0 to 2-fold increases in cell width.[1,33,34] There is a marked increase in size of the small blood vessels as well.[1]

Mitoses are rarely seen in myometrial cells after the initial period of pregnancy. The increased edema is accompanied by an increased transparency of cell cytoplasm.[1] There is an increase in the number of myofibrils per cell. The connective tissue, particularly that associated with uterine arteries, becomes more dense and abundant. Estradiol, progesterone, and relaxin probably influence this process.[35-38] Hypertrophy of fibrous and elastic tissue occurs throughout the uterus.

The increase in growth of uterine cells during pregnancy involves differentiation to give rise to some elongated fusiform cells.[39] A 16-fold increase in uterine muscle mass and a doubling of nuclear volume has been estimated in the human uterus during pregnancy.[40] More recently, histological evidence for a swelling of all diploid nuclei during pregnancy and for polyploidy in a small number of myometrial nuclei from uteri of pregnant women has been reported.[41] Polyploidy appeared before distention, became a factor at 16 weeks, and persisted after pregnancy.

Wathes and Porter[20] have made the interesting observation that distention is associated with a small increase in gap junction formation in the postparturient rat uterus in the absence of estrogen. Estrogen treatment also causes a slight increase, but only the combination of distention plus estrogen produces numbers of gap junctions equivalent to those present in the parturient uterus. Distention was only applied for 48 to 70 hr and involution had occurred over this period, so it is not possible to directly equate sensitivity to estrogen in the distended state in the postpartum rat uterus with the hypertrophic state. However, the suggestion that distended, hypertrophic muscle is more sensitive to an estrogenic stimulus with respect to gap junction formation than nondistended tissue becomes important, since the relationship between the number of gap junctions and the rate in rise in intrauterine pressure suggests that coordination of myometrial activity requires gap junction formation.[20] The response of the uterus to distention in the presence of hormones, which involves hypertrophy, can then be seen as preparing the uterus to respond near term to estrogens with the formation of gap junctions needed for efficient delivery of the fetus.

## B. Biochemical Changes

A clear indication of uterine hypertrophy is an increase in the RNA/DNA ratio. This has been noted in a number of species both in naturally occurring pregnancy and in stretch-induced hypertrophy in experiments with pregnant and nonpregnant animals.[2,12,13,25,33] In pregnancy, these changes are accompanied by increases in phospholipid and total nitrogen per DNA of the order of 5 to 7-fold.[3,33,42-45]

Specific proteins of the contractile apparatus have been shown to increase with uterine hypertrophy.[33] In early studies, Csapo et al.[46] found increased crude actomyosin in the pregnant rabbit uterus compared to the nonpregnant uterus. The same findings pertain in other species.[2] Estrogens also increased the actomyosin content of the uterus after its decline following ovariectomy.[46] Needham and Williams[47] later showed that there were many other proteins in the crude actomyosin fractions including salt-soluble collagen, but confirmed that actomyosin was indeed increased with the hypertrophy accompanying pregnancy. Michael and Schofield[48] found a positive relationship between actomyosin content and tension during pregnancy, and between the number of fetuses and actomyosin content. The same parameters correlated less well with hormonal treatment, suggesting that distention was the primary factor in the hypertrophy.

A number of studies suggest that the control of contractile elements in the uterus is related to the phosphorylation of myosin light chains as proposed for control of other smooth muscles.[49] Phosphorylation of rat uterine myosin light chains is increased during contraction[50,51] and decreased by isoproterenol[51] and relaxin-induced[52,53] relaxation. Rat,[52,53] ovine,[54] rabbit,[55,56] and porcine[57] uterine extracts contain myosin light chain kinase. Purified rat[44] and porcine[57] uterine myosin light chain kinase can be phosphorylated by cAMP-dependent protein kinase. Myosin light chain kinase is a calcium-calmodulin-dependent enzyme in the uterus.[52,53,57,58] Phosphorylation of this enzyme in the absence of calcium-calmodulin incorporates 2 mol of phosphate per mole of enzyme and decreases the affinity of the enzyme for calcium-calmodulin.[57,58] The same decrease in affinity is observed in crude myosin light chain kinase extracts from tissues treated with relaxin,[53] suggesting that this mechanism may be operating *in situ*.

Matsui et al.[55,56] have recently measured myosin light chain kinase, cAMP-dependent protein kinase, and calmodulin activity throughout pregnancy in the rabbit. Both myosin light chain kinase and calmodulin activities are increased per uterus compared to values in ovariectomized or nonpregnant rabbits. However, changes in specific activity are less marked. Over the period of day 10 to day 28, myosin light chain kinase specific activity per mg protein declines slightly but remains elevated compared to that in nonpregnant controls while calmodulin activity remains constant. A more marked decline in cAMP-dependent protein kinase activity was noted over this period, prompting the authors to suggest that by the end of term, the uterus is biochemically equipped to favor contraction over relaxation. While these changes in proteins which affect the uterine contractile apparatus accompany the hypertrophy associated with pregnancy, it remains to be established whether they would occur in experimentally-induced conditions. Matsui et al.[56] have presented some evidence suggesting that estrogens and progesterone increase the specific activity of uterine myosin light chain kinase and calmodulin.

The collagen content of the uterus increases during pregnancy, althouth the non-collagen/collagen protein ratio also increases.[3,33,59,60] In addition there is morphological evidence of increased amounts of collagen accompanying uterine hypertrophy.[1] Early studies on collagen were relatively crude, relying on the amount of insoluble protein. However, estimations of collagen based on hydroxyproline measurements suggest that collagen cannot account for all of the insoluble protein.[59,61,62] Increases in elastin have been implicated. In the human uterus, pregnancy is accompanied by a 7-fold increase in collagen and a 5- to 6-fold increase in elastin content.

Collagen content increases during pregnancy but not in the sterile horn of unilaterally pregnant rats.[63,64] The presence of a wax implant also is associated with an increase in collagen. These data are consistent with a requirement for distention as a stimulus and suggest that an increase in collagen content is part of the hypertrophic response of the muscle. The suggested effects of hormones on collagen turnover during pregnancy[36-38] may play a permissive or secondary role.

Montfort and Perez-Tamayo[60] noted no morphological modification of collagen fibers in the pregnant rat or human myometrium, simply more separation by amorphous material. The collagen content per horn was related to the number of fetuses per horn, substantiating earlier suggestions that distention plays a role in this response.

An increase in energy supply accompanies the uterine hypertrophy during pregnancy. ATP content, easily hydrolyzable ATP phosphate, and creatine phosphate all increase during pregnancy.[33,45] The amount of energy immediately available to uterine muscle compares favorably with that of other smooth muscles but is lower than found in skeletal muscle.

The concentrations of a number of enzymes involved in glycolysis increase during pregnancy.[33,45] While a number of enzymes have been measured in the pregnant state, there are few studies of progressive changes that could be related to uterine hypertrophy. It is clear, however, that the metabolic capability of the tissue increases during pregnancy.

## C. Biophysical Changes

The effect of hormones on the hypertrophic response of the uterus to the stimulus of distention have already been mentioned. Progesterone increases the threshold for stretch-induced hypertrophy by mechanisms which are not understood.[15,16] Although progesterone reduces the resistance of the myometrium to stretching, the increased tension generated by the growing conceptus compensates for the diminished tone. Reynolds[1] argues that since the length to which the uterus must be stretched to induce growth is doubled by progesterone treatment of ovariectomized rabbits, the stimulus for hypertrophy is increased tension rather than increased length of the muscle fibers in this species.

Conrad et al.[65] examined passive stretch relationships in human uterine muscle. Despite problems posed by definition of resting length (pregnant muscle strips demonstrate more passive recoil) and tension (due to differences in degree of stress relaxation), it was clear that pregnancy (and the accompanying hypertrophy) altered passive length-tension properties. The tension generated per unit of elongation was significantly lower in the strips from pregnant as opposed to nonpregnant women. This was true when comparing comparable lengths of muscle in vitro and in vivo. Similar findings have been reported in the rabbit.[16] The increase in distensibility was attributed to an effect of progesterone on the connective tissue of the uterus.

## D. Electrochemical Changes

Measurements of ion distribution and resting potential in the uterus have been subject to controversy because of the difficulty of estimating extracellular space and compartmentalization of ions.[4,66,67] The myometrium possesses considerably higher concentrations of sodium and chloride than do skeletal or cardiac muscles. Membrane potential in myometrium depends on $Na^+$ and $Ca^{++}$ permeability. Regardless of the methodological problems, it is clear that the nonpregnant uterus of many species has a less negative resting potential than the pregnant uterus.[1,4,10,66,67] Estrogens generally increase the resting potential, while progesterone has been reported to either increase or have no effect on it.

Hormones also have marked effects on action potentials in the uterus.[4,66,67] Estrogen treatment is associated with increased electrical excitability whereas progesterone diminishes the frequency of electrical activity measured by the sucrose gap method. The effect of progesterone predominates throughout most of pregnancy, although this varies between species. Since ionic composition does not change markedly in the uterus during pregnancy, shifts in resting and action potential have been attributed to changes in ionic permeability.[66,67] Casteels and Kuriyama[68] have evidence of an increase in $K^+$ permeability during pregnancy which they attribute to the action of progesterone. The relationship of these changes to the uterine hypertrophy occurring during pregnancy remain to be defined.

Hormones also have dramatic changes on the formation of gap junctions between myometrial cells.[19-21] Alterations in these parameters have been studied near the time of parturition but not as a function of stage of uterine hypertrophy.

## III. PARTIAL REVERSAL OF HYPERTROPHY

The reversal of hypertrophy triggered by experimental distention has not been studied extensively. However, the myometrial hypertrophy accompanying pregnancy is rapidly reversed in the postpartum period. Involution involves a reduction in the size but not the number of myometrial cells.[69,70] The degradation of uterine proteins contributes to an increase in urinary nitrogen. Removal of the uterus results in a substantial decrease in urinary nitrogen.[69]

Involution is also characterized by a marked increase in collagenase activity and a breakdown of uterine collagen.[35,36,60] Estradiol, progesterone, and relaxin inhibit this breakdown to varying degrees.[36-38]

Despite the remarkable degree of involution taking place in the postpartum period, the uterus never reverts completely to its original state. Uterine arteries remain widened and thick-walled, and deposits of connective tissue are still observed around the blood vessels.[1,71] The musculature remains significantly enlarged compared to that of a nonpregnant uterus.[71] These changes offer at least a theoretical advantage in subsequent pregnancies.

## IV. SUMMARY AND CONCLUSIONS

Myometrial hypertrophy can occur naturally as a consequence of the hormonal and distention stimuli associated with pregnancy. In this state, hypertrophy is associated with an increase in size of the muscle cells, including an increase in RNA, protein, actomyosin, ATP, and glycolytic enzymes per cell, and an increase in resting potential. Accompanying these changes are increased vascularity, increased connective tissue, and increased distensibility. Hormonal influences fine tune the response to adapt the uterus to the purpose of carrying an enlarging conceptus while maintaining relative contractile quiescence. At the time of parturition, hormones (endocrine and paracrine) convert this hypertrophied tissue into a coordinated, efficient contractile organ well equipped for the task of the delivery of the conceptus. Following delivery, uterine hypertrophy is rapidly but not completely reversed.

Myometrial hypertrophy can also be produced in the nonpregnant state by treatment with estrogens or chronic distention. Although uterine weight changes are similar, there are qualitative and quantitative differences between the hypertrophy generated by these stimuli in comparison with pregnancy. From the information available, it seems logical to conclude that the combined influences (direct and indirect) of a number of hormones during pregnancy influence the nature of the uterine response to the hypertrophic stimuli of estrogens plus distention.

Despite many studies, the precise mechanisms by which hormones, distention, and the combination of the two, trigger uterine hypertrophy are still not known. The biochemical properties of the hypertrophied muscle per se are not well defined. With more recent development of techniques for quantitating uterine proteins, an increased understanding of the components which comprise and regulate the contractile apparatus in this tissue, and more refined methods for studying electrophysiology, it should be possible to gain new insights into the phenomenon of uterine hypertrophy. More precise comparisons could then be made between the properties of muscle which hypertrophied in response to distention, hormonal treatment, or pregnancy. Hopefully such studies will enable elucidation of the mechanisms involved.

## ACKNOWLEDGMENTS

The helpful comments and suggestions of Dr. G. T. Ross and R. K. Creasy are gratefully acknowledged.

## REFERENCES

1. Reynolds, S. R. M., Ed., *Physiology of the Uterus,* Hafner Publishing, New York, 1965, chap. 13-20, 30-32.
2. Wynn, R. M., Ed., *Biology of the Uterus,* Plenum Press, New York, 1977.
3. Finn, C. A. and Porter, D. G., *The Uterus,* Publishing Sciences Group, Acton, Mass., 1975, chap. 11-17.
4. Marshall, J. M., Physiology of the uterus, in *The Uterus,* Norris, H. J., Hertig, A. T., and Abell, M. R., Eds., Williams & Wilkins, Baltimore, 1973, 89.
5. DeMattos, C. E. R., Kempson, R. L., Erdos, T., and Csapo, A., Stretch-induced myometrial hypertrophy, *Fertil. Steril.,* 18, 545, 1967.
6. Lee, A. E., Cell division and DNA during continuous oestrogen treatment, *J. Endocrinol.,* 55, 507, 1972.
7. Lee, A. E., Effects of oestrogen antagonists on mitosis and $^3$H-estradiol binding in the mouse uterus, *J. Endocrinol.,* 60, 167, 1974.
8. Lee, A. E., Rogers, L. A., and Trinder, G., Changes in cell proliferation rate in mouse uterine epithelium during continuous oestrogen treatment, *J. Endocrinol.,* 61, 117, 1974.
9. Martin, L., Finn, C. A., and Trinder, G., Hypertrophy and hyperplasia in the mouse uterus after oestrogen treatment: an autoradiographic study, *J. Endocrinol.,* 56, 133, 1973.
10. Kuriyama, H. and Suzuki, H., Changes in electrical properties of rat myometrium during gestation and following hormonal treatments, *J. Physiol.,* 260, 315, 1976.
11. Allen, E., Smith, G. M., and Gardner, W. U., Accentuation of the growth effect of theelin on genital tissues of the ovariectomized mouse by arrest of mitotis with colchicine, *Am. J. Anat.,* 61, 321, 1937.
12. Csapo, A. I. and Corner, G. W., The effect of estrogen on the isometric tension of rabbit uterine strips, *Science,* 117, 162, 1953.
13. Csapo, A. I., Erdos, T., deMattos, C. R., Gramss, E., and Moscowitz, C., Stretch-induced uterine growth, protein synthesis and function, *Nature,* 207, 1378, 1965.
14. Reynolds, J. R. M. and Kaminester, S., Failure to obtain in immature rabbits uterine growth by chronic distention, *Proc. Soc. Exper. Biol. Med.,* 36, 257, 1937.
15. Reynolds, J. R. M. and Allen, W. M., The effects of progestin on growth of the uterus in response to chronic distention, *Anat. Rec.,* 69, 481, 1937.
16. Currie, W. B., Uterine excitability and distensibility influenced by treatment in vitro with progesterone, *Animal Reprod. Sci.,* 2, 225, 1979.
17. Burack, E., Wolfe, J. M., and Wright, A. W., Effects of the administration of estrogen on the connective tissues of the genital tract of the rat, *Endocrinology,* 30, 335, 1942.
18. Kelly, R. E. and Verhage, H. G., Hormonal effects on the contractile apparatus of the myometrium, *Am. J. Anat.,* 161, 375, 1981.
19. Garfield, R. E., Kannan, M. S., and Daniel, E. E., Gap junction formation in myometrium: control by oestrogens, progesterone, and prostaglandins, *Am. J. Physiol.,* 238, C81, 1980.

20. Wathes, D. C. and Porter, D. G., Effect of uterine distention and oestrogen treatment on gap junction formation in the myometrium of the rat, *J. Reprod. Fertil.*, 65, 497, 1982.
21. Puri, C. P. and Garfield, R. E., Changes in hormone levels and gap junctions in the rat uterus during pregnancy and parturition, *Biol. Reprod.*, 27, 967, 1982.
22. Pollak, A., Gellman, M., and Nebel, L., Persisting structural alterations in the uterus and ovaries of rats induced by intrauterine devices, *J. Reprod. Fertil.*, 33, 129, 1973.
23. Chaudhuri, G., Inhibition of aspirin and indomethacin of uterine hypertrophy induced by an IUD, *J. Reprod. Fertil.*, 43, 77, 1975.
24. Honore, L. H., Menorrhagia, diffuse myometrial hypertrophy and the intrauterine device: a report of fourteen cases, *Acta Obstet. Gynecol. Scand.*, 58, 283, 1979.
25. Lerner, L. J., Hilf, R., Turkheimer, A. R., Michel, I., and Engel, S. L., Effects of hormone antagonists on morphological and biochemical changes induced by hormonal steroids in the immature rat uterus, *Endocrinology*, 78, 111, 1966.
26. Yudaev, N. A. and Pokrovskij, B. V., Mechanism of androgen action on the uterus: influence on the RNA biosynthesis and relationship to the estrogen action, *Endocrinol. Exp.*, 6, 131, 1972.
27. McBee, A. and Stachura, I., Ovarian hilar tumour with coexistent polycystic ovary syndrome and myometrial hypertrophy, *Gynecol. Oncol.*, 8, 370, 1979.
28. Solomon, J., Cocchia, M. A., and DiMartino, R., Effect of delta-9-tetrahydrocannabinol on uterine and vaginal cytology of ovariectomized rats, *Science*, 195, 875, 1977.
29. Williams, J. T. and Kinney, T. D., Myometrial hypertrophy (so-called fibrous uteri), *Am. J. Obstet. Gynecol.*, 47, 380, 1944.
30. Cook, C. B. and Wooster, L. D., Idiopathic uterine hypertrophy, *South. Med. J.*, 61, 246, 1968.
31. Allen, E., Reactions of the genital tissues to estrogens, *Cold Spring Harbor Symp. Quant. Biol.*, 5, 104, 1937.
32. DeClue, J. W. and King, T. M., Observations on uterine growth affected by mechanical distention, *Am. J. Obstet. Gynecol.*, 105, 517, 1969.
33. Hamoir, G., Biochemistry of the myometrium, in *Biology of the Uterus*, Wynn, R. M., Ed., Plenum Press, New York, 1977, 377.
34. Stieve, H., Uber die Neubildung von Muskelzellen in der Wand der schwangeren menschlichen Gebarmutter, *Zentralbl. f. Gynaekol.*, 53, 2706, 1929.
35. Woessner, J. F., Jr., Catabolism of collagen and noncollagen protein in the rat uterus during postpartum involution, *Biochem. J.*, 83, 304, 1962.
36. Woessner, J. F., Jr., Total, latent and active collagenase during the course of post partum involution of the rat uterus, *Biochem. J.*, 180, 95, 1979.
37. Tansey, R. R. and Padykula, H. A., Cellular responses to experimental inhibition of collagen degradation in the post partum rat uterus, *Anat. Rec.*, 191, 287, 1978.
38. Frieden, E. H., Adams, W. C., and Heider, M., Stimulation of amino acid incorporation into uterine proteins by relaxin, *Fed. Proc.*, 42, 1873, 1983.
29. Jeener, R., Acides nucleiques et phosphatases au cours de phenomenes de croissance provoques par l'oestradiol et la prolactine, *Biochim. Biophys. Acta*, 2, 439, 1948.
40. Straus, G., Histoplanimetrische Untersuchungen am menschichen Uterus, *Arch. Gynaekol.*, 207, 572, 1969.
41. Van der Heijden, F. L. and James, J., Polyploidy in the human myometrium, *Z. Mikrosk. Anat. Forsch. Leipzig*, 89, 18, 1975.
42. Wakid, N. W. and Needham, D. M., Effect of estradiol injection on the cytoplasmic fractions of the myometrium in the ovariectomized rat, *J. Biophys. Biochem.*, 10, 136, 1961.
43. Wakid, N. W. and Needham, D. M., Cytoplasmic fraction of the rat myometrium II. Localization of some cellular constituents in the pregnant and ovariectomized states, *Biochem. J.*, 76, 95, 1960.
44. Afting, E. G., Becker, M. L., and Elce, J. S., Proteinase and proteinase-inhibitor activities of rat uterine myometrium during pregnancy and involution, *Biochem. J.*, 177, 99, 1978.
45. Brinkworth, R. I. and Masters, C. J., Effect of pregnancy on the turnover of lactic dehydrogenase and its isozymes in mouse tissues, *Mech. Ageing Dev.*, 8, 69, 1978.
46. Csapo, A., Actomyosin formation of estrogen action, *Am. J. Physiol.*, 162, 406, 1950.
47. Needham, D. M. and Williams, J. M., The proteins of the dilution precipitate obtained from salt extracts of pregnant and nonpregnant uterus, *Biochem. J.*, 89, 534, 1963.
48. Michael, C. A. and Schofield, B. M., The influence of the ovarian hormones on the actomyosin content and the development of tension in uterine muscle, *J. Endocrinol.*, 44, 501, 1969.
49. Adelstein, R. S. and Eisenberg, E., Regulation and kinetics of the action-myosin-ATP interaction, *Ann. Rev. Biochem.*, 49, 921, 1980.
50. Janis, R. A., Moats-Staats, B. M., and Gualtieri, R. T., Protein phosphorylation during spontaneous contraction of smooth muscle, *Biochem. Biophys. Res. Commun.*, 96, 265, 1980.

51. Janis, R. A., Barany, K., Barany, M., and Sarimiento, J. G., Association between myosin phosphorylation and contraction of rat uterine smooth muscle, *Mol. Physiol.*, 1, 3, 1981.
52. Nishikori, K., Weisbrodt, N. W., Sherwood, O. D., and Sanborn, B. M., Relaxin alters rat uterine myosin light chain phosphorylation and related enzymatic activities, *Endocrinology,* 111, 1743, 1982.
53. Nishikori, K., Weisbrodt, N. W., Sherwood, O. D., and Sanborn, B. M., Effects of relaxin on rat uterine myosin light chain kinase activity and myosin light chain phosphorylation, *J. Biol. Chem.*, 258, 2468, 1983.
54. Lebowitz, E. A. and Cooke, R., Phosphorylation of uterine smooth muscle myosin permits actin-activation, *J. Biochem. (Tokyo)*, 85, 1489, 1979.
55. Matsui, K., Higashi, K., Fukanaga, K., Maeyama, M., and Miyamoto, E., Dissociation of alterations in levels of myosin light chain kinase and cyclic AMP-dependent protein kinase in rabbit myometrium during pregnancy, *Endocrinol. Jpn.*, 29, 675, 1982.
56. Matsui, K., Higashi, K., Fukanaga, K., Miyazaki, K., Maeyama, M., and Miyamoto, E., Hormone treatments and pregnancy alter light chain kinase and calmodulin levels in rabbit myometrium, *J. Endocrinol.*, 97, 11, 1983.
57. Higashi, K., Fukanaga, K., Matsui, K., Maeyema, M., and Miyamoto, E., Purification and characterization of myosin light chain kinase from porcine myometrium and its phosphorylation and modulation by cyclic-AMP-dependent protein kinase, *Biochim. Biophys. Acta,* 747, 332, 1983.
58. Nishikori, K., Burroughs, M., and Sanborn, B. M., cAMP-dependent phosphorylation of rat uterine myosin light chain kinase activity partially mimics the effect of relaxin treatment, 65th Annual Meeting of the Endocrine Society, June 1983, 291.
59. Harkness, M. L. and Harkness, R. D., The collagen content of the reproductive tract of the rat during pregnancy and lactation, *J. Physiol. (London)*, 123, 492, 1954.
60. Montfort, I. and Perez-Tamayo, R., Studies on uterine collagen during pregnancy and puerperium, *Lab. Invest.*, 10, 1240, 1961.
61. Needham, D. M. and Cawkwell, J. M., Some properties of the actomyosin-like protein of the uterus, *Biochem. J.*, 63, 337, 1956.
62. Needham, D. M. and Williams, J. M., Salt soluble collagen in extracts of the uterus muscle and foetal metamyosin, *Biochem. J.*, 89, 546, 1963.
63. Harkness, M. L. and Harkness, R. D., The distribution of the growth of collagen in the uterus of the pregnant rat, *J. Physiol.*, 132, 492, 1956.
64. Cullen, B. M. and Harkness, R. D., Collagen formation and changes in cell population in the rat's uterus after distention with wax, *Quant. J. Exp. Physiol.*, 53, 33, 1968.
65. Conrad, J. T., Johnson, W. L., Kuhn, W. K., and Hunter, C. A., Passive stretch relationships in human uterine muscle, *Am. J. Obstet. Gyn.*, 96, 1055, 1966.
66. Kao, C. Y., Electrophysiological properties of the uterine smooth muscle, in *Biology of the Uterus,* Wynn, R. M., Ed., Plenum Press, New York, 1977, 423.
67. Daniel, E. E. and Lodge, S., Electrophysiology of myometrium, in *Uterine Contraction-Side Effects of Steroidal Contraceptives,* Josimovich, J. B., Ed., John Wiley & Sons, New York, 1973, 19.
68. Casteels, R. and Kuriyama, H., Membrane potential and ionic content in pregnant and nonpregnant rat myometrium, *J. Physiol. (London)*, 177, 263, 1965.
69. Wilson, J. R., The puerperium, in *Obstetrics and Gynecology,* Wilson, J. R., Carrington, E. R., and Ledger, W. J., Eds., C. V. Mosby, St. Louis, 1983, 585.
70. Walker, J., MacGillvray, J., and Macnaughton, M. C., *Combined Textbook of Obstetrics and Gynecoloy,* Churchill Livingstone, Edinburgh, 1976, 397.
71. Beker, J. C., Aetiology of eclampsia, *J. Obstet. Gynaecol. Br. Emp.*, 55, 756, 1948.

Chapter 6

# THE POSSIBLE ROLE OF SYMPATHETIC ADRENERGIC NERVES IN THE HYPERTROPHIC RESPONSE OF VASCULAR SMOOTH MUSCLE*

Rosemary D. Bevan

## TABLE OF CONTENTS



* This manuscript is supported by grants USPHS HL32383 and HL32985.

## I. INTRODUCTION

This chapter addresses some of the interactions between peripheral sympathetic nerves and vascular smooth muscle (VSM), particularly the way in which the nerves may influence hypertrophy of the blood vessel wall. The sympathetic outflow from the brain stem, modulated by many reflex homeostatic mechanisms and also by influences from higher brain centers, is the major motor pathway to the circulation. It affects almost all blood vessels containing smooth muscle through postganglionic adrenergic neurons that secrete norepinephrine (NE). In some mammalian vascular beds other types of nerves have been identified which are frequently dilator and also sensory, some of which are peptidergic, but their role is not well defined and there is much species variation in the neurotransmitters and neuromodulators involved.[1,2] The relationship of sympathetic nerves to the arterial side of the circulation has been studied much more than to the venous, although many of these capacitance vessels are regulated by adrenergic nerves.

In this book, hypertrophy is defined as an increase in the amount of smooth muscle present in the wall of hollow organs and includes an increase in cell size, cell number, and extracellular constituents of which the major ones are collagen and elastin and also proteoglycans and glycosoaminoglycans synthesized by smooth muscle cells. It implies an increase over the "normal" condition for age and sex. In the case of the vascular "organ", a vast network throughout the body, this definition would include increases in length and number of vessels containing smooth muscle. Hypertrophy is a normal response of muscle to an increased stretch and/or work load. It can be induced also by growth factors and hormones. This broad definition also encompasses the vascular response to injury, but this aspect is not considered here.

This chapter is concerned with the way in which sympathetic nerves may influence this hypertrophic response. An important emphasis is that because of the heterogeneity of the structure and adrenergic functional features of the vasculature, conclusive evidence for such an effect is hard to find and even were it found in one vascular bed, simple extrapolation to others is at present unjustified.

Reference to reviews of literature on specific topics mentioned in the chapter is made whenever possible.

## II. COMPONENTS OF THE SYMPATHETIC INFLUENCE RELATED TO HYPERTROPHY

Activation of peripheral sympathetic nerves may affect a blood vessel in four major ways by: (1) altering the level of active tone, (2) increasing passive wall tension by elevation of intraluminal pressure, (3) maintaining a trophic influence, and (4) increasing cell metabolism (Figure 1).

At rest the spontaneous low level discharge from the brain stem maintains a level of vasomotor tone, particularly of the outermost smooth muscle cells of innervated blood vessel walls. Epinephrine and norepinephrine (NE) are also released from the adrenal medulla and acting as circulating hormones exert an effect on the entire cardiovascular system, preferentially influencing the inner smooth muscle cells of the vessel wall. General activation of the sympathetic system increases heart force and rate and total peripheral resistance, reduces venous capacitance, and thereby in turn increases venous return and cardiac output. The result is an elevation of blood pressure, both systolic and diastolic. All vascular beds are not innervated to the same degree by sympathetic nerves and this combined with other regional neuroeffector differences means that the increase in cardiac output is differentially distributed. At the same time, the level of

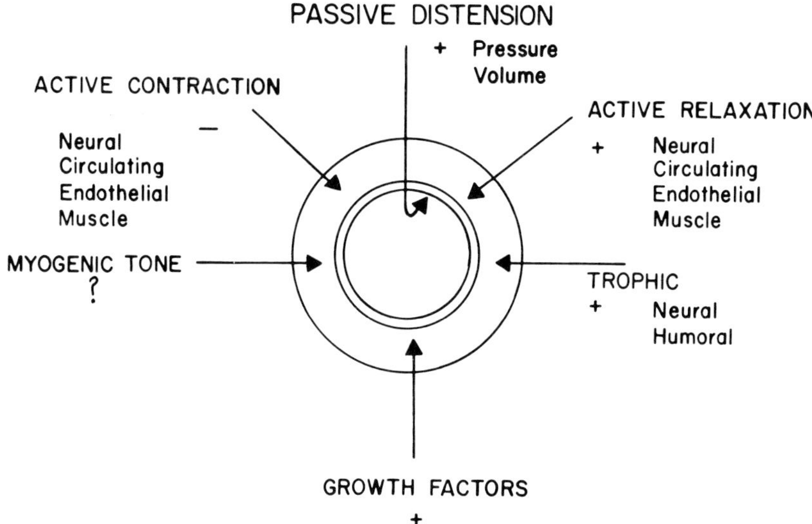



FIGURE 1.   Some factors influencing wall "Hypertrophy".

circulating catecholamines increases, which reinforces or opposes the effect of sympathetic nerve activation depending on the amount of activation and the number and type of receptors present on specific vascular effector tissues.

Nerve impulses, by releasing NE and other substances from the terminal varicosities of sympathetic nerves, exert moment-to-moment control of the state of contraction or "tone" of the smooth muscle. Variations in many components of the neuroeffector mechanism account for some of the differences, both quantitative and qualitative, in the nature of the immediate response to activation of the sympathetic nervous system. These differences occur not only between vascular beds, but also as blood vessels successively branch and get smaller.[3-5]

Many other actions of the sympathetic nervous system and circulating catecholamines have been identified, such as regulation of secretion of a number of hormones and peptides and an influence on intra- and extravascular fluid volume and capillary permeability, which can also affect the vascular smooth muscle cell. In addition, catecholamines mobilize substrate from stored depots in the body and in cells as an energy source for metabolism.[6] Very little is known concerning the effects of catecholamines on smooth muscle cell energy metabolism, although this may be an important influence on the ability of the cell to respond to environmental demands and could affect cell mechanisms involved in hypertrophic responses.

Variations in neuroeffector mechanisms may also modify the long term regulatory or trophic action of sympathetic nerves on the vasculature which have been identified.[7] The term "trophic" is derived from a word meaning nutritional and in the context of this chapter we are interested in such effects of sympathetic nerves on the blood vessel wall. In the interactions of motor nerves with skeletal muscles three components of this long-term regulatory effect have been identified, (1) nerve impulse regulation of activity, (2) a specific influence of the neurotransmitter other than that associated with changes in tone, and (3) the release of unidentified trophic factors. Knowledge of trophic influences on vascular smooth muscle is not as advanced as in skeletal muscle but there is evidence to suggest that they occur and affect growth and development of the vasculature and proliferation and differentiation of the smooth muscle.

Among other substances known to be present and released from the nerve terminals are ATP and neuropeptide Y, which are thought to be involved in nerve-muscle interactions. NE acts on the cell membrane to initiate contraction through $\alpha$-adrenoceptors and possibly other low-affinity sites and dilation through $\beta$-adrenoceptors, by changing its permeability to ions and/or by initiating complex membrane and intracellular processes. Purinergic substances and neuropeptide Y may act by modulating adrenergic transmission or else separately on their specific systems. Many other types of receptors are present on smooth muscle cells, among them are those for other amines, hormones, peptides, and growth factors. Some initiate contraction or relaxation and others complex membrane and intracellular events not immediately related to changes in tone. This makes it difficult to identify a unique influence of sympathetic nerves on smooth muscle cell metabolism related to such events as cell proliferation, hypertrophy, or synthesis of extracellular material other than that which occurs secondarily to other events.

## III. SYMPATHETIC NERVE REGULATION OF VASCULAR TONE

In order to understand ways in which sympathetic nerves might influence hypertrophic responses it is necessary to know about the anatomical and functional relationship of the nerves to the smooth muscle cells.

Considerable heterogeneity occurs in different sections of the vasculature not only in its nervous control but also in the structure of the wall and the function of the muscle. Blood vessels vary in structural design, particularly in their adventitia and media.[8,9] The walls of blood vessels are composed of three layers: the inner endothelium, a muscle layer, and an outer adventitial layer consisting of connective tissue in which the nerves and vasa vasorum are located. The aorta and a number of its major branches are called "elastic" due to alternating membranes of elastic tissue and smooth muscle cells and a collagen and elastic fiber framework. The elastic laminae diminish rapidly in number, and with the exception of the innermost layer disappear in branches of these vessels which are then termed muscular. With diminishing diameter the proportion of elastin and collagen fibers to the circularly oriented smooth muscle is progressively reduced.[10] A reduction in fibrous proteins occurs not only in the muscle layer but also in the adventitia and this affects the relationship of the nerves to the muscle layer. In veins which have thinner walls, there is greater variation in the orientation of the smooth muscle. These differences taken together with a vast array of distinctive functional properties and the complex influence of the local environment[5,11,12] make it exceedingly difficult to define the role that sympathetic nerves may play in the hypertrophic response.

### A. Distribution and Density of Nerves

In the adult mammal, terminal sympathetic nerve axons, when visualized by specific catecholamine fluorescence, form a mesh-like network in the adventitia of the vessel wall. However, only terminal varicosities bare of Schwann cells processes, adjacent to the outermost muscle cells of the media, are thought to be involved in neurotransmission (Figure 2). It is exceptional for the nerves to penetrate the smooth muscle layer.[2,13] Preganglionic axons contact a number of postganglionic neurons, thus ensuring widespread responses. A single postganglionic axon innervates many smooth muscle cells. Since the entire nerve supply to the arteries of a regional vascular bed and other structures may accompany the major supply artery (e.g., mesenteric, basilar and internal carotid arteries, renal artery), the nerves around a larger artery may give a false impression of the density of its functional innervation. There are pronounced regional and species variations in density of adrenergic innervation. For example, in the rabbit, the

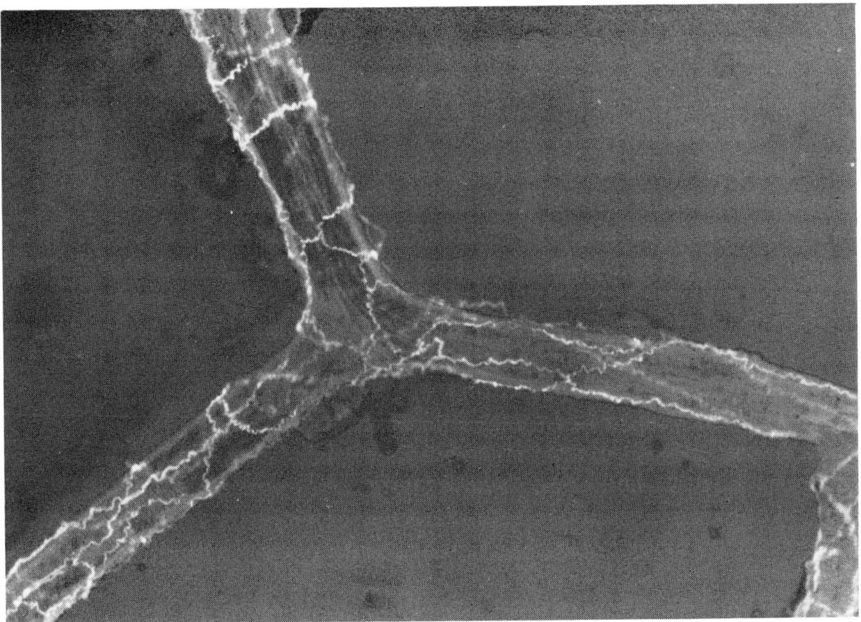

A

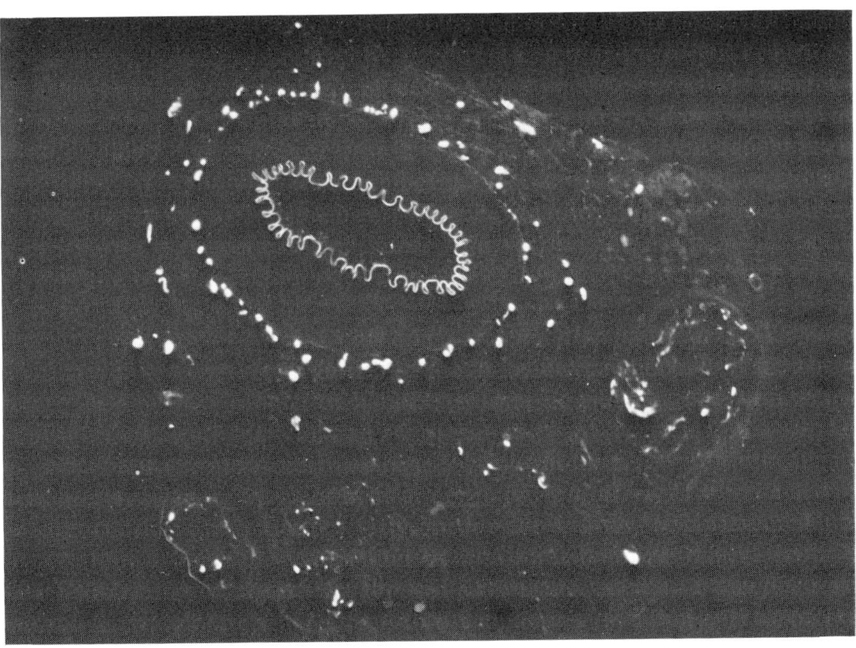

B

FIGURE 2. Typical sympathetic innervation pattern of arteries. Specific catecholamine fluorescence induced by glyoxylic acid method. (A) whole mount preparation of small arteries in margin of rabbit ear showing the network of varicose nerve fibers. (B) transverse section of the rabbit central ear artery. Adrenergic nerve terminals are confined to the adventitia and are primarily at the adventitial-medial junctional zone. The internal elastic lamina is defined by autofluorescence. Smaller branches of the artery and vein are also present but sectioned obliquely.

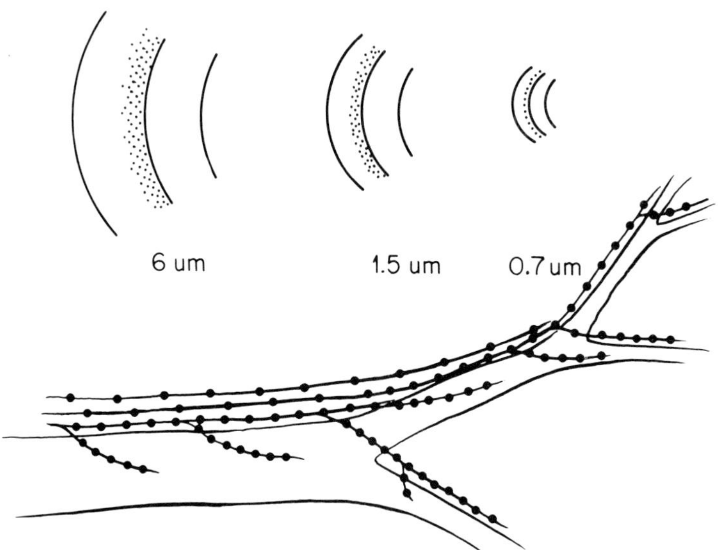



FIGURE 3.   Idealized model of sympathetic innervation of the rabbit central ear artery and a main side branch assuming that all supplying neurons accompany the main artery. The major arterial branch has a wall thickness one-half of its parent vessel. Adrenergic varicosities are shown as black circles. Figures refer to mean distances between varicosities with bare area (i.e., not enclosed by a Schwann cell) and closest smooth muscle cells. Of the areas devoid of Schwann cell covering, 90 to 95% are oriented toward the media. A completely bare varicosity is rarely encountered. Note that the outer axons around the parent vessel innervate more peripheral parts of the arterial bed. Based on data of Dr. Reed Rowan. (From Bevan, J. A., *Fed. Proc.*, 43, 1365, 1984. With permission.)

density of innervation does not parallel the size of the contractile response of a vascular bed to nerve activity. Skin arteries and veins are well innervated, with the density in the veins increasing the more dependent veins. Sympathetic nerve stimulation elicits a marked contractile response. This is not so in the intrapulmonary and cerebral arterial vasculature. Most deep veins of the limbs are not innervated. In some arterial beds the innervation density increases as arteries get smaller, as in the splanchnic bed and in the arteriovenous anastomoses of the rabbit ear. Although it is clear that a vessel must be innervated if it is to respond to changes in sympathetic function, the size and type of the effector response depends upon other features of the effector system such as receptor type, density, local propagation, etc.

## B. Size of Neuromuscular Cleft

As arterial channels diminish in size, the adrenergic varicosities become progressively closer to the smooth muscle cells of the tunica media as there is less intervening connective tissue, fibroblast processes, and Schwann cell ensheathment. Many observations from different vessels confirm this concept (Figure 3).[14-16] Relevant varicosities are considered to be those in the vicinity of smooth muscle cells devoid of Schwann cell cover. Neuromuscular separation is greatest in elastic arteries and conversely is closest to the outermost smooth muscle cells when varicosities are present in small muscular arteries and arterioles. Unlike most other synapses, the distance between the varicosity and the nearest smooth muscle cell is rarely less than 60 nm even in arterioles. These variations have a significant effect on neural control.[17,3] This is discussed in Section G.

## C. Sympathetic Nerve Activity

Recent studies indicate that sympathetic nerve impulses to different organs vary non-uniformly in their pattern and rate of firing under the various conditions which activate the system. These include the afferent discharge from baroreceptors, chemoreceptors, and other visceral afferents, control from other brain centers, the "defense" reaction, hypothalamic stimulation and thermoregulation,[18] and possibly local reflexes mediated through the ganglia. Recording sympathetic nerve activity in human skin and muscle nerves, Wallin and co-workers have demonstrated not only differences in the pattern and role of impulses to these organs, but also between individuals to the same organ.[19] In the adult at rest, the average frequency is low, about 1 to 3 Hz[20] and is usually irregular, occurring in bursts of activity.[19] An irregular pattern of impulses has been shown to affect contractile responses of rat mesenteric arteries and veins differently from a uniform sequence.[21] The extent of neurogenic tone differs in vascular beds.

## D. Presynaptic Control Mechanisms

Neural and many humoral substances have been identified as modulating norepinephrine (NE) release through activation of receptors on the nerve terminal and consequently can influence the transmitter concentration in the neuromuscular cleft. Perhaps the most important mechanism is one through which NE influences its own release through a presynaptic $\alpha_2$ negative feedback system.[22]

## E. Postsynaptic Effects of Sympathetic Nerve Activity

When the adrenergic nerves of vessels with their innervation on the outside of the media are stimulated in vitro under isometric conditions, the maximum contraction obtained even in those vessels with dense innervation is usually only 50 to 70% of the maximum obtainable response that the vessels can attain to other constrictors[3] This is due to a number of factors, the most important being the transmitter diffusion gradient, the metabolism of NE as it diffuses through the wall, activation of $\beta$-adrenoceptors to cause vasodilation, and in addition, the production of endogenous vasodilating substances such as prostaglandins or factors from endothelium. Consequently, the inner vascular smooth muscle cells are less affected by NE released by nerve stimulation. These are the cells that are preferentially influenced by circulating vasoactive substances and the endothelium. However, these factors have not been systematically studied in small arteries and arterioles, where spacial separation between nerves and endothelium is diminished.

NE interacts at smooth muscle membranes with two major classes of receptors: contraction is mediated through $\alpha$-adrenoceptors and relaxation by $\beta$-adrenoceptors. The number, proportion, and position of adrenoceptors differs in blood vessels and in species.[3]

*$\alpha$-Adrenoceptors* — There are marked variations in $\alpha$-receptor regulation of the vasculature. In a study of rabbit arteries, it was found that the sensitivity of the smooth muscle to NE can vary over several magnitudes in different parts of the vascular tree. Sudden changes of sensitivity occur at specific sites along the main branches of the aorta and distal to these sites there are further changes.[23] These changes can be related to a great extent to changes in agonist affinity for the $\alpha$-adrenoceptor.[24] In the pulmonary arterial tree, as arteries become smaller there is progressive diminution in their response to $\alpha$-adrenoceptor agonists relative to other agonists.[25] In both the smaller pulmonary and cerebral arteries the response is limited by receptor number.[26] In the vascular bed of the rabbit ear, however, there is no change with diminution in size of artery, with NE eliciting a maximum response in vessels of all sizes.[27] The characteristics of the $\alpha$-adrenoceptor defined pharmacologically are different in different ves-

sels.[12] Alpha-receptors are classified into two groups, $\alpha_1$ and $\alpha_2$, which appear to have different mechanisms for signal transduction. $\alpha_1$ are associated with increased turnover of phosphoinositol and changes in membrane permeability to ions, whereas $\alpha_2$ are inhibitory to adenylate cyclase in a number of tissues examined.[28,29] Recently low affinity binding sites for NE mediating a contractile response have been identified in all arteries tested in the rabbit except the pulmonary.[30] These may be similar to the proposed gamma receptors of Hirst and Nield.[31] Gamma receptors have been defined as junctional receptors in arterioles through which NE can cause local depolarization when applied iontophoretically. The effects are resistant to $\alpha$-adrenoceptor antagonists. They may or may not be involved with excitatory junctional potentials and there is some debate regarding the physiological significance of these potentials. Subsequent intracellular events are not known.

$\beta$-Adrenoceptors — These are linked to adenylate cyclase and membrane hyperpolarization. Although a systematic assessment of their contribution to adrenergic nerve mediated responses has not been carried out in the vasculature, in certain regions of the vasculature, they appear to be dominant, e.g., small coronary arteries, cerebral arteries of rat and pig, arterioles of skeletal muscle, and adipose tissue. Species, regional, and age variations occur.

## F. Transmitter Disposition

The action of NE following release from the varicosities adjacent to the outermost layer of vascular smooth muscle is terminated in smaller vessels by a neuronal uptake mechanism. The NE not taken up into the varicosities diffuses through the vascular wall where it is metabolized and into the adventitial capillaries and hence to the circulation where it may influence smooth muscle at distant sites. This contrasts with the larger arteries where although NE is also taken up into the nerve terminals by the same process, its action on the tunica media is largely terminated by uptake into the smooth muscle cells and subsequent metabolism.

## G. Differences Between Large and Small Arteries

In large conduit arteries because of the wide neuroeffector cleft, neuronally released NE has a widespread action associated with its diffusion into the vascular wall. Because of this and because there is little additional electrical spread of excitation, the response of the artery wall is relatively small and slow. The smooth muscle cells respond to the local concentration of NE. Its effect is terminated by NE uptake into the smooth muscle cell. In well innervated small arteries which have a much narrower cleft, the effect of released NE is localized and excitation is electrically propagated through muscle cells. Because of this and because the NE effect is terminated by neural uptake, vessel tone follows much more closely changes in sympathetic activity. These extremes are illustrated by the electrophysiologic events recorded in vessels of different size. Kuriyama and Kitamura[32] have concluded that when the cleft is narrow as in small arteries, excitatory junction potentials summate into an action potential and in addition, there is a graded depolarization of the membrane. In medium sized vessels with wider clefts, only graded depolarization occurs, and in the largest vessels, usually no membrane potential change is recorded.

A feature of small arteries which is not seen in the larger ones is the presence of myogenic as well as neurogenic tone.[33] Mellander and Johansson[34] considered myogenic tone the most important determinant of vascular resistance at rest. It is an important component of autoregulation as it is stretch dependent and occurs independently of nerves or circulating vasoactive substances.[35] It has not been determined exactly at what branching order in the arterial tree myogenic tone is found or whether it occurs abruptly, although it seems to be first seen in the smaller muscular arteries. There may well be regional differences. As this is a non-neurogenic response to vascular load, its presence adds another complicating factor to the determination of any specific role for the sympathetic nervous system in hypertrophy of the blood vessel wall.

## H. Summary

Many examples could be cited of differences of structural characteristics and functional properties of large, medium, and small blood vessels in different vascular beds including, in addition to those already discussed, endothelial characteristics,[36] level of metabolism, and other types of nerves. Particularly striking is the variation between species. Hemodynamics is undoubtedly a primary cause of structural variations along the length of the vasculature but functional differences must be determined by local environmental factors where cells of different origin are in close contact during embryological development. In the microvasculature, local metabolic and hormonal influences are extremely important as regulating factors, especially for vasodilation, while intrinsic myogenic and neurogenic tone may coexist. The major motor innervation, the sympathetic outflow, through post ganglionic (adrenergic) nerves permits a coordinated circulatory response to varying circumstances while maintaining blood pressure. It exerts a tonic effect on much of the vasculature. Under stressful conditions when there is increased sympathetic activity, neurogenic tone is increased especially in the mesentery, skin, and muscle. Blood flow is maintained to vascular beds of essential organs because of modifying local factors which are varied in nature, e.g., a limited number of $\alpha$-adrenoceptors. $\beta$-receptor predominance, vasodilator nerves or other strong vasodilator influences including those originating from the endothelium and the relative importance of myogenic tone.

## IV. SYMPATHETIC ACTIVITY AND INTRAVASCULAR PRESSURE

The sympathetic nervous system affects vascular smooth muscle through its influence on the blood pressure and hence the local intravascular pressure. This is accomplished by regulating the caliber of arteries and veins, cardiac output, plasma volume, and levels of circulating vasoactive substances such as angiotensin.

According to the "Laplace Law" the tension on the wall of a blood vessel is defined as:

$$T = P \times r/w$$

where T = tension, P = pressure, r = radius, and w = wall thickness. Elevation of pressure would passively distend the wall and lead to a greater radius and increased wall tension. There would be a reduction in the ability of the smooth muscle to shorten. Hypertrophy of the wall occurs when increased wall tension is sustained and is seen typically proximal to a stenosis, e.g., in coarctation of the aorta. The complex cellular events that occur in hollow smooth muscle organs are detailed in other chapters of this book. Distal to a constriction, a reverse process occurs in association with a diminished wall tension.[37,38] Whether these structural changes are caused by chronic alterations in cell tension or other mediators is an area of active investigation.

In the vasculature dynamic changes in pressure and flow are constantly occurring due to many factors, including the level of sympathetic activity. It is evident that sympathetic activity in a vascular bed where $\alpha$-adrenoceptors predominate, would tend to reduce vascular diameter and thus wall tension through a local action on the vessel wall. Opposing these effects are the local consequences of a raised arterial pressure associated with general sympathetic activity that tend to increase diameter and wall tension (Figure 1). Whether or not hypertrophy occurs as a result of these factors depends on their relative size — and there are many variables, including innervation density, receptor population, elasticity of the vessel wall, and amount of increase in pressure.

In the mature cardiovascular system, elevations in intravascular pressure and an active increase in smooth muscle cell tension would be expected to occur simultaneously in association with a generalized increase in sympathetic nerve activity in the mesentery, skin, and skeletal muscle beds. If the overall result is a diminution in artery diameter, the sympathetic activity would result in a "protective effect" in vascular beds with a rich sympathetic vasoconstrictor innervation. The smaller arteries and larger arterioles which are especially well innervated and have most neurogenic tone would be protected. Vessels further downstream would also be protected from the elevated intravascular pressure. "Upstream", where there is less neurogenic constriction, vessels would distend and tend to hypertrophy. In this way sympathetic activation of smooth muscle in the resistance vessels where neurogenic tone, when it occurs, is greatest would be an initiating factor in hypertrophy. If neurogenic tone to a vascular bed is removed, then protection of "downstream" vessels would be lost and increased wall tension on the smaller vessels would tend to promote distal hypertrophy when the blood pressure rises. Thus in any particular vascular bed, whether hypertrophy occurs in response to increased pressure or not depends on the interplay between passive distension and the degree of neurogenic constriction. A trophic effect of the nerve, either due to activation of the muscle or release of a specific factor (see below) would be a further complicating factor which may cause muscle hypertrophy. In many vascular beds another factor, intrinsic myogenic tone found in small arteries and arterioles, may reinforce neurogenic tone.

## V. HYPERTENSION AND SYMPATHETIC NERVES

Chronic hypertension is associated with hypertrophy of the media of the vascular wall, particularly of the smaller arteries and arterioles. The elevated total peripheral resistance present in various animal models is not uniformly distributed in the body or along the arterial tree. Chronic hypertension can be induced in many different ways and the mechanisms involved have been studied in great depth in the rat. Each form of hypertension has distinctive features. A relationship between sympathetic nervous system hyperactivity and vascular hypertrophy in initiating and/or maintaining sustained arterial pressure elevation has been proposed.[39-41] Considerable evidence implicates central and peripheral mechanisms elevating sympathetic vasomotor tone and their participation in the development and maintenance of sustained hypertension in some human and a number of experimental animal models of hypertension.[42,43]

### A. The Spontaneously Hypertensive Rat (SHR)
In this genetically hypertensive rat, hypertension develops at an early age. Numerous studies comparing the SHR to the normotensive Wistar Kyoto (WKY) strain indicate that the elevated pressure is associated with increased peripheral resistance, sympathetic nervous system hyperactivity, and structural and functional abnormalities of the vascular wall. As much of the past evidence is cited by Gray[44] and Mulvany,[45] mainly recent evidence is referenced here.[42,43] Arterial wall hypertrophy (detailed and discussed in previous chapters), reduced compliance of veins and changes in microvascular patterns exist. In addition, abnormalities of smooth muscle function are present,

which includes alteration in excitation-contraction coupling, increased sensitivity to agonists including NE, changes in membrane properties with increased permeability to ions, and alterations in myogenic tone. Abnormalities in membrane properties are not confined to vascular smooth muscle and are probably also present in adrenergic nerves. The relationship between the two is not clear as several studies indicate that sympathetic nerve activity influences membrane properties.[85,86,111]

## 1. Sympathetic Hyperactivity

There is no general agreement as to the time of appearance of the sympathetic hyperactivity during growth, as interstrain differences complicate time sequence studies. A study by Smith et al.[46] on the neurogenic contribution to the elevation of blood pressure indicates that, that in comparison to the WKY control, increases of sympathetic activity, arterial pressure, and basal vascular tone occur in parallel and are detected from birth. Others place the onset of elevated pressure after 4 weeks of age. SHR differs from WKY in certain aspects of the sympathetic control of the vasculature, specifically in (a) accelerated development,[46] (b) adrenergic hyperinnervation identified before 2 weeks of age[47,48] and present in many adult arteries,[49-53] (c) increased sympathetic nerve activity demonstrated in renal, cervical, cardiac and splanchnic nerves[54-57,46] (d) increased NE content,[58] (e) loss of accommodation in sympathetic neurons,[59] and (f) greater release of NE in response to nerve stimulation.[53,60] Increased heart rate and increased cardiac output are early features of the SHR rat and are probably indicative of increased sympathetic cardiovascular control. Although the primary genetic alterations are not known, the SHR is the best documented example of increased sympathetic nerve influence on the vasculature at an early age, although it must be emphasized that abnormalities in the smooth muscle and other systems are also present.

## 2. Development of Hypertrophy

Hypertrophy of large and middle-sized arteries has been observed within the first few weeks of life in cerebral, mesenteric, renal, carotid, and tail arteries.[61-63] This early postnatal period is one of rapid growth of the vasculature associated with intense proliferative activity of vascular smooth muscle cells. In the rabbit, with comparable postnatal development, this has been documented in the larger arteries but has not been studied in the smaller arteries and arterioles. Proliferation appears to occur in waves.[7] It is possible that the early observations on vascular hypertrophy in the SHR are in fact due to accelerated development, or that development is simply out of phase with the WKY strain. However, hypertrophy is found in the adult, and noted to be in arteries over 75 to 100 microns in diameter in the mesenteric circulation in tissues fixed for microscopy.[49,64] The concurrent sympathetic hyperinnervation and hyperplasia or increased size of smooth muscle cells of the arterial vasculature could indicate elevated levels of a nerve growth factor produced by the smooth muscle, or a trophic effect of nerves on muscle development, or both. These two possibilities probably could not be distinguished by observations of timing alone. Target organ influence on pattern and density of sympathetic innervation has been shown by transplantation of tissues into the anterior eye chamber and in tissue culture.[65-68] Although the well characterized nerve growth factor (NGF) content of effector organs is increased after sympathetic denervation, it is now recognized that many other growth factors exist. The interpretation of what is taking place is complicated by the concurrent elevation of arterial pressure due to both increased nerve activity and the level of non-neurogenic peripheral resistance.[46] Contributing to the latter could be myogenic tone, circulating vasoactive substances, and functional and structural changes in the vasculature.

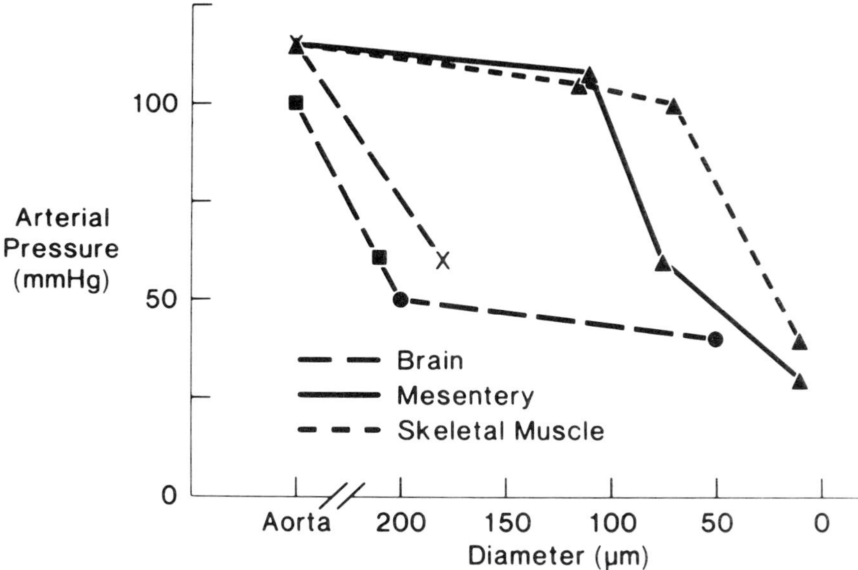



FIGURE 4.    Gradient in pressure between aorta and small arteries in brain, mesentery, and skeletal muscle. All data are from anesthetized cats. (From Baumbach, G., Busija, D., Werber, A., and Heistad, D., Role of Autonomic Innervation in Modification of Cerebral Vascular Responses in *Neurotransmitters and the Cerebral Circulation,* L. E. R. S., Vol. 2, MacKenzie, E. T., et al., Eds., Raven Press, New York, 1984. With permission.)

### 3. Regional Changes: Mesenteric, Skeletal, and Skin Vasculature

Detailed in vivo and in vitro studies of structural and functional changes in resistance vessels in a variety of vascular beds are only available for rat models of hypertension. Measurements of pressure and flow correlated with lumen diameter and wall thickness have been made in the microcirculation of many SHR and WKY vascular beds in vivo under general anesthesia and also when perfused. These have been complemented by morphologic studies. In comparison to the normotensive WKY, the adult SHR despite elevated arterial pressures has a near normal capillary pressure and blood flow.[69-71] This is accomplished by different mechanisms in different vascular beds which presumably reflect their varied properties. The different gradients in pressure between aorta and small arteries in brain, mesentery, and skeletal muscle has been measured in anesthetized cats and is illustrated in Figure 4. In the adult microvasculature of three skeletal muscles of the rat — the spinotrapezius, gracilis, and cremaster,[69,71,72] and also in mesentery[73] and skin,[74] the major changes were a diminished number of arterioles of lumen size less than 30 μm and of capillaries. Hypertrophy of the wall was not present in the smaller arterioles, but occurred upstream from vessels of this size.[51] These features were already present at 5 to 6 weeks of age in skeletal muscle[69] and became accentuated with increasing age with continuing elevation of blood pressure. The age at which the architectural pattern is altered is not known although Prewitt et al.[71] did not find rarefaction of the smallest arterioles in the gracilis muscle of 5-week-old rats, but vasoconstriction was present.

The age at which the vasomotor influence on different vascular beds becomes fully mature has not been defined. Adrenergic hyperinnervation and hyperactivity has been shown to develop very early in the postnatal growth period in the mesenteric vasculature of SHR. It is of interest that in a comprehensive study of the development of sympathetic innervation to proximal and distal arteries of the rat mesentery in a normotensive rat strain, Hill et al.[75] observed that the innervation pattern and functional integrity to the distal vessels matured slightly earlier than to the proximal. The innervation was denser in the smaller vessels.[75] Adult levels were achieved by about 16 days of age. In the rat mesentery all the vessels of the arterial tree with the exception of precapillary arterioles constrict in response to nerve stimulation.[76] An abrupt doubling of nerve activity was recorded about 3 weeks of age in the normal Sprague-Dawley rat,[77] and subsequently declined to adult levels. This hyperactive period appears to coincide with the onset of the marked increment in arterial pressure difference between SHR and WKY. Cervical ganglionic neurotransmission was functional by the first postnatal day in both SHR and WKY,[46] but in the SHR the neurally mediated component of blood pressure showed accelerated development for t first 6 weeks of postnatal life and then was not different from WKY.

Morphometric measurements at the light and electron microscopy level of the mesenteric arterial tree down to first order arterioles at 3 to 4 weeks of age showed hyperplasia of muscular arteries and first order arterioles, but not the superior mesenteric artery. Retrograde hypertrophy of the vasculature up to and including the superior mesenteric artery occurred at 10 to 12 weeks and 18 to 20 weeks of age: hypertrophy of endothelial cells, increased elastin and both hyperplasia and hypertrophy of smooth muscle cells, and increased nerve density in arteriolar vessels was present.[51] A similar sequence probably occurs in the gracilis muscle as in vivo microvascular studies indicated the rarefaction of small arterioles and capillaries proceeded from a stage of capillary rarefaction at 6 to 8 weeks of age, through one of predominantly functional arteriolar rarefaction, and by 12 weeks of age there was arteriolar loss. In both spinotrapezius and cremaster muscle of the rat, constriction is maintained during nerve stimulation down to the smallest precapillary arterioles,[78,79] although the smallest precapillary vessels relax after a short period[78] presumably local influences. As hypertrophy of the wall of the thoracic aorta of the SHR occurs and this structure is not innervated by sympathetic nerves, it would appear that other factors are involved in the hypertrophic response and one obvious candidate is elevated pressure. The exact role of the peripheral sympathetic nervous system could only be unequivocally defined by prevention of innervation of the vasculature, but it does appear to exert a protective action on the microvasculature of those beds that are highly innervated. Excessive neurogenic control associated with a dominance of alpha-adrenoceptor activation of smooth muscle may contribute to the rarefaction of the microvasculature.

### 4. Regional Changes: Cerebral and Jejunal Vasculature

In contrast to the three regional beds discussed above, in the cerebral vasculature of adult SHR, normal flow and capillary pressures were achieved by active constriction of all arteriolar sizes and by hypertrophy of the wall. When arterial pressure was reduced, the arterioles dilated, probably reflecting the intrinsic myogenic response. Numbers of vessels were not different in SHR and WKY.[80] In the cerebral vasculature of the rat, which autoregulates, both large and small arteries exhibit intrinsic myogenic tone and all sizes contribute appreciably to the control of blood flow and resistance.[80] Rat cerebral arteries are innervated by adrenergic nerves but $\beta$-adrenoceptors are dominant, at least in the larger ones.[81] Pial arteries do not constrict to NE (personal communication; Halpern). The microvasculature of the wall of the jejunum showed slightly different features, but essentially the same characteristics as the cerebral vasculature of the SHR. There was little alteration in vessel numbers. This part of the mesenteric vasculature is undoubtedly influenced by gastrointestinal hormones and local factors.

In summary, it is conceivable that in those vascular circuits which have a strong adrenergic vasoconstrictor influence, the pattern of the microvasculature during growth and in the developing phase of hypertension is altered in the SHR, resulting in fewer smaller arterioles and capillaries of possibly increased internal diameter. Hypertrophy develops upstream. In vasculatures in which the sympathetic vasoconstrictor influence is small, the evidence indicates that the pattern of the microcirculation is not altered; there is no reduction in the microvascular bed, but the vessels exhibit hypertrophy. Changes occur on the venous side also, but their distribution is not as well documented.

### B. Renal Hypertension in the Rat

It is interesting to compare the changes in the microcirculation in the SHR with those in other forms of hypertension. In a renal form induced by unilateral nephrectomy and constriction of the artery to the remaining kidney, a number of investigators have noted a different sequence of events.[71] A progressive increase in wall/lumen ratio and structural diminution of lumen size of small arterioles together with a progressive rarefaction of arterioles and capillaries was observed at 4 and 8 weeks after initiation. Pressure measurements indicated resistance was also increased upstream.[82] A more extensive functional and structural impairment of lumen diameter and hypertrophy of the vascular wall including vessels as small as precapillary arterioles occurred, which may be related to the abrupt elevation of arterial pressure and increased plasma angiotensin levels. These changes differ from those in the SHR, where in the same vascular bed, hypertrophy was not seen in the smallest arterioles whose maximum internal diameter was equal to or increased. Infusion of angiotensin has been shown to increase protein synthesis and DNA levels in the vasculature, necrosis and proliferation of smooth muscle cells, and fibrosis in the vascular wall.[83,84] However, another factor may also be involved in an explanation of the differences seen. In the SHR, hypertension occurs during a rapid period of growth, whereas in the experimentally induced forms of hypertension, the surgical procedures are initiated at a later age and pressure rises during a different phase of maturation.

In the SHR we have a unique example of adrenergic hyperinnervation of the arterial side of the circulation. Although this may also occur in veins, and it would be of interest to analyze functional and structural alterations in vessels not exposed to the elevated pressure, there is insufficient information available to do so. However, an important influence of the sympathetic nervous system on the altered membrane potential in the SHR has been observed in both arterial and venous smooth muscle.[85,86] There is evidence that the sympathetic nervous system can exert an influence on the normal developmental pattern of a vascular bed in which there is marked neurogenic tone, specifically that of the rabbit ear.[87] This will be discussed below.

## VI. GROWTH OF THE VASCULATURE

Blood vessels initially consist of a single layered tube of endothelium derived from mesenchyme. Although the major conduit vessels, e.g., the aorta and branchial arch derivatives are genetically programmed, the peripheral vasculature develops by a process of endothelial sprouting and extensive remodeling to form arteries, capillaries, and veins. Principles involved in the development of the vasculature were defined by Thoma[88] who observed the development of chick embryos. The presence of endothelium, blood circulation and external mechanical factors, such as longitudinal tension associated with body and organ growth are important factors in the differentiation of the vasculature. Circulating or locally derived growth factors may also be involved.

The smooth muscle of the media is derived from undifferentiated mesenchyme which condenses around the endothelium to form arteries and veins. Sympathetic nerves have not been reported to be present in the vessel wall before the endothelium and media are defined. Subsequent development has been mostly studied in the rat. The adrenergic nerves grow along the vasculature and the plexus increases in density, complexity, and varicosity number at the same time that the arterial walls are increasing in thickness and length — a process that involves cell proliferation and differentiation and formation of elastin, collagen, and extracellular matrix. The possibility exists that nerves have a different relationship to the vasculature during growth. The ultrastructural appearance of sympathetic nerves as they grow into the vascular wall has been studied by Todd and Tokita[80] in muscular arteries of the Wistar rat. Initially the varicosities of nerve processes contain mainly empty vesicles and are not enclosed in Schwann cells. Subsequently, vesicles with typical dense cores appear. Both clear and dense core vesicles increase in number and nerve processes become partially enclosed in Schwann cells. The density of varicosities increases and peaks during the second postnatal week in this strain and then decreases as the rate of growth of the wall exceeds the increase in varicosities. Receptors are present on the smooth muscle before functional connections are established,[90,91] although it is possible that the nerves release vasoactive substances spontaneously or with nerve activity.

There is no definitive evidence on the relationship of the sympathetic nerves to the major proliferative phase of the larger arteries in the first few postnatal weeks. However, in the rat the aorta and some of the larger arterics develop in the absence of an adrenergic innervation. It was, however, noted by Bevan in 1975[92] that fewer smooth muscle cells were seen synthesizing DNA in the chronic sympathetically denervated ear artery in rabbits 6 weeks of age. This does not necessarily indicate a specific action on cell proliferation, but could indicate a nonspecific trophic action of sympathetic nerves facilitating the developmental program of cell proliferation or differentiation. Many studies describe the early embryology and development of vascular beds (for references see 93 and 94). There are comparatively few observations of the postnatal development of the pattern of a vascular bed. Morris[94] has described this process in the rabbit ear by study of vascular corrosion casts and scanning electron microscopy. The arrangement of the major arteries was determined before birth, even down to the arteries passing to the lateral margin of the ear which are about 100 $\mu$m in diameter in the adult. Despite a growth in length of the ear cast from birth to maturity of approximately 14 times, there was no change in the number of arteries arising from the central ear artery (CEA). However, the distance between these major arteries increased indicating proportional growth between the branches. In the lateral margins of the ear, as the area supplied by the lateral marginal artery increased, it branched more extensively. Arterio-venous anastomoses (AVA), which are densely innervated by sympathetic nerves, are present in large numbers in the lateral ear margins. They formed at a steady rate during growth of the ear. In the normal development of this vascular bed, the pattern of the arteries is fixed very early in development and only the microvessels appeared to increase in number with further growth.

Surgical denervation of the young rabbit ear enhanced proliferation of AVA in comparison to the contralateral sham-operated ear.[87] This effect was not seen after denervation of adult ears. Combined sympathetic and sensory denervation has been observed to produce an acute but temporary increase in blood flow and temperature of the adult ear, but this has not been measured at either age (see Grant[95]). However, local blood flow does not effect the modeling of the developing vasculature.[96,97] It is not known why the age related difference occurred. Although the vasculature is innervated by sympathetic fibers at the younger age, the extent of neurogenic tone is not known. However, it would appear that the absence of the nerves contributed to the increase of AVA. Plasticity is certainly retained in these small vessels, at least in capillaries and small veins, as can be demonstrated in wound healing and inflammation and elegantly described by Clark et al.[98] Many factors, both physical and chemical, influence the formation of new vessels.[99] New growth appears to occur where endothelial sprouts are unimpeded by other cells. Larger vessels, perhaps because they have two defined concentric layers in addition to endothelium, have more limited options and adapt to changes in pressure and flow by remodeling lumen size and by changes in wall thickness and length. This has been shown to occur in normal growth, hypertension, and changes in blood flow.[37,38,11] Quite different effects were seen in the rabbit car following denervation of larger arteries and is discussed in the following section.

## VII. TROPHIC ACTIONS OF SYMPATHETIC NERVES ON THE VASCULATURE

In skeletal muscle the motor innervation has been clearly shown to exert a long term regulatory or "trophic" effect on functional and structural properties, in addition to controlling contractile responses. It is necessary for postnatal growth and the maintenance of differentiation of the mature phenotype. The frequency-pattern of nerve impulses and chemicals derived from nerves determine the characteristics of skeletal muscles. The motor nerve regulates and maintains membrane properties and initiates complex intracellular events including protein synthesis. Other means of maintaining skeletal muscle activity, such as electrical stimulation or stretch only partly compensate for the nerves.[100-102]

In various intact target tissues, sympathetic nerves have been identified as promoting cell division and growth by influencing DNA, RNA, and protein synthesis,[103] mitotic rate,[104] cell proliferation, and cell size.[105] In some tissues the influence has been to maintain the differentiated state.[106] Contractile ability of smooth muscle has been shown to be dependent on maintenance of nerve traffic and maintenance of normal structure and function of an artery.[107,108]

It has been shown that in the ear vasculature, sympathectomy in young animals produces greater alteration in structure and function than in the adult.[109,110] It is difficult to analyze the effects of sympathetic nerves on vascular smooth muscle (VSM), which is a component of a tubular structure constantly subject to changes in tension and is in close apposition to blood which contains many vasoactive substances and growth factors.

Considerable evidence exists that sympathetic nerves exert a trophic influence on membrane properties of vascular smooth muscle cells and are responsible for the long-recognized phenomenon of postsynaptic, nonspecific increased sensitivity to agonists and ions.[111,112] This change is found in both the young and mature and is associated with impaired activity of $Na^+K^+ATPase$, an electrogenic pump and a decreased resting membrane potential.[113,114] Thus the sympathetic nervous system influences the distribution of ions across the cell membrane and the transport of nutrients into the cell. However, the relationship of this property to the intracellular events related to hypertrophy is not known.

Trophic effects of sympathetic nerves on the vasculature have been mostly determined by denervation or decentralization techniques. These avoid some of the difficulties of interpretation that widespread impairment of sympathetic function presents. This approach cannot identify the earliest influence of the nerves and some of the effects observed may be due to substances released from degenerating nerve terminals or denervated muscle. Loss of nerves can, however, indicate the cell processes influenced by them. Evidence for trophic actions will be presented from two different vascular beds.

In other tissues conflicting evidence suggests that sympathetic nerves (a) stimulate cell proliferation, (b) inhibit cell proliferation, and (c) are important for maintaining differentiation of cells. These effects could be mediated in a nonspecific manner by increasing cell activity by the pattern of nerve traffic, by release of trophic substances, by activation of an $\alpha$ or $\beta$ or possibly a low affinity site or a combination or by changing the membrane potential. Because of all the variable factors in sympathetic nerve muscle relationships, it would be surprising if a single mechanism could define the influence of nerve on a smooth muscle cell that would lead to or inhibit hypertrophic responses. Different intracellular pathways are activated by these events.

## A. The Vasculature of the Rabbit Ear

Only sympathetic vasomotor and substance P-like containing nerves which are probably sensory have been identified in the major vessels. Substance P itself does not alter the contractile response of these arteries in vitro (unpublished observation). As previously mentioned, following total denervation of the ear during the early postnatal period, there was a 2.5 fold increase in arterio-venous anastomoses (AVAs) in the microcirculation, but in adult rabbits the number did not differ from those in the contralateral ear (Figure 5a).[87] This is consistent with the observation of Rusterholz and Mueller,[110] who demonstrated that chronic sympathetic denervation by superior cervical ganglionectomy at 4 weeks of age, when the rabbit ear is still rapidly growing, lowered vascular resistance during constant flow perfusion at maximum relaxation in comparison to the contralateral normally innervated ear. In the adult rabbit denervation did not alter perfusion pressure (Figure 5b). It is of additional interest that when intravascular pressure was elevated by aortic coarctation in the growing rabbit, no difference in perfusion pressure was found between denervated and innervated ear vasculature.[125] A quite different effect has been found in the major artery to the ear (CEA). After chronic denervation, in comparison to an identical segment of the contralateral ear, the luman diameter, cross-sectional area of the media and wall/lumen ratio was reduced (Figure 6).[87,115] It appeared that the artery had failed to grow as well as the innervated side and contained fewer smooth muscle cells. The extracellular composition of the wall was also altered and it was "stiffer". These structural differences were age related being most pronunced the younger the age denervated (Figure 7). Branco et al.[106] have noted that after removal of the superior cervical ganglion (SCG) from young adult rabbits, structural alterations in the outermost 2 to 3 layers of the media of the central ear artery (CEA) consisted of clusters of larger dedifferentiated smooth muscle cells adjacent to degenerating nerve terminals and an increase in fibroblasts and extracellular material. It is not clear whether these changes are caused by substances released from the degenerating terminals or to the absence of a trophic effect of the nerves. However, neural influences on collagen metabolism have been demonstrated in skeletal muscle where increased turnover and deposition of collagen Types 1 and 3 and fibronectin occurred following denervation.[116] Fibroblast proliferation occurs immediately following denervation of skeletal muscle[117] and has also been noted by Branco et al.[106] in the ear vasculature. Fronek et al.[118] also noted a significant increase in collagen content of the rabbit aorta after chronic sympathectomy by 6-hydroxy dopamine, concluding that the sympathetic nerves influenced the metabolism of the vessel wall.

Other changes occurring in rabbit arterial vasculature following removal of the vascular innervation are impairment of endothelial-dependent relaxation mediated by acetylcholine without a concomitant change in the ability of the vascular smooth muscle to relax,[119] and also an increased level of myogenic tone in small arteries.[120] These effects might be related to a change in blood flow, but this has not been measured. In

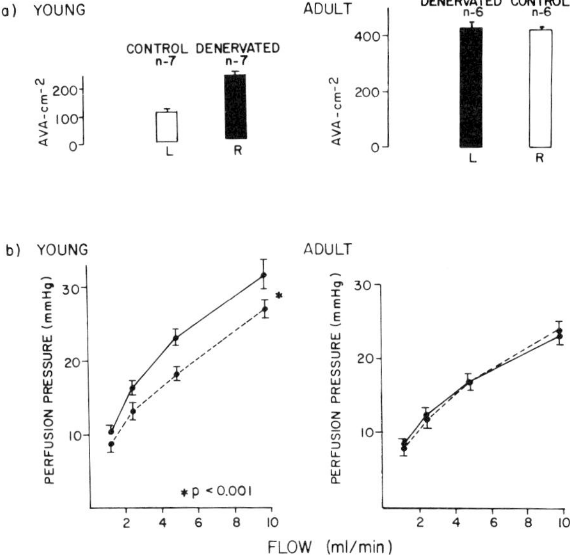


FIGURE 5.    (A) Effects of surgical denervation on density of arterio-venous anastomoses in a defined segment of lateral margin of rabbit ear. L = left ear, R = right ear. Control values from sham-operated ears and denervated values from contralateral ears of the same animals and obtained from vascular corrosion casts by Dr. J. Morris, 4 weeks after denervation of the right ear at 10 to 12 days of age in young animals and 4 weeks after surgery in adult animals (16 weeks of age). From Morris, J. L. and Bevan, R. D., *Am. J. Anatomy,* 176, 497, 1986. With permission.) (B) Flow-pressure curves obtained at maximal dilation for innervated and ears denervated by superior cervical ganglionectomy from rabbits 4 weeks of age, perfused 9 weeks later (left, n = 9) and for rabbits denervated at 16 weeks of age and perfused 10 weeks later (right, n = 10). Vetical bars represent ± SE. (From Mueller, S. M. and Rusterholz, D. B., Trophic Influence of Sympathetic Nerves on the Peripheral and Cerebral Vasulature, in *Cerebral Blood Flow: Effects of Nerves and Neurotransmitters,* Vol. 14, Heistad, D. D. and Marcus, M. L., Eds., Elsevier Science Publishers, Amsterdam, 1982. With permission.)

this vascular bed, adrenergic nerves thus appear to regulate many aspects of structure and function and promote the normal development of the vascular wall associated with growth.

## B. The Cerebral Vasculature

In the innervation of the cerebral vasculature a variety of putative neurotransmitters including peptides have been described. A trophic effect of the sympathetic innervation has been detected. In the rabbit, unilateral superior cervical ganglionectomy caused sympathetic denervation most marked in the ipsilateral middle cerebral artery. In the normal rabbit following chronic sympathectomy, the middle cerebral artery with branches weighed signifcantly less than their matching counterparts on the contralateral side.[109] Aubineau et al.[121] have demonstrated that after chronic unilateral sympathectomy, resting blood flow in the ipsilateral cortex was decreased by a mean of 17% compared to the contralateral cortex. In the stroke prone spontaneously hypertensive rat (SHR-SP), unilateral sympathetic denervation at 3 weeks of age significantly atten-

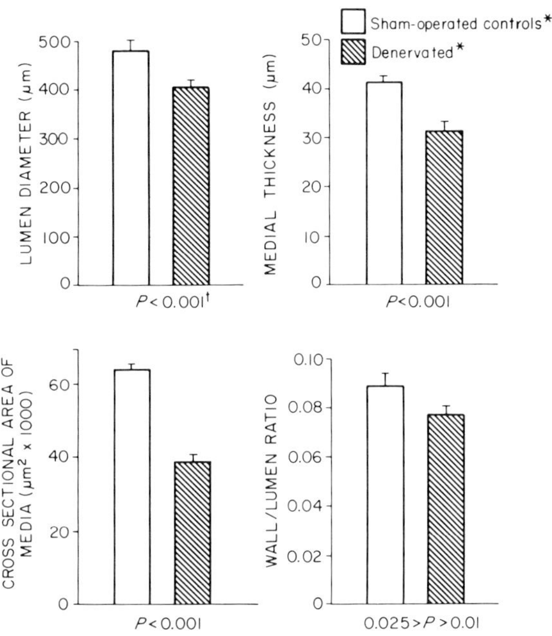


FIGURE 6. Effect of chronic complete denervation by resection of all nerve trunks and superior cervical ganglion on lumen diameter, medial thickness, cross-sectional area of media, and wall/lumen ratio of rabbit central ear artery. Denervation was carried out at 10 days of age and examined 4 weeks later. Paired comparisons were made between denervated and contralateral innervated sides. (From Bevan, R. D., *Hypertension*, 6(Suppl. III), III-19, 1984. With permission from the American Heart Association.)

uated the development of vascular hypertrophy in intraparenchymal vessels.[122] Evidence for a protective action of sympathetic nerves on the cerebral vasculature in SHR-SP hypertensive rats has been provided by Sadoshima et al.[123] who found that cerebral strokes affected almost exclusively the hemisphere ipsilateral to the sympathectomy in the SHR-SP. In addition, the blood brain barrier was better protected from the hypertension in the normally innervated hemisphere of the SHR with vascular hypertrophy. It may be concluded from these studies that structural modifications of cerebral vessels occurred in these rats in the absence of the sympathetic innervation. A protective action of neurogenic tone preventing hypertrophy associated with an increase in wall tension has been detected in a superficial limb vein.[106] Hyperplasia of vascular smooth muscle cells has also been noted in the absence of sympathetic nerves in the radial and arteries of the chicken wing.[124] It is likely that neuro-trophic influence does not necessarily play a causative role in hypertrophy of smooth muscle, but it may facilitate the response to other factors.

## C. Conclusion

The anatomical distribution of the sympathetic innervation of the vasculature indicates its importance in circulatory control and the hemodynamic effects following activation have been extensively described in the literature.

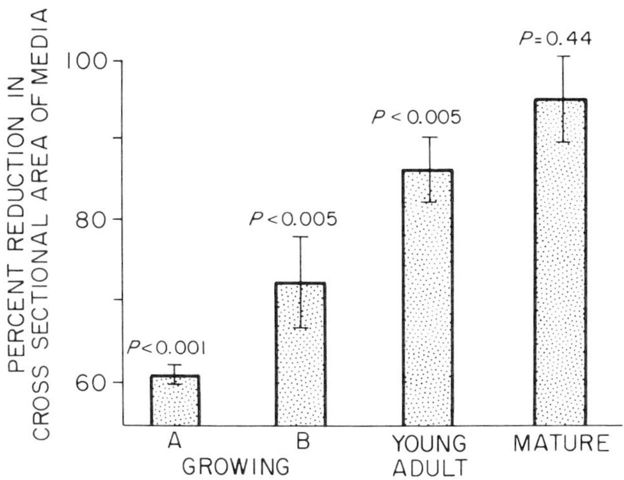


FIGURE 7. Comparison of percentage reduction of cross-sectional area of medial smooth muscle layer of denervated ear arteries of the rabbit compared with their controls in 4 different age groups. In growing group A, total denervation of ear was performed at 10 days and examined 4 weeks later (n = 5). Excision of superior cervical ganglia only was performed 8 weeks prior to measurement in other groups. In group B, age at denervation was 4 weeks (n = 7); young adult 9 to 11 weeks (n = 13); mature over 4 months (n = 7). Paired observations in control and denervated arteries in each age group were compared by paired t-test. p < 0.05 is significant. Vertical bars indicate SEM. Reproduced by permission. (From Bevan, R. D., *Hypertension*, 6(Suppl. III), III-19, 1984. With permission from the American Heart Association.)

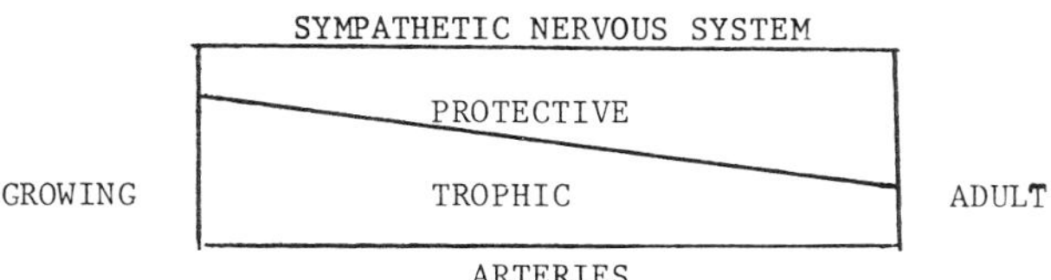


This diagram represents functions of sympathetic nerve action relevant to vessel wall hypertrophy. By limiting distension of the vascular wall through a vasoconstrictor action, it has a local "protective" effect limiting the hypertrophic response of the vascular wall to a generalized increase in intravascular pressure. In resistance vessels, the "protective" action is complicated by the presence of myogenic tone and such vasodilator influences as $\beta$-adrenoreceptor mediated effects, metabolites, endothelial-derived vasodilation, and vasodilator nerves. These would tend to modify the protective action. The relative importance of these factors differs in vascular beds and along the vascular network and between species. In addition, a "trophic" action of sympathetic nerves associated with nerve activity can be discerned which influences the development and maintenance of vascular structure and function and is particularly prominent during growth. However, there is not enough known concerning the trophic effects of sympathetic nerves to determine the mechanisms involved and their role in hypertrophy. There is also evidence from the specialized thermoregulatory vascular bed of the rabbit ear, which has marked sympathetic vasoconstrictor tone, that these nerves may be a regulatory influence during growth on the number of vessels in the microcirculation.

# REFERENCES


1. Bevan, J. A. and Brayden, J. E., Non-adrenergic vascular neural vasodilator transmitter mechanisms, *Circ. Res.*, in press.
2. Burnstock, G., Griffith, S. G., and Sneddon, P., Autonomic nerves in the precapillary vessel wall, *J. Cardiovasc. Pharmacol.*, 6(Suppl. 2), S344, 1984.
3. Bevan, J. A., Bevan, R. D., and Duckles, S. P., Adrenergic regulation of vascular smooth muscles, in *Handbook of Physiology, Section 2*, Vol. 2, Bohr, D. F., Somlyo, A. P. and Sparks, H. V., Eds., American Physiological Society, Baltimore, 1980, 515.
4. Rosell, S., Neuronal control of microvessels, *Ann. Rev. Physiol.*, 42, 359, 1980.
5. Bell, D. R., Webb, R. C., and Bohr, D. F., Functional bases for individualities among vascular smooth muscles, *J. Cardiovasc. Pharmacol.*, 7(3), S1, 1985.
6. Landsberg, L. and Young, J. B., Catecholamines and the adrenal medulla, in *Williams Textbook of Endocrinology*, Foster, D. W. and Wilson, J. D., Eds., 7, 891, 1985.
7. Bevan, R. D., Trophic effects of peripheral adrenergic nerves on vascular structure, *Hypertension*, 6, III-19, 1984.
8. Rhodin, J. A. G., Architecture of the vessel wall, *Handbook of Physiology*, Section 2, Vol. 2, Bohr, D. F., Somlyo, A. P., and Sparks, H. V., Eds., American Physiological Society, Baltimore, 1980, 1.
9. Gabella, G., An introduction to the structural variety of smooth muscles, *Vascular Neuroeffector Mechanisms, 4th Int. Symp.*, Bevan, J. A. et al., Eds., Raven Press, New York, 1983, 13.
10. Mathieu-Costello, O. and Fronek, K., Morphometry of the amount of smooth muscle cells in the media of various rabbit arteries, *J. Ultrastructure Res.*, 91, 1, 1985.
11. Bevan, J. A., Some bases of difference in vascular response to sympathetic activity — variations on a theme, *Circ. Res.*, 45, 161, 1979.
12. Bevan, J. A. and Bevan, R. D., Patterns of $\alpha$-adrenoceptor regulation of the vasculature, Proc. of the 5th Meeting on Adrenergic Mechanisms, *Blood Vessels*, 21, 110, 1984.
13. Burnstock, G. and Costa, M., *Adrenergic Neurons: Their Organization, Function and Development in the Peripheral Nervous System*, Chapman and Hall, London, 1975.
14. Cowen, T., An ultrastructural comparison of neuromuscular relationships in blood vessels with functional and "non-functional" neuromuscular transmission, *J. Neurocytol.*, 13, 369, 1984.
15. Rowan, R. A. and Bevan, J. A., Quantitative ultrastructural measurement of innervation density, neurotransmitter vesicles, and neuromuscular cleft width in the rabbit central ear artery and its main side branch, *Blood Vessels*, in press.
16. Devine, C. E. and Simpson, F. O., The fine structure of vascular sympathetic neuromuscular contacts in the rat, *Am. J. Anat.*, 121, 153, 1967.
17. Bevan, J. A. and Su, C., Variation of intra- and perisynaptic adrenergic transmitter concentrations with width of synaptic cleft in vascular tissue, *J. Pharmacol. Exp. Ther.*, 190, 30, 1974.
18. Barman, S. M., Gebber, G. L., and Calaresu, F. R., Differential control of sympathetic nerve discharge by the brain stem, *Am. J. Physiol.*, 247, R513, 1984.
19. Wallin, G. and Fagius, J., The sympathetic nervous system in man — aspects derived from microelectrode recordings, *Trends in Neurosci.*, 9(2), 63, 1986.
20. Folkow, B., Impulse frequency in sympathetic vasomotor fibres correlated to the release and elimination of the transmitter, *Acta. Physiol. Scand.*, 25, 49, 1952.
21. Nilsson, H., Ljung, B., Sjablam, N., and Wallin, B. G., The influence of the sympathetic impulse pattern on contractile responses of rat mesenteric arteries and veins, *Acta. Physiol. Scand.*, 123, 303, 1985.
22. Westfall, T. C. and Meldrum, M. J., Alterations in the release of norepinephrine at the vascular neuroeffector junction in hypertension, *Ann. Rev. Pharmacol. Toxicol.*, 25, 621, 1985.
23. Bevan, R. D., Yang, J. S.-J., and Bevan, J. A., Embryology and functional variation in the arterial system. in *Vascular Neuroeffector Mechanisms*, Vol. IV, 4th Int. Symp. on Vascuar Neuroeffector Mechanisms, Japan, Bevan, J. A., Maxwell, R. A., Fujiwara, M., Shibata, S., Toda, N., Mohri, K., Eds., Raven Press, New York, 1983, 209.

24. Oriowo, M. A. and Bevan, J. A., Differing sensitivity of large rabbit arteries to norepinephrine; correlation between sensitivity and agonist dissociation constant, *Fed Proc.*, 45(3), 559, 1986.
25. Su, C., Bevan, R. D., Duckles, S. P., and Bevan, J. A., Functional studies of the small pulmonary arteries, *Microvasc. Res.*, 15, 37, 1978.
26. Laher, I. and Bevan, J. A., Alpha-adrenoceptor number limits response of some rabbit arteries to norepinephrine, *J. Pharm. Exp. Ther.*, 233, 290, 1985.
27. Owen, M. P., Quinn, C., and Bevan, J. A., Phentolamine-resistane neurogenic constriction occurs in small arteries at higher frequencies, *Am. J. Physiol.*, 249, H404, 1985.
28. Langer, S. Z., Presynaptic regulation of the release of catecholamines, *Pharmacol. Rev.*, 32, 337, 1980.
29. Alabaster, V. and Davey, M., Precapillary vessels: effects of the sympathetic nervous system and of catecholamines, *J. Cardiovasc. Pharmacol.*, 6, S365, 1984.
30. Laher, I., Kayhal, M. A., and Bevan, J. A., NE sensitive, phenoxybenzamine-resistant receptor sites associated with contraction in rabbit arterial but not venous smooth muscle: possible role in adrenergic neurotransmission, *J. Pharmacol., Exp. Ther.*, 237(2), 364, 1986.
31. Hirst, G. D. S. and Neild, T. O., Evidence for two populations of excitatory receptors on arterial smooth muscle, *Nature*, 283, 767, 1980.
32. Kuriyama, H. and Kitamura, K., Electrophysiological aspects of regulation of precapillary vessel tone in smooth muscles of vascular tissues, *J. Cardiovasc. Pharmacol.*, 7(Suppl. 3), S119, 1985.
33. Johnson, P. C., The myogenic response, in *Handbook of Physiology,* Section 2, Vol. 2, Bohr, D. F., Somlyo, A. P., and Sparks, H. V., Eds., American Physiological Society, Baltimore, 1980, 409.
34. Mellander, S. and Johansson, B., Control of resistance, exchange, and capacitance functions in the peripheral circulation, *Pharmacol. Rev.*, 20(3), 117, 1968.
35. Hwa, J. and Bevan, J. A., Stretch-dependent (myogenic) tone in rabbit ear resistance arteries, *Am. J. Physiol.*, 250, H87, 1986.
36. Vanhoutte, P. M. and Miller, V. M., Heterogeneity of endothelium-dependent responses in mammalian blood vessels, *J. Cardiovasc. Pharm.*, 7(3), S12, 1985.
37. Guyton, J. R. and Hartley, C. J., Flow restriction of one carotid artery in juvenile rats inhibits growth of arterial diameter, *Am. Physiol. Soc.*, 248, H540, 1985.
38. Langille, B. L. and O'Donnell, F., Reductions in arterial diameter produced by chronic diseases in blood flow are endothelium-dependent, *Science*, 231, 407, 1986.
39. Folkow, B., Hallback, M., Lundgren, Y., Sivertsson, R., and Weiss, R., Importance of adaptive changes in vascular design for establishment of primary hypertension, studied in man and in spontaneously hypertensive rats, *Circ. Res.*, 32(Suppl. 1), I-2, 1973.
40. Folkow, B., Physiological aspects of primary hypertension, *Physiol. Rev.*, 62, 347, 1982.
41. Abboud, F. M., The sympathetic system in hypertension. State-of-the-art review, *Hypertension*, 4(Suppl. II), II-208, 1982.
42. Brody, M. J., Barron, K. W., Berecek, K. H., Faber, J. E., and Lappe, R. W., Neurogenic mechanisms of experimental hypertension, in *Hypertension,* 2nd ed., Genest, J., Kuchel, O., Hamet, P., and Contin, M., Eds., McGraw-Hill, New York, 1983, 117.
43. Dietz, R., Schomig, A., and Rascher, W., Pathophysiological aspects of genetically determined hypertension in rats, with special emphasis on stroke-prone spontaneously hypertensive rats, in *Handbook of Hypertension,* Vol. 4, de Jong, W., Ed., Elsevier, New York, 1984, 256.
44. Gray, S. D., Spontaneous hypertension in the neonatal rat, *Clin. Exper. Hyper. — Theory and Practice,* A6(4), 755, 1984.
45. Mulvany, M. J., Do resistance vessel abnormalities contribute to the elevated blood pressure of spontaneously-hypertensive rats? *Blood Vessels,* 20, 1, 1983.
46. Smith, P. G., Poston, C. W., and Mills, E., Ontogeny of neural and nonneural contributions to arterial blood pressure in spontaneously hypertensive rats, *Hypertension,* 6(1), 54, 1984.
47. Scott, T. M. and Pang, S. C., The correlation between the development of sympathetic innervation and the development of medial hypertrophy in jejunal arteries in normotensive and spontaneously hypertensive rats, *J. Autonomic Nerv. Syst.*, 8, 25, 1983.
48. Henrich, H., Hertel, R., and Assman, R., Structural differences in a mesentery microcirculation between normotensive and spontaneously hypertensive rats, *Pflugers Arch.*, 375, 153, 1978.
49. Ichijima, K., Morphological studies on the peripheral small arteries of spontaneously hypertensive rats, *Jpn. Circ. J.*, 33, 785, 1969.
50. Lee, T. J.-F. and Saito, A., Altered cerebral vessel innervation in the spontaneously hypertensive rat, *Circ. Res.*, 55, 392, 1984.
51. Lee, R. M. K. W., Forrest, J. B., Garfield, R. E., and Daniel, E. E., Morphometric study of structural changes in the mesenteric blood vessels of spontaneously hypertensive rats, *Blood Vessels,* 20, 57, 1983.

52. Lee, R. M. K. W., Vascular changes at the prehypertensive phase in the mesenteric arteries from spontaneously hypertensive rats, *Blood Vessels,* 22, 105, 1985.
53. Cassis, L. A., Stitzel, R. E., and Head, R. J., Hypernoradrenergic innervation of the caudal artery of the spontaneously hypertensive rat: An influence upon neuroeffector mechanisms, *J. Pharmacol. Exp. Ther.,* 234(3), 792, 1985.
54. Judy, W. V., Watanabe, A. M., Henry, D. P., Besch, H. R., Murphy, W. R., and Hockel, G. M., Sympathetic nerve activity. Role in regulation of blood pressure in the spontaneously hypertensive rat, *Circ. Res.,* 38(Suppl. II), II-21, 1976.
55. Thoren, P. and Ricksen, S.-E., Recordings of renal and splanchnic sympathetic nervous activity in normotensive and spontaneously hypertensive rats, *Clin. Sci.,* 57, 197s, 1979.
56. Juskevich, J. C., Robinson, D. S., and Whitehorn, D., Effect of hypothalamic stimulation in spontaneously hypertensive and Wistar-Kyoto rats, *Eur. J. Pharmacol.,* 51, 429, 1978.
57. Schramm, L. P. and Barton, G. N., Diminished sympathetic silent period in spontaneously hypertensive rats, *Am. J. Physiol.,* 236, 147, 1979.
58. Head, R. J., Cassis, L. A., Robinson, R. L., Westfall, D. P., and Stitzel, R. E., Altered catecholamine contents in vascular and nonvascular tissues in genetically hypertensive rats, *Blood Vessels,* 22, 196, 1985.
59. Yarowsky, P. and Weinreich, D., Loss of accommodation in sympathetic neurons from spontaneously hypertensive rats, *Hypertension,* 7, 268, 1985.
60. Ekas, R. D., Jr. and Lokhandwala, M. F., Sympathetic nerve function and vascular reactivity in spontanesouly hypertensive rats, *Am. J. Physiol.,* 241, R379, 1981.
61. Nordborg, C. and Johansson, B. B., Cerebral arterial morphometry in young adult stroke-prone spontaneously hypertensive rats, in *Hypertensive Mechanisms,* Rascher, W., Clough, D., and Ganten, D., Eds., Schattauer Verlag, Stuttgart, 1982, 165.
62. Gray, S. D., Early postnatal differences between WKY and SHR in anatomical and physiological parameters, *Fed. Proc.,* 41, 1589, 1982.
63. Karr-Dullien, V., Bloomquist, E. I., Beringer, T., and El-Bermani, A-W., Arterial morphometry in neonatal and infant spontaneously hypertensive rats, *Blood Vessels,* 18, 253, 1981.
64. Lee, R. M. K. W., Forrest, J. B., Garfield, R. E., and Daniel, E. E., Ultrastructural changes in mesenteric arteries from spontaneously hypertensive rats. A morphometric study, *Blood Vessels,* 20, 72, 1983.
65. Malmfors T., Furness, J. B., Campbell, G. R., and Burnstock, G., Reinnervation of smooth muscle of the vas deferens transplanted into the anterior chamber of the eye, *J. Neurobiol.,* 2, 193, 1971.
66. Ebendahl, T., Olson, L., and Seiger, A., The level of nerve growth factor (NGF) as a function of innervation, *Exp. Cell Res.,* 148, 311, 1983.
67. Chamley, J. H. and Campbell, G. R., Tissue culture: interaction between sympathetic nerves and vascular smooth muscle, Proc. 2nd Int. Symp. on Vascular Neuroeffector Mechanisms, Odense, Denmark, July 29 to August 1, 1976, 10.
68. Korsching, S. and Thoenen, H., Treatment with 6-hydroxydopamine and colchicine decreases nerve growth factor levels in sympathetic ganglia and increases them in the corresponding target tissues, *J. Neurosci.,* 5(4), 1059, 1985.
69. Zweifach, B. W., Kovalcheck, S., DeLano, F., and Chen, P., Micropressure-flow relationships in a skeletal muscle of spontaneously hypertensive rats, *Hypertension,* 3, 601, 1981.
70. Harper, S. L. and Bohlen, H. G., Microvascular adaptation in the cerebral cortex of adult spontaneously hypertensive rats, *Hypertension,* 6, 408, 1984.
71. Prewitt, R. L., Chen, I. I. H., and Dowell, R. F., Microvascular alterations in the one-kidney, one-clip renal hypertensive rat, *Am. J. Physiol.,* 246, H728, 1984.
72. Hutchins, P. M. and Darnell, A. E., Observation of a decreased number of small arterioles in spontaneously hypertensive rats, *Circ. Res.,* 34, 35(Suppl. I), I-161, 1974.
73. Hertel, R. and Henrich, H., Difference of catecholamine fluorescence in walls of microvessels of different branching order in the mesentery preparation of SHR and WKY, *Pflugers Arch.,* Suppl. 384, R6, 1980.
74. Haack, D. W., Schaffer, J. J., and Simpson, J. G., Comparisons of cutaneous microvessels from spontaneously hypertensive, normotensive Wistar-Kyoto, and normal Wistar rats, *Proc. Soc. Exp. Biol. Med.,* 164, 453, 1980.
75. Hill, C. E., Hirst, G. D. S., and Van Helden, D. F., Development of sympathetic innervation to proximal and distal arteries of the rat mesentery, *J. Physiol. (London),* 338, 129, 1976.
76. Furness, J. B., Arrangement of blood vessels and their relation with adrenergic nerves in the rat mesentery, *J. Anat.,* 115, 347, 1973.
77. Mills, E. and Smith, P. G., Functional development of the cervical sympathetic pathway in the neonatal rat, *Fed. Proc.,* 42, 1639, 1983.

78. Marshall, J. M., The influence of the sympathetic nervous system on the microcirculation of skeletal muscle, *J. Physiol. (London)*, 258, 118, 1976.
79. Gootman, P. M., Baez, S., and Feldman, S. M., Microcirculatory responses to central nerve stimulation in the rat, *Am. J. Physiol.*, 225, 1375, 1973.
80. Harper, S. L., Bohlen, H. G., and Rubin, M. J., Arterial and microvascular contributions to cerebral cortical autoregulation in rats, *Am. J. Physiol.*, 246, H17, 1984.
81. Winquist, R. J. and Bohr, D. F., Characterization of the rat basilar artery in vitro, *Experientia*, 38, 1187, 1982.
82. Meininger, G. A., Harris, P. D., Joshua, I. G., Miller, F. N., and Wiegman, D. L., Microvascular pressures in skeletal muscle of one-kidney one-clip renovascular hypertensive rats, *Microcirculation*, 1(3), 237, 1981.
83. Fernandez, D. and Crane, W. A. J., New cell formation in rats with accelerated hypertension due to partial aortic constriction, *J. Pathol.*, 100, 307, 1970.
84. Engler, E., Matthias, D., and Becker, C. H., Pathomorphological reactions of myocardium and intramural vessels and rats in the course of hypertension induced by depot angiotensin. Autoradiographic, light and electron microscopic investigations, *Exp. Pathol.*, 18, 37, 1980.
85. Abel, P. W. and Hermsmeyer, K., Sympathetic cross-innervation of SHR and genetic controls suggests a trophic influence on vascular muscle membranes, *Circ. Res.*, 49, 1311, 1981.
86. Willems, W. J., Harder, D. R., Contney, S. J., McCubbin, J. W., and Stekiel, W. J., Sympathetic supraspinal control of venous membrane potential in spontaneous hypertension in vivo, *Am. J. Physiol. Soc.*, 243, C101, 1982.
87. Morris, J. L. and Bevan, R. D., Proliferation of arteriovenous anastomoses in the developing rabbit ear is enhanced after denervation, *J. Anat.*, 176, 497, 1986.
88. Thoma, R., Uber die Histomechanik des Gefasssystems und die Pathogenese der Angiosklerose, *Virchows Arch. (Pathol. Anat.)*, 204, 1, 1911.
89. Todd, M. E. and Tokito, M. D., An ultrastructural investigation of developing vasomotor innervation in rat peripheral vessels, *Am. J. Anat.*, 106, 195, 1981.
90. Su, C., Bevan, J. A., and Assali, N. S., Development of neuroeffector mechanisms in the carotid artery of the fetal lamb, *Blood Vessels*, 14, 12, 1977.
91. Ljung, B. and Stage D., Postnatal ontogenetic development of neurogenic and myogenic control in the rat portal veins, *Acta Physiol. Scand.*, 94, 112, 1975.
92. Bevan, R. D., Effect of sympathetic denervation on smooth muscle cell proliferation in growing rabbit ear artery, *Circ. Res.*, 37, 14, 1975.
93. Pallie, W. and Phil, D., Embryology of the human arterial system (arteriogenesis), in *Structure and Function of the Circulation*, Vol. 1, Schwartz, C. J., Werthessen, N. T., and Wolf, S., Eds., Plenum Press, New York, 1980, 21.
94. Morris, J. J. and Bevan, R. D., Development of the vascular bed in the rabbit ear: scanning electron microscopy of vascular corrosion casts., *Am. J. Anat.*, 171, 75, 1984.
95. Grant, R. T., Further observations on the vessels and nerves of the rabbit's ear, with special reference to the effects of denervation, *Clin. Sci.*, 2, 1, 1935.
96. Clark, E. R., Studies on the growth of blood-vessels in the tail of the frog larva — by observation and experiments on the living animal, *Am. J. Anat.*, 23, 37, 1918.
97. Clark, E. R. and Clark, E. L., Microscopic observations of the extraendothelial cells of living mammalian blood vessels, *Am. J. Anat.*, 66, 1, 1940.
98. Clark, E. R., Hitschler, W. J., Kirby-Smith, H. T., Res, R. O., and Williams, R. G., General observations on the ingrowth of new blood vessels into standardized chambers in the rabbit's ear, *Anat. Rec.*, 50, 129, 1931.
99. Ryan, T. J. and Barnhill, R. L., Physical factors and angiogenesis, in *Development of the Vascular System*, Ciba Foundation Symp. 100, Nugent, J. and O'Connor, M., Eds., Pitman, London, 1983, 80.
100. Guth, L., "Trophic" influences of nerve on muscle, *Physiol. Rev.*, 48, 645, 1968.
101. Gutmann, E., Neurotrophic relations, *Ann. Rev. Physiol.*, 38, 177, 1976.
102. McArdle, J. J., Molecular aspects of the trophic influence of nerve on muscle, *Prog. Neurobiol.*, 21, 135, 1983.
103. Srinivasan, R. and Chang, W. W. L., Effect of neonatal sympathectomy on the postnatal differentiation of the submandibular gland of the rat, *Cell Tissue Res.*, 180, 99, 1977.
104. Klein, R. M. and Torres, J., Analysis of intestinal cell proliferation after guanethidine-induced sympathectomy, *Cell Tissue Res.*, 195, 239, 1978.
105. Muir, T. C., Pollock, D., and Turner, C. J., The effects of electrical stimulation of the autonomic nerves and of drugs on the size of salivary glands and their rate of cell division, *J. Pharmacol. Exp. Ther.*, 195, 372, 1975.

106. Branco, D., Teixeira, A. A., Azevedo, I., and Osswald, W., Structural and functional alterations caused at the extraneuronal level by sympathetic denervation of blood vessels, *Naunyn-Schmiedeberg's Arch. Pharmacol.,* 326, 302, 1984.

107. Bevan, R. D. and Tsuru, H., Long-term influence of the sympathetic nervous system on arterial structure and reactivity: Possible factor in hypertension, in *Disturbances in Neurogenic Control of the Circulation,* Abboud, F. M., Fozzard, H. A., Gilmore, J. P., and Reis, D. J., Eds., Clin. Phys. Series, American Physiological Society, Bethesda, Md., 1981, 153.

108. Smith, P. G., Role of the sympathetic nervous system in functional maturation of Muller's smooth muscle in the rat, *J. Pharmacol. Exp. Ther.,* 235, 330, 1985.

109. Bevan, R. D., Tsuru, H., and Bevan, J. A., Cerebral artery mass in the rabbit is reduced by chronic sympathetic denervation, *Stroke,* 14, 393, 1983.

110. Rusterholz, D. B. and Mueller, S. M., Sympathetic nerves exert a chronic influence on the intact vasculature that is age related, *Ann. Neurol.,* 11, 365, 1982.

111. Fleming, W. W., The trophic influence of autonomic nerves on electrical properties of the cell membrane in smooth muscle, *Life Sci.,* 22, 1223, 1978.

112. Westfall, D. P., Lee, J. J-F., and Stitzel, R. E., Morphological and biochemical changes in supersensitive smooth muscle, *Fed. Proc.,* 34(10), p. 1985, 1975.

113. Aprigliano, O. and Hermsmeyer, K., Trophic influence of the sympathetic nervous system on the rat portal vein, *Circ. Res.,* 41, 198, 1977.

114. Abel, P. W., Urquilla, P. R., Goto, K., Westfall, D. P., Robinson, R. L., and Fleming, W. W., Chronic reserpine treatment alters sensitivity and membrane potential of the rabbit saphenous artery, *J. Pharmacol. Exp. Ther.,* 217, 430, 1981.

115. Bevan, R. D. and Tsuru, H., Functional and structural changes in the rabbit ear artery following sympathetic denervation, *Circ. Res.,* 49, 478, 1981.

116. Salonen, V., Lehto, M., Kalimo, H., Penttinen, R., and Aro, H., Changes in intramuscular collagen and fibronectin in denervation atrophy, *Muscle & Nerve,* 8, 125, 1985.

117. Murray, M. A. and Robbins, N., Cell proliferation in denervated muscle: identity and origin of dividing cells, *Neuroscience,* 7, 1823, 1982.

118. Fronek, K., Bloor, C. M., Amiel, D., and Chvapil, H., Effect of longterm sympathectomy on arterial wall in rabbits and rats, *Exp. Mol. Pathol.,* 28, 279, 1978.

119. Mangiarua, E. I. and Bevan, R. D., Altered endothelium-mediated relaxation after denervation of growing rabbit ear artery, *Eur. J. Pharmacol.,* 122, 149, 1986.

120. Mangiarua, E. I., Joyce, E. H., and Bevan, R. D., Denervation increases myogenic tone in a resistance artery in the growing rabbit ear, *Am. J. Physiol.,* 250, H889, 1986.

121. Aubineau, P., Reynier-Rebuffel, A. M., Bouchaud, C., Jousseaume, O., and Seylaz, J., Long-term effects of superior cervical ganglionectomy on cortical blood flow of non-anesthetized rabbits in resting and hypertensive conditions, *Brain Res.,* 338, 13, 1985.

122. Hart, M. S., Heistad, D. D., and Brody, M. J., Effect of chronic hypertension and sympathetic denervation on wall/lumen ratio of cerebral vessels, *Hypertension,* 2, 419, 1980.

123. Sadoshima, S., Busija, D., Brody, M., and Heistad, D., Protection against stroke by sympathetic nerves, in *Cerebral Blood Flow: Effect of Nerves and Neurotransmitters,* Heistad, D. and Marcus, M. L., Eds., Elsevier/North Holland, New York, 1982, 309.

124. Hartley, L. and Campbell, G. R., Sympathetic denervation of arteries in the wing of the chicken, *Neurosci. Lett.,* 11, S47, 1983.

125. Mueller, S. M. and Rusterholz, D. B., The trophic influence of sympathetic nerves on rabbit ear vasculature is absent in coarctation hypertension, *Artery,* 11(5), 345, 1983.

126. Baumbach, G. L. and Heistad, D. D., Effects of sympathetic stimulation and changes in arterial pressure on segmental resistance of cerebral vessels in rabbits and cats, *Circ. Res.,* 52, 527, 1983.

127. Fronek, K. and Zweifach, B. W., Microvascular pressure distribution in skeletal muscle and effect of vasodilation, *Am. J. Physiol.,* 228, 791, 1975.

128. Kontos, H. A., Wei, E. P., Navari, R. M., Levasseur, J. E., Rosenblum, W. I., and Patteson, J. L., Jr., Responses of cerebral arteries and arterioles to acute hypotension and hypertension, *Am. J. Physiol.,* 234, H371, 1978.

129. Shapiro, H. M., Stromberg, D. D., Lee, D. R., and Wiederhelm, C. A., Dynamic pressures in the pial arterial microcirculation, *Am. J. Physiol.,* 221, 279, 1971.

Chapter 7

# MECHANICAL CONTROL OF SMOOTH MUSCLE GROWTH

## John R. Guyton

## TABLE OF CONTENTS

## I. GENERAL PRINCIPLES

### A. Levels of Feedback Control

Since the chief function of smooth muscle is mechanical — i.e., to form containing walls in hollow organs that either hold or propel luminal contents under pressure — one may suspect that the chief influence on smooth muscle growth is also mechanical. The functional capacities and hence the sizes of smooth muscles throughout the body are well matched to mechanical requirements. We can consider for a moment the possibility that evolution has provided genes that simply target the growth of individual smooth muscles to meet the average functional requirements for those muscles in the species. Under this concept there would be no feedback control to match smooth muscle growth to varying mechanical requirements within the individual organism. However, much evidence indicates that this is not the case. What the genes actually provide is not a set of growth targets, but rather a set of mechanisms of feedback control of growth, whereby smooth muscle cells, sometimes in conjunction with other cells, are able to sense mechanical stress and respond with appropriate growth or atrophy.

The first part of this review consists of this introduction, an operational definition of tissue growth, and a brief sketch illustrating how classical mechanics may be applied to the problem of smooth muscle growth. Considerable evidence favors the idea of a fundamental feedback loop of tensile or stretch stimulus and growth response, acting probably at the level of the smooth muscle cell, regardless of the organ or tissue within which the smooth muscle is found. The second part of this review will cite the evidence for this feedback loop. For orderly growth, growth under special circumstances, and especially the attainment of correct lumen dimensions, neighboring cell types and neurohormonal signals from distant sites play important roles in modulating smooth muscle growth. These roles tend to differ according to the specific organ in which the smooth muscle is found. These diverse heterocellular responses are sometimes stimulated mechanically, an example being the effect of blood flow on circumferential growth of the arterial media. The third part of this review will consider mechanical aspects of heterocellular responses resulting in lumen size determination and smooth muscle growth. A most interesting question is how any cell, smooth muscle or othewise, may sense mechanical stress and respond by generating signals for growth. A few data from in vitro studies are available to address this question, and these studies will be covered in the fourth part of this review.

### B. Operational Definition of Growth

It will be helpful to think of smooth muscle growth in a functional, mechanical sense, and we shall define it, for purposes of this review, as an increase in the mass of stress-bearing structures. For the most part, these structures are the linear polymers comprising actin and myosin, along with their associated proteins, and the linear, cross-linked extracellular polymers of elastin and various collagens. Other aspects of tissue growth, including cell division and the synthesis of DNA, RNA, protein, membranes, and other products, will be regarded as subordinate to the chief goal of providing increased strength and sometimes contractility to the tissue. The issue of hyperplasia versus hypertrophy, though interesting in itself and sometimes pertinent to this review, will not be a major focus of discussion. Much more to the point is the question of the nature of growth stimulation, whether the pattern of resultant growth is hyperplasia or hypertrophy. It seems possible that the same mechanical stimulus — for example, tensile stress — may elicit hyperplasia in a smooth muscle tissue during fetal and neonatal life and hypertrophy in the same tissue during adult life.[1,2] If this is so and can be proven, it will be interesting, of course, to discover why the pattern of

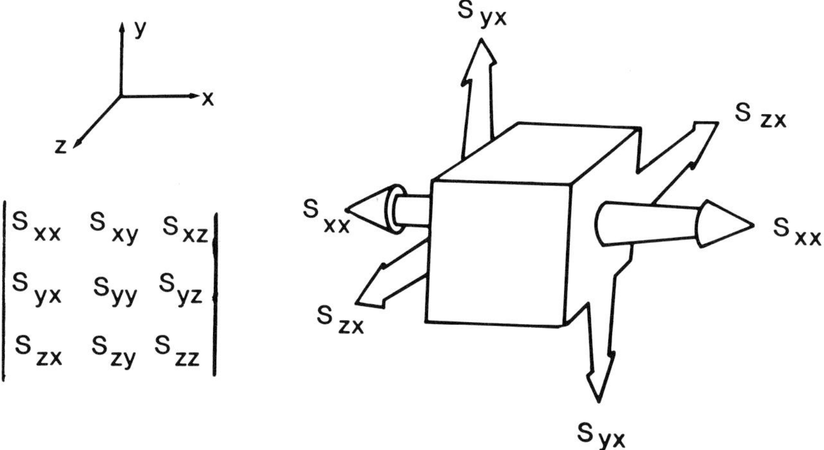



FIGURE 1. The stress tensor. All nine components in matrix form are represented at lower left. Three of these components, each represented as a combination of balanced forces acting upon opposite sides of the cube, are shown at right. Stress is determined as force/area, as the size of the cube approaches zero.

growth is different. Yet this question should not overshadow what seems, at least to this author, to be a more interesting problem, that is, the nature of the cellular transducing mechanism that converts tensile stress into a signal governing the production of tension-bearing structures. This review will not provide the answer. The studies cited will only form a background, from which one may hope to formulate testable hypotheses.

## C. Tissue Stress and Strain

Texts on classical mechanics, particularly the fields of viscoelasticity and rheology, should be consulted for more complete treatment of the points to follow.[3-5] Mechanical properties of the vascular wall, studied extensively because of the relation to hemodynamics, have been well reviewed by Dobrin[6] and Milnor.[7]

### 1. Stress

Stress is the totality of forces acting to deform a tissue, divided by appropriate measurements of tissue area. In general mathematical terms, stress is expressed as a tensor field, which means that at every point within the tissue the stress tensor is specified. The nature of the stress tensor acting at a given point can be grasped with the aid of Figure 1, which shows a small cube of tissue, the dimensions of which can be imagined to approach the limit of zero. Both the magnitude of forces acting on the cube and the area of a cube face will approach zero, but the ratio force/area, which defines the components of stress, will remain finite. The force acting upon each face of the cube can be resolved into three orthogonal components. If the cube is imagined to be small enough, then the forces acting on any face will be balanced by equal and opposite forces acting on the opposite face, as illustrated by the diagram. Thus the stress tensor has nine components, specified by the three axes normal to the cube faces and three components of balanced forces for each pair of faces. These components are conveniently represented in matrix form, as shown on the left-hand side of Figure 1. As long as matter can be represented as a continuum, a reasonable assumption for practical work and certainly for consideration of biologic tissue, then it can be shown that the stress tensor is symmetric. This means that $S_{xy} = S_{yx}$, $S_{xz} = S_{zx}$, and $S_{yz} = S_{zy}$. Thus stress is fully specified, when 6 independent scalar quantities are specified at each point within a tissue. It is helpful to relate the stress components to commonly recognized physical quantities. The diagonal components of the stress matrix, $S_{xx}$, $S_{yy}$, $S_{zz}$, comprise tensile stress. A negative value for tensile stress denotes compressive stress. These stresses cause elongation or shortening of the tissue. The symmetric nondiagonal components, $S_{xy}$, $S_{xz}$, $S_{yz}$, are termed shear stress, and they cause angular deformations of the tissue. A special situation applies to a fluid medium at rest. In this case the shear stress components are all zero. Furthermore, $S_{xx} = S_{yy} = S_{zz} = -P$, where P denotes pressure.

It is important at times to distinguish between stress evaluated at the surface of a piece of tissue, readily referred to as stress "upon" the tissue, and stress evaluated throughout the tissue interior, or stress "within" the tissue. In the former case the stress is defined by the tensor field existing at those points comprising the tissue surface, and in the latter the entire tensor field is considered, including both interior and surface points. In considering stress upon a tissue, one may treat the tissue as a "black box", without *a priori* assumptions about interior structure. The deforming effect of stress may still be measured reproducibly and analyzed in a macroscopic sense, that is, according to a chosen mathematical model for the behavior of the system as a whole. To analyze stress within a tissue, however, some type of assumptions about the interior mechanical structure must be made. Mathematical analyses developed for use in materials science can have hidden assumptions of homogeneity, isotropy, and solidity of materials, and these assumptions may not apply in biology, as soon as the extrapolation from macroscopic to microscopic stress is attempted. We shall return to this point later.

Most analyses of stress and response involve simple one-component stress models applied to situations of favorable tissue geometry. Stress will often be expressed in cylindrical or polar coordinates in preference to orthogonal coordinates, when the former can make the stress field mathematically homogeneous. For hollow organs in many situations, circumferential tensile stress may be the predominant stress component within the organ wall, and only this component may need to be determined. However, it is worth remembering that the complex geometry and stress patterns commonly found in smooth muscle tissues may lead to unique growth patterns. For example, Robertson[8] has attributed the growth of smooth muscle pads surrounding the orifices of branching arteries to stress patterns produced as the pulsating main vessel alternately pushes and tugs on the branch. Other examples of obviously complex stress patterns will occur wherever smooth muscle-containing organs deviate from ideal cylindrical or spherical shapes.

### 2. Strain

Strain is defined as the extent of deformation produced by stress and is generally expressed as a fractional change in dimensions such as volume or length, or as an angular shape change. The term "stretch" is best understood as the linear or areal strain induced by externally applied tensile stress. Volume strain of smooth muscle is usually considered negligible — ie., if the tissue is stretched, it becomes thinner in exact proportion, because volume is conserved. In a limited sense, experimental verification for this assumption has been obtained.[9] However, the experimental analysis considered the arterial wall to be a solid, rather than porous, structure, and thus neglected the movement of water into or out of the total space enclosed by the tissue. Especially in considering phenomena that may take hours or days to develop, the question of volume strain may have to be reexamined, since the water content of the tissue may change.[10]

Experimentally, a predetermined stress or a predetermined strain may be applied to the tissue being studied. The alternate variable is then measured. The term "mechanical loading" of the tissue can be applied to either strategy or a combination of the two. In the living organism, likewise, either stress or strain, or both, may be appear to be applied to tissues. For the vascular wall, especially the arterial wall, two predetermined variables, blood pressure and blood flow, are closely related to tensile stress and shear stress, respectively. The bladder wall, on the other hand, must adapt itself to a predetermined strain specified by urine volume.

### 3. Elasticity, Viscosity, and Plasticity

A tissue may exhibit elastic behavior, in which case the deformed tissue, upon the removal of stress, will eventually return to its original resting shape and dimensions. If either the deformation or the return to resting state does not occur promptly with the application of stress, but requires an appreciable time to develop, then the tissue is said to exhibit viscoelastic behavior. The combination of shock absorbers and springs in an automobile suspension constitutes a viscoelastic system. Finally, a deformation that never returns to the original state after removal of stress indicates a plastic response of the tissue.

Whether tissue strain should be considered elastic, viscoelastic, or plastic depends greatly upon the time frame and degree of stress considered. The stress imposed upon every smooth muscle-containing organ will have both steady and time-dependent components. Viscoelasticity is determined from the strain response only to time-dependent stress occurring within the experimental period. The pulse pressure in arterial blood is a natural, cyclical stress often employed in the analysis of viscoelastic properties of arteries in vivo. It is important to remember, when attempting to correlate measured mechanical behavior to the in vivo physiologic state, that the experimental design will often neglect the steady or long-term components of stress. Tissue growth is a long-term process, and its correlation with short-term mechanical behavior may be inaccurate. One approach to circumvent this problem can be to measure short-term biochemical phenomena which themselves may correlate with growth — e.g., rates of amino acid uptake.[12,13]

Isolated smooth muscle strips or rings are often stretched to specific increments of resting length over several hours prior to studies of contractile behavior. Tensile stress in the muscle is initially high, but diminishes gradually to a quasi-steady state suitable for experimentation. This poorly understood plastic phenomenon is termed "stress relaxation". A similar process, termed "creep", occurs after a predetermined, constant stress is applied to the tissue — e.g., by hanging a weight on the muscle strip. The tissue stretches slowly, reaching a fairly constant length after several hours. Creep and stress relaxation are especially important problems for the investigator who would like to study tissues in vitro over a period of days rather than hours, because these processes actually continue at a slow rate. The conditions have not yet been found under which smooth muscle can be maintained indefinitely under steady tension in cell culture or organ culture, and yet most smooth muscle in vivo exists under precisely this condition. Approaches to the application of mechanical stimulation to cultured cells and organs will be discussed in the last part of this review, but, in fact, none accomplish the goal of producing steady, measurable tension in the cell layer or tissue. Cyclic tension and cyclic or steady shear are other matters and may be more amenable to experimentation.

### 4. Active Vs. Passive Tension

Up to this point, we have treated smooth muscle tissue as a passive mechanical system. For some situations, such as the modeling of pulse wave velocity in arteries, a passive viscoelastic model can be entirely appropriate. Ruegg[13] has pointed out that an active force-generating response of muscle to increased strain (myogenic response) may actually mimic passive viscous behavior. In attempting to discern the pathway of growth response to smooth muscle to mechanical stimulation, however, passive models generally seem inadequate. An increase in tension must be identified as active or passive, because in the first case tissue length is decreased, and in the second it is increased. A question that will come up continually is whether the growth response of smooth muscle may be related to active tension, passive tension, or both. There is not yet a single instance in which this issue can be regarded as settled. Some suggestive data are available, however, and the question will be addressed repeatedly in this review.

*5. Microscopic Mechanics — Importance of Tensile Stress*

In order to discover the cellular transducer that converts stress or strain to a signal for growth, stress and strain will have to be understood at the microscopic level and even at the molecular level, at least in qualitative terms. The extent to which tissues are nonhomogeneous and anisotropic must be considered carefully, as one attempts to deduce the state of microscopic stress from macroscopic measurements. Two examples from everyday life may be considered. The first demonstrates that shear stress, considered macroscopically, can be experienced by component structures as a combination of tensile and compressive stress. Consider an ordinary screen door that hangs by hinges on one side and has a rod with a turnbuckle fastened diagonally as a brace. The stress produced by gravity is primarily a shear stress upon the door as a whole. Yet the components of the door accomodate this stress through a combination of tensile stress in the diagonal bracing rod and compressive stress in the parts of the frame below the brace. In specimens of smooth muscle, at the cellular level, not only diagonal bracing, but also criss-cross diagonal patterns are so common that a developmental mechanism for responding to shear stress in this manner must be suspected. At any rate, what is most likely experienced at the cellular level is a combination of tensile and compressive stress. The second example from everyday life is that of a balloon which is being compressed by two hands. Macroscopically the stress on the balloon is compressive. This stress is separable into components of increased pressure (compression) of the gas within the balloon and increased tension in part of the wall of the balloon. In fact, the solid, flexible wall of the balloon will experience increased tensile stress in some form almost regardless of the nature of the applied stress. Although a balloon model may not be precisely accurate, smooth muscle cells are known to be surrounded not only by the plasma membrane, but also by a sheath of basement membrane and collagen fibers.[14,15] Thus even compressive forces may be experienced as tension by certain cellular and extracellular components. Resistance to compression is readily provided by the incompressibility of water contained within microscopic analogues of balloons. Ion pumps, tissue gels, and osmotic forces are responsible for the long-term maintenance of tissue incompressibility, or turgor.

Further insight into the necessity of regarding stress in microscopic terms can be gained by considering the effects of atmospheric pressure on thick-walled hollow vessels. On at least two occasions[16,17] it has been argued that the contribution of atmospheric pressure to wall tension in such vessels may have considerable impact. It is pointed out that the general equation for total circumferential wall tension (T) per unit length in a cylinder is

$$T = P_1r_1 - P_2r_2 \tag{1}$$

where $P_1$ and $P_2$ are internal and external pressures, and $r_1$ and $r_2$ are internal and external radii, respectively. By this formula, T will be negative, if $P_2/P_1 > r_1/r_2$, a condition that indeed is fulfilled for small arteries and some other smooth muscle-containing organs. This leads to the statement that "the conclusion is inescapable, the [small] arterial wall must be under compression."[17] This would seem to invalidate the simple use of transmural pressure difference in determining wall tension, and yet the simplistic approach, namely,

$$T = (P_1 - P_2) \cdot r_1 \tag{2}$$

probably correlates better with smooth muscle tissue growth than the first, "general" equation.[18] The fallacy in applying the general equation arises in the extrapolation from the macroscopic level to the microscopic or molecular level of mechanics. Compressive, isotropic stress, within the vessel wall due to hydrostatic (atmospheric) pressure, can be borne by all of the molecules comprising the arterial wall, including fluid water and nontension-bearing cellular organelles. In contrast, tensile stress affects a distinct minority of the mass and space in the vessel wall. Tensile stress can be conceived as concentrated along the fibrils of actomyosin, collagen, and elastin, and transmitted through correspondingly small regions of cellular membranes. The signals for tissue growth presumably are generatd in these or similar small regions. Thus it seems possible that the hydrostatic pressure component of tissue stress may be largely or entirely neglected. In a thick-walled hollow organ, the transmural pressure difference, not the absolute pressures multiplied by radii, determines the relevant tensile component of stress, and the perferred expression for wall tension is Equation 2.

It is appropriate to ask to what extent and with what kind of confidence mechanics at the microscopic level can ever be deciphered. No single approach will suffice for this task. A combination of light and electron microscopy, physiologic and physical studies, pharmacologic intervention, and biochemical purification or selective removal of tissue components will all have to be correlated. The studies of Glagov and co-workers[15,19] on the microscopic and ultrastructural architecture of arterial wall at various degrees of distention represent good examples of this eflla[20] has reviewed the structural attachments that function for smooth muscle as tendons do for skeletal muscle — i.e., to allow the transmission of force. The necessity of correlating mechanical function to chemistry and structure continues to be an intriguing challenge.

## II. GROWTH RESPONSES TO TENSILE STRESS OR TO STRETCH

### A. Myometrium

Perhaps the best documented relationship of smooth muscle growth to stretch occurs in uterine smooth muscle, known as the myometrium.[21-23] Myometrial growth during pregnancy is interesting, because most of it occurs during a time when active, cyclical force development is suppressed. Considering the extraordinary forces necessary for parturition, it is obvious that the combination of mechanical and hormonal influences on myometrial growth must provide a contractile reserve able to generate, over a relatively brief period of time, forces that are well beyond the previous experience of the tissue.[24,25] Estrogen, progesterone, and stretch play crucial roles in bringing about this type of smooth muscle growth.

### 1. Mechanical Activity During the Menstrual Cycle

During fetal life, mechanical forces probably aid in shaping the organ and determining lumen dimensions, but these forces are unknown. The next major phase of uterine growth occurs just prior to menarche, presumably under the influence of estrogen. Menarche also marks the onset of increased uterine mechanical activity, which has been characterized by the measurement of intrauterine pressure via balloon-tipped catheter. Baseline pressure is 10 to 15 mm Hg; superimposed upon this are frequent cyclical rises in pressure of minimal amplitude (less than 10 mm Hg) during most of the cycle, but attaining 70 mm Hg or more during menstruation.[26] The high pressure may function not only to expel luminal contents, but also to help close bleeding vessels. Whether wall tension may play any role in myometrial growth at or before menarche is unknown; an understanding of the temporal relationship between intrauterine pressure development and myometrial growth would be a first step in determining such a role.

### 2. Uterine Growth and Mechanical Activity during Pregnancy

Most of the description in this paragraph is taken from the classic review of uterine growth by Reynolds.[21] During pregnancy the human uterus increases from a nongravid weight of 30 to 60 gm to 1000 to 1400 gm at term, and much of the mass increase can be attributed to the myometrium. Smooth muscle cells replicate rapidly early in pregnancy, leading to smaller, more numerous cells, while the uterus as a whole increases little in size. The uterine lumen during this time is minimally distended by the small trophoblast and amniotic volume, so that mechanical stimulation of growth is minimal. From some time in the first trimester until term, almost no further hyperplasia is seen. Hypertrophy during this time assumes massive proportions, with human myometrial cells doubling in width and increasing 10-fold in length to attain final lengths greater than 0.5 mm. Myometrial wall thickness has been measured carefully in the mouse. At mid-pregnancy wall thickness is 50% greater than in the nongravid uterus; at term, however, it decreases to a dimension about one half that of the nongravid uterus. This attenuation of wall thickness is implicit in the time course for uterine mass growth given by Reynolds. Late in pregnancy, mass growth levels off, while fetal expansion continues to distend the uterus.

Chemical measurements of tension-bearing proteins in the uterine wall show total mass and compositional changes during pregnancy. Actomyosin concentration in rabbit myometrium almost doubles during the latter two thirds or pregnancy, a period when stretching of the wall by fetal growth may be important.[27] Uterine collagen mass rises 4- to 5-fold during pregnancy in the rat; but because of an even greater increase in other tissue components, collagen concentration decreases slightly.[28] Elastic components of the uterus also appear to hypertrophy, but as Reynolds has pointed out, this effect is confined to the corpus uteri. Connective tissue in the uterine isthmus is relatively underdeveloped, and, presumably, the isthmus thereby may stretch more easily. This fact may aid in the dropping of the fetus at near term, and in cervical effacement and dilatation.[21]

The most significant aspect of mechanical activity of the myometrium, from early pregnancy until near term, is a lack of cyclic tension development. At 14 weeks in human pregnancy, cyclic pressure increments in the uterine lumen average only 3 mm Hg, while resting pressure remains at pregestational levels. A large, unused capacity for tension development is shown by the fact that pressures of 60 to 70 mm Hg can be generated, at the same gestational time, if abortion is induced by hypertonic saline.[24] To understand the tensile stress perceived by individual myometrial cells during pregnancy, two physiological aspects in addition to the low intrauterine pressure must be considered. The first is increasing lumen size, which leads to increasing wall tension, even while pressure remains constant.[25] Nevertheless, the tension developed and experienced by the muscular wall as a whole remains only a small fraction of the capacity for tension development. The second consideration may explain this partially. Pregnancy is characterized not only by decreased electrical activity, but also by a lack of propagation of action potentials among myometrial cells.[29] The contraction of one cell, triggered by an action potential, may serve only to stretch other cells with which it is linked in series. Tensile stress and strain at the cellular level thus may run through cycles that are not manifested by changes in intrauterine pressure. However, it has not been possible so far to assess quantitatively the extent of forces affecting individual cells.

During the latter stages of pregnancy, a gradual development of synchronous, propagating contractions is seen. At 36 weeks of pregnancy, the human uterus displays average 18 mm Hg cyclic increments in intraluminal pressure. From 36 to 40 weeks, spontaneous contractions increase in frequency, but these are not accompanied by a corresponding increase in myometrial mass. At the onset of labor, contractile force development increases abruptly, with the development of 45 mm Hg increases in pressure.[24] Because of vastly increased lumen size, wall tension with these contractions may be 10-fold or more greater than that developed in the nonpregnant, menstruating uterus.[25]

### 3. Effects of Estrogen and Progesterone

The growth-promoting effects of estrogen on the myometrium have long been recognized.[21] Estrogen administration to an ovariectomized animal leads rapidly to an increase in RNA synthesis, especially ribosomal RNA, and subsequently to increased protein synthesis.[30] Hypertrophy rather than hyperplasia of the myometrial cells apears to result, if estrogen is given alone.[21,31] In the ovariectomized animal without estrogen, the resting membrane potential appears to be too low to allow the generation of action potentials, and the muscle is therefore quiescent. Estrogen raises the resting membrane potential and stimulates action potential generation and contractile activity.[26] Although estrogen appears to stimulate RNA and protein synthesis directly, additional hypertrophic effects might result from the stimulation of mechanical activity or from the increase in endometrial volume causing distension of the surrounding myometrium.

Studies of progesterone effects have often employed estrogenic stimulation either simultaneously or sequentially.[21,30] With previous estrogen stimulation, progesterone can cause marked mitotic activity in the myometrium.[21] It is possible that progesterone is responsible for the hyperplasia of myometrial cells seen early in pregnancy. Progesterone also has major effects on electromechanical activity of myometrial cells in many species, inducing a hyperpolarization of the cell membrane.[26] The development and propagation of action potentials is inhibited in the rabbit uterus. The disintegration of electrical activity is associated with a lack of effective force development.[29] These effects of progesterone are quite dependent upon the species studied; in the rabbit and probably several other species progesterone does appear to be responsible for the reduced contractile activity of the uterus during pregnancy.[26]

### 4. Effect of Distension

The factor of uterine distension as a potent growth stimulus has received much attention from Reynolds[21] and from Csapo and co-workers.[22,23] A common observation, established early in this work, is that in animals with a duplex uterus, such as the rabbit, rat, or mouse, in a unilateral pregnancy the gravid horn gains much more wall mass than the sterile horn. Furthermore, the mass of a horn correlates well with the number of fetuses it holds. To confirm the distension-growth relationship experimentally, Reynolds inserted oversized cylindrical paraffin pellets into the uterine lumen. In adult ovariectomized rabbits, a uterine horn can be stimulated to grow in mass by more than 150% by this technique. Much of the increase can be attributed to the myometrium. An increase in mitotic figures is seen in both rabbit and rat myometrium stimulated in this manner, and cellular hypertrophy also occurs.[21,32] Interestingly, the growth response to distention is absent in sexually immature rabbits, suggesting that previous exposure to estrogen and progesterone with the attendant maturing of the uterus is necessary. In the adult ovariectomized rabbit treated with estrogen, Reynolds found the distension-growth response to be much reduced. Progesterone-treated rabbits, on the other hand, exhibited a growth response similar in magnitude to untreated rabbits, but required about twice as much uterine distension to produce optimal growth. Reynolds[21] considered this outcome to favor a role for wall tension in uterine growth, since with progesterone-induced relaxation a greater degree of stretch would be required to produce the same tension.

The results of Csapo and co-workers[22,23] do not agree on all points with those just given. These workers used inflatable balloons to produce chronic distension of one horn of the rabbit uterus. Postpartum involution of the uterus could be prevented by the maintenance of distension. In nonpregnant, ovariectomized rabbits, distension was considered to produce myometrial hypertrophy without hyperplasia.[22] However, it is not clear that hyperplasia could really be ruled out by the lack of increased mitoses as late as 5 to 15 days after the stimulus, and no detailed results are given for the mitotic counts. In estrogen-treated animals, considerable endometrial growth occurred in both distended and control horns, while myometrial growth was substantial only in the distended horn. Progesterone treatment was associated with myometrial hypertrophy in both the distended and control horns. Based on these results, the authors suggested that myometrial hypertrophy could result either from progesterone or from stretch. Stretch rather than active tension was favored as a mechanical stimulus, since spontaneous, cyclic contractions were almost absent with progesterone treatment. However, little attention was given to baseline, resting intrauterine pressures in this study, and published tracings do not show a reduction in baseline pressure with progesterone.[22,23,29] A constant baseline pressure in concert with increasing lumen dimensions would give rise to increased steady wall tension. Thus the steady component of wall tension is a possible, though unproven, factor in myometrial growth.

Specific tension-bearing proteins also appear to increase in response to uterine distension. Actomyosin content was found to correlate highly with the number of fetuses per rabbit uterine horns; this effect was especially evident in unilaterally pregnant rabbits.[27] A very similar result was obtained for the collagen content of the two horns of the rat uterus.[28] Marked increases, up to 3-fold, in total mass of collagen, with modestly decreasing collagen concentration in the uterine wall, were also found in an experiment in which rat uteri were distended with paraffin wax.[32]

### 5. Strain Vs. Stress as a Growth Signal

In attempting to deduce the nature of the mechanical stimulus to myometrial growth, we must first admit that a clear answer is not yet available. However, active, cyclic tension development, particularly the propagating muscular activity that can effectively raise intrauterine pressure, does not appear to be linked to growth stimulation. This leaves two candidate variables that can be named at present — stretch of the wall and resting wall tension. These variables can be identified as strain and stress, respectively, and they are obviously closely related. Resting wall tension is a largely unexplored physiologic factor, dependent upon resting luminal pressure, which is mentioned but not well documented in several publications. Resting luminal pressure and wall tension actually must operate in the context of a homeostatic system that includes (1) growth of fetus and placenta, (2) the development of fluid pressure via the secretion of amniotic fluid, (3) an unknown mechanism for regulating the volume of this fluid, and (4) the support of constant tension by the uterine wall as it stretches and grows. It would be most interesting to find ways to distinguish primary from secondary effects within this homeostatic system.

## B. Smooth Muscle of the Urinary Tract

### 1. Ureters

The ureters serve a purely mechanical function of transporting urine from kidneys to bladder, but they are not simply passive conduits. Ureteral smooth muscle cells near the renal pelvis have a pacemaker function, generating slow waves of electrical excitation. These signals propagate downward toward the bladder in the form of groups of action potentials, which correlate with peristaltic contraction.[33] Pressure and diameter changes in canine ureters, in normal and pathologic states, have been studied by Rose and colleagues.[34,35] These studies exemplify the type of data useful for estimating a potential mechanical stimulus to smooth muscle growth. Normally, relatively modest intraluminal pressure cycles peaking at 15 to 23 mm Hg are generated from a baseline of 8 mm Hg resting pressure. These pressure cycles propagate downward toward the bladder, driving boluses of urine before them. The ureteral lumen is completely closed between boluses. From measurements of outside diameter in surgically exposed ureters, histologic measurements of wall thickness, and pressure determinations, Rose et al.[34] estimated circumferential tensile stress in the ureteral wall, giving values averaging $1.3 \times 10^4$ dynes/cm$^2$ for baseline and $3.4 \times 10^4$ dynes/cm$^2$ for peak tensile stress. Lack of consideration of tissue shrinkage during histologic preparation may have made these values artifactually high by perhaps 20 to 30%. After acute ureteral occlusion, pressure waves become more frequent, and their amplitude increases 3-fold or more. Within 5 to 20 min, the ureter reaches a state of steady distension and pressure without contractile cycles. Lumen pressure in the obstructed ureter continues to increase and is maintained at 1 hr at 40 to 50 mm Hg, presumably through the force of glomerular filtration. A steady high wall tension of about $10 \times 10^4$ dynes/cm$^2$ was estimated at this point. The relative contributions to wall tension of active smooth muscle contraction versus passive stretch in this case were unclear. In chronically obstructed ureters, both radius and length of the ureter were increased. Contractile waves were present, but irregular, and did not effect lumen closure. Lumen pressures were no greater than those recorded in normal ureters (about 20 mm Hg), presumably because of adaptive and pathologic changes in the formation of urine in the kidney. Wall tension estimated for chronically obstructed ureters varied greatly with a mean of $10 \times 10^4$ dynes/cm$^2$.

An experimental technique for producing partial obstruction of the ureters was used by Gee and Kiviat,[36] who noted a 4-fold increase in the area of the muscularis layer on proximal transverse sections 8 weeks after partial obstruction. The number of smooth muscle nuclear profiles doubled over the same time. Interestingly, large amounts of collagen and large elastic fibers in the muscularis layer, unusual in sham-operated ureters, were demonstrated histologically in experimental ureters. An experimental model of longitudinal ureteral stretching in dogs has been described by Crooks and co-workers.[37] While the authors focused primarily on its potential value in surgical repair, the model did provide an impression of muscular hypertrophy and increased extracellular fibrous tissue secondary to longitudinal stretch.

### 2. Urinary Bladder

The function of the urinary bladder is considerably more complex than that of the ureters. Smooth muscle in the bladder wall, known as the detrusor muscle, is subject to a unique degree of voluntary control. Sensory fibers in the bladder wall detect distention of the bladder and make afferent connections with the sacral spinal cord, which is able to trigger the excitation of parasympathetic ganglia. The detrusor muscle is richly innervated by parasympathetic, cholinergic terminals, which cause contraction. Other neurotransmitter terminals are also present, but their function is unclear. The spinal reflex loop is subject to inhibition by signals arriving from the cerebral cortex; otherwise, urination might occur at any time or place.[38]

The rather complex nervous control of the bladder allows a unique type of functional behavior — that is, distention from zero to 400 to 500 m*l* luminal volume with a minimal increase in filling pressure. At almost any time, however, voluntary mechanisms can lead to detrusor contraction and relaxation of the urethral sphincter, with micturition occurring at relatively modest bladder intraluminal pressures of 15 to 30 mm Hg. With partial urethral obstruction, pressure in the bladder must rise considerably higher, 35 to 75 mm Hg, to effect voiding. This condition is associated with thickening of the detrusor muscle, such that hypertrophic trabeculations appear on the luminal surface.[39] In a study of partial obstruction induced by partial ligation of the urethra in rats, Mattiasson and Uvelius[40] found a 7-fold increase in bladder weight. This corresponded to detrusor hypertrophy sufficient to allow high pressure (70 mm Hg) to develop in the distended bladder. Control bladders could develop similar pressures with maximal electrical stimulation of parasympathetic nerves, but this was true only at low lumen volumes. Interestingly, the hypertrophic bladders were ineffective at low volumes, because of an altered force-length relationship in the muscle. At optimal length, the maximal stimulated active stresses developed in control and hypertrophic muscles were very nearly equal, averaging $8 \times 10^5$ dynes/cm². A different approach to induction of detrusor hypertrophy, which again may point toward wall tension as a determining factor, was developed by Peterson and co-workers.[41] The bladder was injected transurethrally with a mixture of warm paraffin and petrolatum jelly, which conformed to lumen shape on cooling. This procedure might be expected to affect bladder volume chronically, without affecting voiding pressure, since urethral function was not impaired. By the LaPlace relationship ($T = Pr$), increased bladder volume would increase wall tension. Striking detrusor hypertrophy resulted, up to a 7-fold increase in weight, but unfortunately, parameters of intravesical pressure and frequency and volumes of voiding were not determined. In another study,[42] in which increased urine volume flow was caused by parabiosis of a normal rat to one with bilateral nephrectomy, bladder hypertrophy was also found. This was felt to be due to increases in both the frequency and volume of voiding.

Various means of interrupting the nervous control of bladder function can provide unique insights on the mechanical control of smooth muscle growth in this organ, suggesting particularly a minimal role for active contractility. Langworthy and Kolb[43] reported in 1938 histologic results of various denervation procedures on the cat urinary bladder, including section of preganglionic parasympathetic sacral roots, posterior sacral roots, and postganglionic sympathetic fibers. Increased bladder volume occurred temporarily after parasympathetic motor denervation, and permanently after sensory denervation. Motor denervation resulted in hypertrophy of the muscle; the most striking cases of hypertrophy were unilateral, obtained in bladders with parasympathetic denervation of the ipsilateral side only. Five bladders with sensory denervation (posterior sacral root section) were examined histologically, two of these having had sympathectomy performed as well. Generally atrophy of the bladder wall was seen; however, these results were qualified by the statement that bacterial cystitis was severe in these animals. After sympathectomy alone, the bladders tended to be smaller than normal, and no hypertrophy of the muscle was evident. The interesting finding of unilateral hypertrophy following unilateral parasympathectomy was confirmed by one other investigator,[44] but not by Carpenter in 1951. Carpenter[45] weighed the cat bladder and found greater than 2-fold increases in weight following bilateral parasympathectomy. The volume of urine remaining in the bladder after daily expression of urine by the investigator's hand pressing on the abdominal wall was thought to be a key parameter in the hypertrophy. The highest bladder weights were obtained after an operation that left intact the pudic innervation of the striated muscle of the external urethral sphincter, leading to urethral resistance and high intravesical pressure. Most recently bilateral parasympathetic denervation has been carried out with rat bladders, which were found to increase 4- to 5-fold in weight.[10] At the time of killing, the denervated bladders were noted to contain up to 7 m$\ell$ urine compared to less than 0.5 m$\ell$ in control bladders. Not surprisingly, muscle strips isolated from the denervated bladders were found to contract optimally at greater lengths, corresponding to the high volumes present in vivo.

These experiments illustrate the potential wealth of information that might be gained from appropriate manipulation of bladder innervation, volume, and pressure. Detrusor growth may occur within 20 days in the cat and within 10 days in the rat. The results thus far seem to favor passive stretch or perhaps passive tensile stress in detrusor hypertrophy.

## C. Vascular Smooth Muscle

Growth of vascular smooth muscle has received much attention, primarily because of the magnitude of the clinical problem of atherosclerosis. Only a small proportion of this work has considered mechanical stimulation of growth. The hypothesis that we wish to consider is that of direct stimulation by wall tension. By this mechanism abnormal smooth muscle growth in arteries and arterioles may be stimulated pathophysiologically by hypertension. Smooth muscle growth in veins exposed to elevated venous pressure or transposed surgically to arterial sites may be stimulated similarly.

Injury to arterial endothelium is postulated to be a key factor in many instances of pathologic smooth muscle growth, and under some circumstances mechanical phenomena such as hypertension or shear stress appear to cause endothelial injury.[46,47] Although endothelial injury or dysfunction cannot be excluded as a physiologic mechanism of smooth muscle growth, the complexity of this topic, its coverage in well-informed recent reviews,[48,49] and the need to cover wall tension as an alternative, attractive, physiologic growth stimulus all place the discussion of endothelial injury largely beyond the scope of this review. On the other hand, a very likely physiologic function of endothelium in long-term lumen diameter regulation in arteries will be covered in a subsequent section.

### 1. Blood Pressure

Mean arterial pressure in man is ordinarily between 80 and 110 mm Hg, and all mammals exhibit pressures close to this range.[18] Whether pulse pressure or pulse rate may contribute to normal arterial morphogenesis independently of mean pressure is unclear. A study on cyclic stretching of cultured smooth muscle,[50] to be described later, would suggest a positive answer, but in vivo data on this question have not yet been obtained. The diurnal variability of hemodynamic parameters may also be important. Systolic hypertension (greater than 150 mm Hg) may occur in every normal individual at some time during the day. A marked decline in systolic and diastolic pressures (15 to 40 mm Hg) generally occurs with sleep.[51]

### 2. Tension-Bearing Components of the Arterial Wall

The acute response of an artery to increasing pressure consists of fairly rapid dilatation until physiologic pressure is reached, then little further diameter increase up to a bursting point which is above physiologically attainable pressures.[6,19,52] Roach and Burton[52] hypothesized that this behavior could be due to the presence in the arterial wall of both elastin, which is easily stretched, and collagen, which resists stretching strongly. The collagen is considered to have a relatively loose arrangement at pressures less than physiologic, leaving elastin to bear most of the tensile load. At physiologic pressures and above, the collagen fibers are pulled taut and dominate the support of tension. This theory was supported by digestion experiments, in which arteries were treated with crude trypsin to degrade elastin, or formic acid to degrade collagen. By present standards, these must be regarded as relatively nonspecific digestions, so that quantitative aspects of tension support remain uncertain. Wolinsky and Glagov[19] fixed rabbit aortic segments at various pressure, with attention to maintenance of in vivo length, and examined the tissue by light and electron microscopy. An important result was that, even at pressures up to 200 mm Hg, collagen fibers in the adventitia did not appear stretched, while collagen fibers in the medial layer were straight and well oriented circumferentially. It was suggested that the adventitia contributes little to tension support at pressures in the physiologic range. In the media, collagen was assigned an important role in tension support, but the participation of elastin in a "two-phase" tension support system was emphasized. The "two-phase" material may have a greater tensile strength than that of the stronger component alone, because the more extensible component (elastin) is able to distribute stressing forces. This allows stresses about structural defects (a major cause of failure in stiff mechanical systems) to be transferred to other locations in the system.

The contribution of active smooth muscle contraction to the maintenance of normal arterial dimensions in vivo has sometimes been considered insignificant, but the steepness of the pressure-diameter curve at physiologic pressure implies that a very substantial smooth muscle contribution to tension could be missed easily. Recently, intracoronary nitroglycerin infusion in conscious, chronically instrumented dogs was shown to produce a 3.4 ± 0.5% diameter increase in coronary arterial diameter, with no change or a slight decrease in perfusing pressure.[53] This study and a similar one[54] show at least some contribution of pharmacologically responsive smooth muscle tone to tension support. To assess the magnitude to this contribution, the passive stress-strain relationship — i.e., the pressure-diameter curve — will have to be assessed in the same experimental preparation under conditions that rule out active contractility. The overwhelming majority of studies of arterial mechanics in vivo have focused on pulsatile behavior. Smooth muscle contraction is much too slow to contribute dynamically to pulsatile diameter changes, but an altered contractile state may contribute in a minor way to the high-frequency stiffness of the wall.[6] A question that appears relatively unexplored is whether pulsatility, particularly in vivo, might alter the long-term state of smooth muscle contraction via myogenic or endothelium-mediated mechanisms. We must conclude that the contribution of tonic smooth muscle contraction to steady-state tension support in arteries is an underdeveloped topic.

If indeed collagen and elastin bear most of the tension in arteries under physiologic conditions, some difficult questions must be answered, including the following: how can the tension of these elements be developed and maintained, especially considering that collagen and elastin are proteins subject to slow, but definable metabolic turnover? Why has evolution provided a potent cellular contractile apparatus in arteries? After all, arteries, except after catastrophic trauma, do not regulate the flow of blood by offering a substantial resistance, but rather function to conduct blood with minimal resistance. There is no doubt that smooth muscle cells, since they synthesize collagen and elastin, largely determine the total support of tension in the arterial media. One can hypothesize further that actomyosin, if not bearing a considerable portion of the tension itself through a series or parallel mechanical arrangement, may at least determine the resistance to stretch of other, more passive elements. However, the mechanisms involved are by no means clear. The possibility that smooth muscle contractility may be involved in long-term diameter regulation in arteries will be considered further in a later section.

### 3. Medial Tension and Thickness

Wolinsky and Glagov[18] have measured medial thickness and diameter in the descending thoracic aorta of ten mammalian species, ranging in size from mice (body weight as low as 28 g) to pigs (up to 200 kg). The measurements incorporated fixation at physiologic pressure and correction for shrinkage artifact. (It should be noted that histologic preparation may induce considerable shrinkage, amounting to 30% in this study, or alternatively tissue expansion, depending on technique. The assumption that shrinkage is isotropic, so that the shrinkage of wall thickness can be estimated from circumferential shrinkage, is widely applied, but has not been tested directly.) From the measurements of Wolinsky and Glagov and their estimates of mean arterial pressure, one may estimate that average circumferential tensile stress in thoracic aorta of various mammalian species falls within the range of 1 to $3.5 \times 10^6$ dynes/cm$^2$. This range is intriguingly narrow, given the 7000-fold range of body weight, and it suggests that wall tensile stress may be a regulated quantity in arteries. The mechanism of such regulation could well be an intrinsic growth response of smooth muscle cells to tension, but these results by themselves could also be explained by evolutionary adaptation. Other observations do address the question of evolutionary targeting versus adaptive response. The postnatal development of large pulmonary and systemic arteries was studied by Leung et al.,[55] in order to assess the effects of physiologically decreased and increased pressures, respectively, on tension-bearing components of these vessels. Differences in total fibrous protein accumulation and elastin accumulation, presumably adaptive responses, paralleled differences in medial tension development.

The thoracic aorta is an example of an elastic artery, a category that includes the aorta, carotids, and proximal portions of their branches. These arteries are characterized by the organization of the media into musculoelastic lamellae, which are bands of smooth muscle cells circumscribed by thick, circumferential elastin plates. Wolinsky and Glagov[18] emphasized the role of the musculoelastic lamella as a functional unit in the wall of elastic arteries. The number of these lamellae in thoracic aortic media varied from 5 in the mouse to 72 in the pig. The constancy of tension per lamellar unit, a maximum 3-fold variation, was a particularly striking feature noted by Wolinsky and Glagov. As might be anticipated from the preceding discussion, the average thickness of each lamella was also relatively constant from species to species. The number of musculoelastic lamellae appears to be determined early in life, near the time of birth.[1] It would be interesting to have estimates of wall tension in the various species at birth, in order to gauge the possibility that the number of lamellae is actually determined in response to this factor.

### 4. Intimal Thickening

Large arteries in adult humans, pigs, horses, and cows all are known to develop, over a period from late in intrauterine life to adulthood, diffusely thickened intimas, composed of modified smooth muscle cells and extracellular fibrous components, covered by a monolayer of endothelium. In the corresponding arteries in small animals such as mice and rats, the intima consists only of the endothelial layer resting almost directly on the internal elastic lamina. In animals of intermediate size (rabbit, cat, dog, and monkey), intimal thickening is largely confined to regions near the origins of the main arteries. Thus there appears to be a correlation between intimal thickening and arterial size in an absolute sense.[56] Furthermore, a relationship between rate of intimal growth and rate of postnatal arterial diameter growth — i.e., the most marked thickening occurring during the first 3 decades of human life — has also been suggested.[57] How may a relationship between arterial size, or rate of increase of size, and diffuse intimal thickening be explained? One possibility would involve a relationship between arterial size and alterations in endothelial function or integrity, but studies to support this notion by comparing arteries of different sizes are lacking. An alternative answer may lie with the fact that total circumferential wall tension is high in large arteries (since blood pressure is relatively constant among species). The number of medial musculoelastic lamellae actually fails to rise proportionately, and the tension per lamella and individual lamellar thickness are increased.[58] Tensile stress for the innermost medial layers may be considerably higher than that affecting external layers, as suggested by Doyle and Dobrin for dog carotid artery.[59] High tensile stress acting on the innermost part of the arterial wall may not only increase the thickness of the medial layers, but also stimulate migration of smooth muscle cells and proliferation in the intima. Thus it seems possible that the postnatal arterial media by itself may be limited in its capacity to make a fully adaptive response to tension, because of the inability to form additional lamellae,[1] and that intimal thickening arises in part from a continuing tensile stimulus.

In certain large muscular arteries of the human fetus, including the coronary, popliteal, and distal brachial arteries, focal cushion-like intimal thickenings composed of longitudinal muscle and elastic tissue have been found.[8,60] The thickenings occur particularly at the mouths of branch vessels. Their longitudinal muscular orientation contrasts with the more circular arrangement of medial muscle cells. These early, focal intimal thickenings progress with age and are found to merge with the musculoelastic layer (deepest layer) of diffuse intimal thickening. Causally, this process has been ascribed to unusual, focal pulsatile stresses. At branch sites, the radial pulsations of the branch are transmitted to the parent vessel in part as longitudinal pulsatile stress. Circularly oriented muscle in the parent vessel media may offer relatively weak resistance to this stress. The popliteal and lower brachial arteries, situated near major joints, appear to be relatively weakly tethered to the surrounding tissue, and thus to have increased pulsatile strain. In adult man, relative intimal thickness reaches its peak in the coronary arteries, with nonatherosclerotic intima eventually exceeding the media in thickness.[56] This fact has long been attributed to the bending and longitudinal stretching of coronaries that accompany cardiac motion. In all of these situations, detailed mechanical studies and correlation with focal intimal thickening have yet to be performed. Furthermore, the possible contributions of endothelial alterations[48,49] due to either pulsatile strain or blood flow shear[61-63] must be considered.

### 5. Effects of Hypertension

Hypertension is well known to cause thickening of the arterial wall, an effect that has long been ascribed to increased wall tension.[64,65] In the media, the musculoelastic lamellae expand in thickness, but do not increase in number.[66] Both the weight of thoracic aortic media and absolute amount of actomyosin were found by Seidel to be increased in spontaneously hypertensive rats (SHR).[67] Hydralazine treatment and lowering of blood pressure in SHR and control rats led to decreased levels of actomyosin, although total medial protein and DNA remained high in these rats with previously established hypertension.[68] Recently studies by Olivetti et al.[69] and Owens et al.[2] have suggested that cellular hypertrophy rather than hyperplasia occurs during hypertension-induced medial growth. The latter authors also found that an increased fraction of smooth muscle nuclei, approximately 20%, were tetraploid and octaploid in hypertensive rats. Wolinsky found absolute increases in collagen and elastin mass in thoracic aorta media of two-kidney, one-clip Goldblatt hypertensive rats; these connective tissue changes were persistent after reversal of the hypertension.[65]

Arterial medial hypertrophy in hypertension is usually considered to be due to increased tensile stress.[65] Hypertension also leads to growth and thickening of the arterial intima.[70] The mechanism could be similar to that outlined earlier for diffuse intimal thickening. However, there is clearly no consensus for tensile stress as the major cause for intimal growth in hypertensive states. Determination of the major mechanism for intimal growth is clinically quite important, since atherogenesis has a clear predilection for regions of thickened intima. One of the best documented effects of hypertension is to cause the margination and diapedesis of blood monocytes into the arterial intima, but not as far as the media.[71] Once inside the tissue, the monocytes become macrophages, and as such they presumably secrete the same growth factor for smooth muscle that macrophages are known to produce and secrete in vitro.[72] Other effects of hypertension are to increase endothelial permeability and to cause a thickened region of basement membrane-like material just beneath the endothelium.[73] These indications of effect on arterial endothelium could relate directly or indirectly to growth of underlying smooth muscle. Endothelium produces both stimulatory and inhibitory factors affecting smooth muscle growth.[74,75] A role for some of these factors in vivo is suspected, but not proven; their possible relationship to hypertension is completely obscure.

Evidence for structural arteriolar changes in hypertension was obtained by Folkow and co-workers[76,77] in plethysmographic studies on forearm blood flow. After a maximum vasodilatory stimulus (ischemia), hypertensive subjects were unable to decrease forearm blood flow resistance as effectively as control subjects. Other studies using a morphologic approach have clearly shown an increased ratio of wall thickness to lumen radius in small arterial resistance vessels[78] and arterioles.[79] In the former vessels, this is thought to represent true tissue growth — that is, an increase in medial cross-sectional area and number of smooth muscle cell layers. For arterioles it is difficult to be certain that comparable vascular sites (order of branching) are being sampled. It has been suggested that a structural narrowing (decreased diameter) of the arteriole rather than increased medial cross-sectional area accounts for the increased wall-to-lumen ratio.[79]

## 6. Veins

Uvelius, Arner, and Johansson[80] reported hypertrophy of smooth muscle in the rat portal vein 5 days after induction of portal hypertension, accomplished by partial ligation of the hepatic branches of the portal vein. Cellular hypertrophy, not hyperplasia, accounted for smooth muscle growth. The rat portal vein contains mostly longitudinal muscle, and this layer as well as the smaller circular muscle layer participated in hypertrophy. The capacity to bear passive longitudinal stress increased, but active longitudinal contractile force did not. The significance of these results is unclear, since the increased stress imposed by portal hypertension is circumferential, and thus at a right angle to the longitudinal direction of experimental force measurement.

Veins used as grafts in the arterial circulation develop intimal thickening and increased contents of protein, actin, myosin, and collagen per unit vessel length.[81,82] Seidel and co-workers found that maximum active contractile stress (force/wall cross-sectional area) in the grafts was reduced compared to normal veins. Total force development was increased, but not in proportion to hypertrophic wall thickening. The same authors pointed out that the vessel hypertrophy could be due to increases in wall tension, oxygen partial pressure, or shear stress.[82]

## D. The Role of Tension or Stretch in Smooth Muscle Growth

Experimental evidence for a causative effect of tension or stretch in determining smooth muscle growth in vivo has been cited for each of the three organ systems reviewed. For the cardiovascular system in particular, the hypothesis of a tension-induced growth response may provide a unifying explanation for a number of diverse observations. The interested reader can also refer to work on gastrointestinal smooth muscle, where similar experiments and observations have been described[14,83-85] (see also the chapter by Gabella in this volume). It must be noted that all of the data thus far cited have been obtained in vivo, and that confirmatory data from pure cell cultures (Section IV of this review) are generally sparse. Thus it is strictly correct to speak of a tension- or stretch-induced growth response only at the tissue level. Yet the universality of the response in diverse smooth muscle tissues, with varying types and degrees of innervation, strongly suggests that the response actually takes place at the level of the individual smooth muscle cell. This issue will require continued attention and very likely a high degree of ingenuity for a satisfactory answer.

Three general observations should be made. First, hormonal and neuronal influences play important modulating roles on the smooth muscle growth response. These roles are especially evident in the uterus and the urinary system. Second, growth seems to correlate best with long-term, steady stress or strain relationships, rather than short-term, easily demonstrated, contractile behavior. To put it in another way, static mechanical properties of tissues appear more relevant to the growth response than dynamic mechanical properties. This may contrast with the general situation in skeletal muscle, where repetitive short-term exercise is obviously accompanied by hypertrophy. It is a mistake, however, to equate long-term phenomena with "passive" behavior and short-term phenomena with "active" behavior of smooth muscle. Although short term active responses are more readily demonstrated as active, there is considerable evidence for a constant baseline level of tone in smooth muscle tissues. Even when it appears that extracellular fibrous proteins bear most of the load, and when tissue stress is altered little after poisoning the smooth muscle cells, one must still face the question of how, in the history of development of the tissue, tension may have been imparted to the collagen and elastin.

Studies of mechanical influences on smooth muscle growth often try to interpret whether stretch or tension is more important in determining the growth. Since stretch is defined as tissue strain, and tension divided by area is identified with stress, it is tempting to ask whether stress or strain determines growth. However, this is fallacious thinking, because stress and strain are not separable phenomena. Stress is always accompanied by strain, and vice versa. When tension, as opposed to stretch, is mentioned as an inducer of growth, the type of tension referred to is generally active contractile tension. Because tissue always has some elasticity, stretch is always accompanied by some type of stress, although it may only be passive. The question of the mode of growth induction may be clarified by referring to Figure 2. This illustrative physiologic model of a smooth muscle cell contains an active contractile component and both series of parallel elastic components. Because of the slow time scale over which growth stimulation is presumed to occur, it may be possible to neglect viscous mechanical behavior. Growth stimulation could be coupled to any of the three mechanical elements in the figure. If it is the parallel elastic component, then stretch could activate growth, regardless of the state of contractility. If the coupled element is the series elastic component, then either passive stretch or active contraction would activate growth. Finally, if some step or structure directly related to contraction were involved, then growth would only follow active contraction. It should be emphasized that such a model is at this point only illustrative. It would be wrong to equate elasticity measured over short

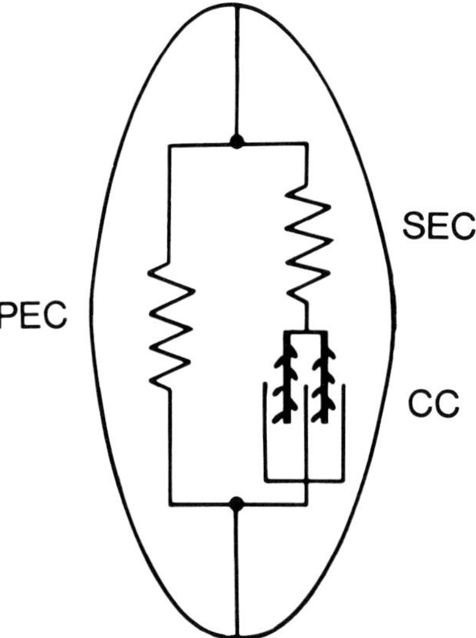



FIGURE 2.   Model of a smooth muscle cell, illustrating possible sites of coupling between mechanical activity and growth stimulation. CC, contractile component; SEC, series elastic component; PEC, parallel elastic component.

time periods with the elastic elements just discussed. Yet it may be possible to gain some understanding of the subcellular compartments involved, by studying mechanical growth control according to such a model.

## III. LUMEN SIZE REGULATION

Smooth muscle function, and therefore growth, has a three-dimensional character that cannot be neglected (Figure 3). In this respect the situation for smooth muscle differs from that for skeletal muscle, where muscle length is determined within a narrow range by bony attachments. The lumen dimensions of a hollow organ obviously become a major determinant of smooth muscle growth, or vice versa. A general scheme for the factors that are involved in three-dimensional smooth muscle growth are shown in Figure 4. We must not forget that the mechanical load, against which smooth muscle operates, is subject to physiologic control. This load, controlled by forces that tend to expand the lumen, must be defined in terms of pressure and volume (or more specifically, volume/time, i.e., either flow or cycles of filling and emptying, depending on whether the organ functions as a conduit or reservoir). For arteries, the expansive forces are generated by cardiac action and controlled by arteriolar constriction. For the ureters and bladder, the mechanical load is created by the combination of glomerular filtration and renal tubular reabsorption. For the uterus, increasing fetal volume is obviously important, but a stable luminal pressure is actually established by the control of amniotic fluid pressure.

The overall response of smooth muscle to its mechanical load must be characterized in terms of both dimension change and generation of tension. Growth of smooth muscle must accommodate both of these aspects, and therefore, growth must be able to respond selectively to either the dimensional or tensile aspects of the mechanical load. Let us consider the area that smooth muscle sheet in the wall of any hollow organ would cover, if it were dissected free, cut open, and laid out flat. Factors that govern lumen size will determine this area. Wall thickness, on the other hand, is related to the

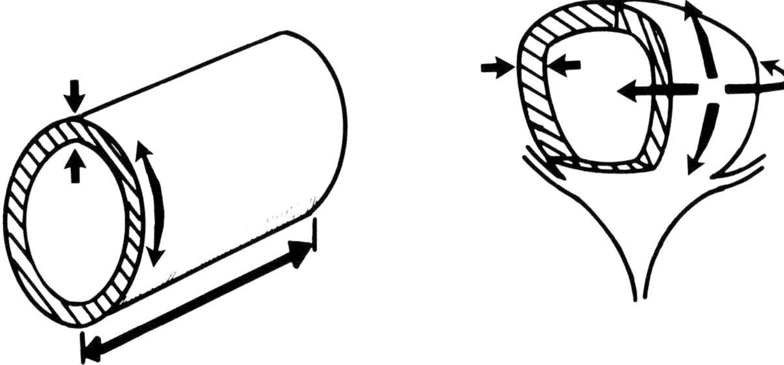

FIGURE 3.    Representations of a blood vessel and the urinary bladder, showing the
dimensions in which smooth muscle growth may occur.

tensile force that must be developed in the smooth muscle sheet. Both area, related to
lumen size, and thickness must be controlled during the course of smooth muscle
growth. As will be discussed below, these two aspects of growth can be interactive, but
they clearly respond to distinct stimuli.

## A. Volume-Dependent Lumen Size Regulation

Organs that function to accommodate cyclically an increasing volume within their
lumen, then to expel it, have a type of lumen size regulation that may be termed
volume-dependent. The pregnant uterus is an obvious example, and it will serve to
reinforce the idea that control of smooth muscle growth in terms of lumen size must
be considered separately from growth control for wall thickness. Growth in wall thickness corresponds to growth of reserve contractile power and is directed toward the
eventual goal of expelling the fetus. Growth in lumen size has as its goal the accommodation of the growing fetus and placenta. Special mechanisms, especially the inhibition of myometrial contraction by progesterone or other agents, are available to help
accomplish this latter goal. However, it must be remembered that there is always a
residual, resting tension in the uterine wall, whether generated actively or passively.
This tension is precisely balanced with the stress imposed by amniotic fluid pressure;
how such a balance is achieved, so that amniotic fluid volume remains under control,
is unclear.

In the urinary bladder, growth of wall thickness appears to be a response to wall
tension, as has been discussed. The factors that lead to luminal expansion are quite
different. Partial urethral obstruction in humans, typically developing over a period of
years, causes high voiding pressure and detrusor hypertrophy, but through most of the
course no increase in average lumen dimensions. Only late in the course of the problem
do marked postvoiding residual urine volume and urinary retention develop. Prior to
this time, the typical clinical picture is one of frequent voiding of small volumes due to
instability of the bladder reflex loop, and because of the frequency of voiding, bladder
lumen dimensions are small. Animal models, unfortunately, fail to mirror this sequence of events, probably because partial urethral obstruction is induced suddenly.[40]
In contrast to the wall thickness increase following urethral obstruction, tremendous
volume dilatation of the bladder is produced after sensory denervation due to diabetes
mellitus in humans[86] or to posterior spinal root section in animals.[43] The key difference
is that in this case, frequency of voiding is altered by the cessation of spinal and cortical
afferent signals that indicate bladder filling. The primary point to be made from these

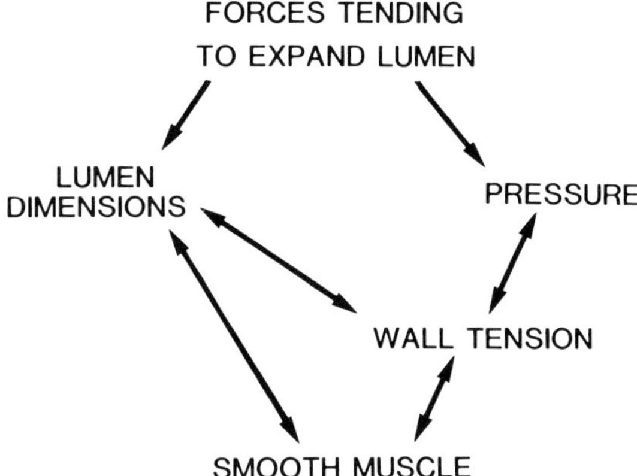


FIGURE 4. Interrelationships between the mechanical load and smooth muscle response.

examples is that factors determining wall thickness growth are clearly separable from those determining lumen expansion.

## B. Dilatation, Increased Wall Tension, and Thickness Growth

Of course, the expansion of lumen size need not be accompanied by true growth (defined as an increased tissue mass) at all. Relaxation and stretching of the muscle sheet will accommodate the expansion of lumen size, but without tissue growth, if the sheet at the same time becomes attenuated in thickness. Yet the muscle sheet, now expanded in area and diminished in thickness, may be required to sustain or develop the same luminal pressure as before. By the LaPlace relationship, wall tension will be increased at least in proportion to the increase in radius of the lumen. We recall that for a cylinder,

$$T = rP \tag{3}$$

where T is tension per unit length, r is radius, and P is transmural pressure, assuming a thin-walled organ for simplicity. This increase in total tension will be applied to a wall that is thinner, in inverse proportion to the area increase. Thus tensile stress in the wall, which is $S_r = T/h$, with h representing wall thickness (again assuming a thin walled organ), will be increased by a factor equal to the increase in radius raised to a power of 2 or more, depending upon the lumen shape. If, as already discussed, smooth muscle does respond to tensile stress by hypertrophy or hyperplasia, it will now be triggered to do so.

The sequence just postulated may remain valid, even if the increase in lumen size is generated by a initial relaxation of the smooth muscle. Relaxation, defined as a decrease in active tension development, actually leads to increased total wall tension by the Law of LaPlace, as long as the organ dilates and luminal pressure remains constant. The putative apparatus for sensing tension for smooth muscle growth regulation need not be the same as, or even operate in series with, the apparatus for generating active contraction. The series-parallel model for contractile-elastic behavior of smooth muscle (Figure 2) may help to clarify this point. Consider that the growth response of smooth muscle may depend upon the stretch of the parallel elastic component. Relaxation of the sliding filaments actually places this component under greater tension. Thus a rational sequence from smooth muscle relaxation to growth can be proposed.

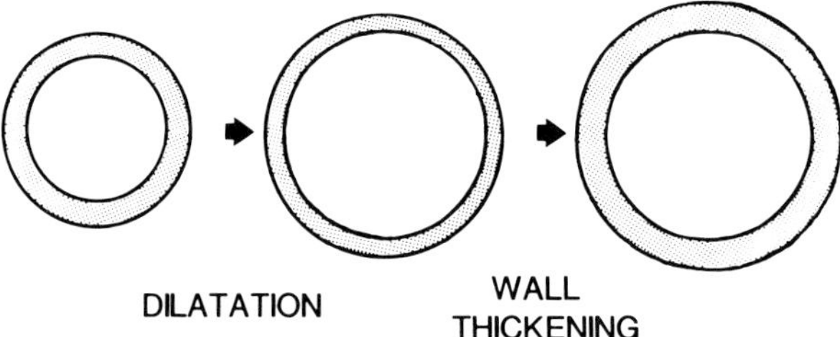



FIGURE 5. Hypothesized two step sequence in growth of smooth muscle with expanding lumen size. Dilatation of vessel lumen occurs first. Increased wall tension then leads to compensatory wall thickening and an increase in tissue mass. While represented in sequence for illustration, the two steps might occur virtually simultaneously.

There seem to be two alternative hypotheses for the way in which the growth of lumen size might be accomplished. The first is simply that cells may enlarge or may proliferate primarily in the directions appropriate to expand the area of the smooth muscle sheet. The second is that stretching of the muscle, due to lumen pressure, combined in some cases with relaxation of the muscle, occurs first (Figure 5). This not only causes thinning of the muscle sheet, but also leads to increased tension in some key cellular structure, triggering a secondary response of growth of wall thickness. The growth of wall thickness compensates for the original thinning and also for increased total wall tension incurred through the mechanism of the LaPlace relationship. In a given instance, growth of the smooth muscle sheet to accommodate lumen expansion may occur by either of these processes, or by a combination of them. Theoretically, the second alternative may be more attractive, in that growth as an increase in tissue mass is viewed as a unitary process responding to a single primary stimulus (wall tensile stress), and during this process tissue will be added in one dimension only, that of wall thickness. The first alternative of direct growth in directions tangential to the lumen surface is not as simple as it first appears, because the growth response of the myocyte to a luminal expansion stimulus would have to be quite differently oriented from the growth response to wall tension alone.

## C. Flow-Dependent Lumen Diameter Regulation in Arteries

The long-term regulation of lumen diameter of arteries is an interesting phenomenon that may serve as an example for lumen size regulation in several tubular organs that function mechanically more as conduits than reservoirs. Even in the artery the mechanism for regulating lumen size is not entirely clear, but much more is known about this phenomenon in the artery than in veins, ureters, gut, and various ducts.

If arterial diameter regulation is considered to operate as a homeostatic system, then one must expect the target of the system to relate to the obvious function of arteries — that is, the conduction of blood to distal vessels. Clearly blood flow is regulated by the microvasculature and not by the diameters of arteries. During development of the organism, every artery attains a diameter adequate to support flow to its field of distribution with minimal pressure drop. How is the balance between arterial diameter and tissue requirements achieved? By far the simplest explanation is that flow itself is the mediator, and that cells of the arterial wall are able to sense and respond to changes in flow rate during development. Thoma in 1893 observed in chick embryos that vessels with the fastest blood flow became the main arteries, while those with slower velocity became atrophic. Combining this with other observations, he proposed that the veloc-

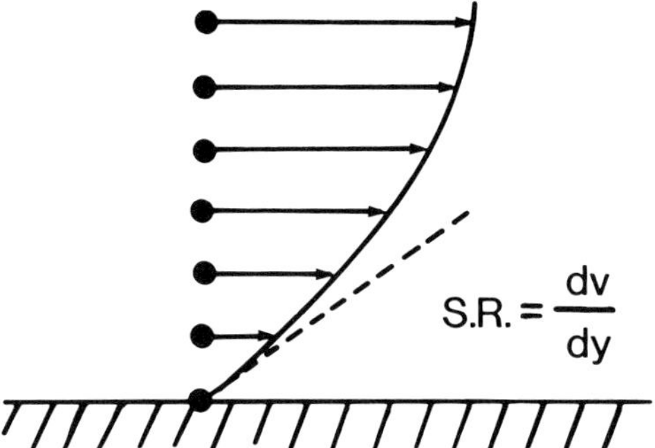



FIGURE 6.    Illustration of wall shear rate. While blood flow velocity at the luminal surface is zero, the first derivative of velocity with respect to position has a finite value, termed
the wall shear rate. It is represented in the figure as a tangent to the velocity profile. Shear
stress is simply equal to the shear rate multiplied by fluid viscosity.

ity of blood flow was the controlling factor in vascular diameter regulation.[87] In the adult organism, one may consider the case of a vascularized tumor, where an artery supplying, but not within, the tumor may expand several-fold or more in diameter. The phenomenon of diameter growth of the artery forming the proximal limb of an arteriovenous fistula is well known and has been attributed to flow.[88] Recently Kamiya and Togawa[89] studied the response of canine carotid arterial diameter to increased flow induced by a distal carotid-to-jugular anastomosis. Turbulence was documented by Doppler technique at the anastomosis, but it was felt to be absent at the site of interest proximal to the anastomosis. Wall shear rates (Figure 6) were calculated by the classical Poiseuille equation[7] and therefore represented rough estimates of shear, neglecting entry effects, pulsatility of flow, and pulsatility of the vessel wall. Nevertheless, at 6 to 8 months the luminal diameter had increased to the point that shear rate estimates had virtually normalized, as compared with the contralateral artery. There were three exceptions where diameter growth did not fully respond, i.e., dogs in which carotid flow increased more than 4-fold. Despite the exceptions, the results support the notion that flow may act directly on the luminal surface to influence arterial growth, since flow is perceived at the luminal surface as fluid shear. Wall shear rate, with pulsatility and other factors properly taken into account, may well be the homeostatic target for arterial lumen expansion. From Evans Blue staining in several chronic and acute dogs, Kamiya and Togawa felt that increased endothelial permeability might mediate the arterial diameter response. Permeability change, however, seems inadequate as a total explanation, since other conditions associated with permeability increase actually lead to decreased arterial diameter.[90]

Recently two groups reported preliminary results on arterial diameter change under conditions of reduced flow. Guyton and Hartley[91] placed silver clips, one flow-restricting and the other completely nonobstructive, on the carotid arteries of weanling rats. After 6 to 12 weeks, during which the rats grew 5 to 6 fold in body weight, measurements of arterial diameters and blood flow velocity (pulsed Doppler technique) were made. Flow-restricted arteries clearly attained smaller diameters than contralateral control arteries, and the degree of flow restriction, as indicated by flow pulsatility indices, correlated inversely with diameter increase. Langille and O'Donnell[92] used ligatures to reduce flow in the adult rabbit carotid by 65%, compared to the unrestricted control side. Internal diameter measurements were obtained from vascular casts. Within 24 hours a diameter decrease of 13% was seen, and by 1 month a diameter decrease of 65% and atrophy of the arterial media were found. These results show that adult arteries retain the ability to atrophy in response to decreased flow over a relatively short period of time. It appeared that the flow-related constrictive response could be prevented by removing endothelium from the rabbit carotid at the time of placing the ligature.

The anatomic position of endothelium at the arterial luminal surface makes it a prime candidate as a mediator of flow- or shear-dependent effects on the arterial wall. Endothelial cells are able to respond to shear stress by elongation in the direction of stress and narrowing in the perpendicular direction.[93,94] Whichever of the two alternative mechanisms for diameter growth is operative, as discussed earlier, the endothelium might mediate the response. The first mechanism would be a direct effect on smooth muscle growth, with the growth oriented to produce either elongation of myocytes or increased numbers in the circumferential direction. Endothelial cells produce both a polypeptide stimulator of smooth muscle cell growth[74] and a heparinlike inhibitor.[75] Perhaps of greater importance are the recent suggestions that endothelium can mediate a flow-dependent dilatation of arteries. Holtz, Giesler, and Bassenge[54] found that infusion of adenosine or dipyridamole was able to increase the diameter of dog coronary arteries, only when flow through the arteries increased as a result of distal arteriolar dilatation. When flow in the artery was held constant by the use of a pneumatic occluder, these agents had no effect on arterial diameter. Thus flow was implicated as the true stimulus for large vessel diameter change. Hull and colleagues[95] have reported in abstract that abruptly opening a femoral artery-to-jugular vein shunt in a dog, causing greatly increased flow, initially led to an arterial pressure and diameter decrease, followed over 20 sec by an arterial dilatation to a diameter greater than control, while pressure remained decreased. Denudation of the femoral artery endothelium prior to opening the shunt abolished the dilatory response.

The capacity of endothelium to produce a vasodilatory effect via the production of prostacyclin is now well known.[96] Weksler and co-workers[97] have noted that mechanical stimulation of endothelium, i.e., rocking or vibration, leads to brief bursts of prostacyclin ($PGI_2$) production, and on this basis, have speculated "that the flow of blood itself [could be] the natural stimulus for $PGI_2$ production in vivo." Recently Frangos and colleagues have found that increases in pulsatile shear can generate sustained increases in prostacyclin production in cultured endothelium.[98] Since prostacyclin is not an effective vasodilator for all arteries, another more general dilating factor, the endothelium-derived relaxing factor of Furchgott,[99] may be more promising as a potential mediator of flow-dependent arterial dilatation. Like prostacyclin, this appears to be an arachidonic acid metabolite, with a half-life recently reported as 6 sec. A novel mode of synthesis has been proposed, involving neither the cyclooxygenase nor the lipoxygenase pathway.[100] Because of the difficulty of working with this factor, there are as yet no reports of the effect of shear rate on its production.

To summarize, current knowledge about long-term arterial diameter regulation clearly shows a relationship to blood flow. A speculative hypothesis about diameter growth of arteries is as follows: increased wall shear rate of blood flow causes the endothelium to secrete a vasoactive substance that relaxes medial smooth muscle and dilates the vessel. Continuing chronic muscular relaxation may eventually lead to further dilatation as fibrous tissue proteins, collagen and elastin, undergo their normal turnover and are remolded according to the state of active contraction of smooth muscle cells. Arterial dilatation causes increased wall tension, and mass growth of tissue ensues.

The hypothesized sequence of dilatation, increased wall tension, and growth may have parallel manifestations in other smooth muscle-containing organs. The ureters are known to respond to increased flow with diameter and mass increase.[101] In man[102] and dog,[103] but not in the rat,[104] increased urine output from the kidney is associated with increases in urinary prostaglandin E2. This prostaglandin is known to relax some types of smooth muscle, but its effect on ureteral muscle is unclear.

# IV. MECHANICAL EFFECTS ON CELL AND TISSUE GROWTH IN VITRO

In vitro culture systems generally have the advantages of dealing with homogeneous cell populations, using defined media, and providing large numbers of replicate specimens for each experiment. Despite these advantages, the study of mechanical effects on cultured cells and tissues remains at a rudimentary level. Perhaps the major obstacle at the present time is that cells in culture cannot be organized sufficiently into a tissue that can either sustain or generate mechanical forces comparable to those found in vivo. Considerable ingenuity has often guided the application of mechanical measurements or stimuli to cultured cells or tissues, but the completed experiment usually can be interpreted in alternative ways. Nevertheless, some progress has been made. It will be helpful to review experiments performed both with smooth muscle and with a variety of other tissues and cells. This brief treatment will focus primarily on what might be termed the bioengineering aspects of the problem — that is, the manner in which mechanical stimulation can be applied to cells and the manner in which mechanical endpoints can be measured.

## A. Effects Observed with Cultured Smooth Muscle

Early work by Lelkes and Karmazsin[105] took advantage of the fact that the chick embryo heart can continue to beat while maintained in organ culture up to 4 weeks. Qualitative histologic comparisons were drawn between cultures in which the heart continued to beat and those in which heart motion ceased. An interesting observation, thus far unique to this type of culture system, was that elastin in the aortic root could be identified histologically only in the cultures with a beating heart.

Two systems for cyclical stretching of cultured smooth muscle cells have been studied. Leung, Mathews, and Glagov[50,106] used a substrate composed of elastin, prepared as the insoluble residue following hot alkali extraction of slices of bovine aortic media. Rabbit aortic smooth muscle cells, subcultured once, were plated at confluencey on the surface of the elastin membrane and cultured for 3 days under mechanically stationary conditions. Cells typically grew in a monolayer at the membrane surface and perhaps one additional layer in the elastin interstices just beneath the surface. A 10% stretch was applied at a rate of 52 c/min. Synthesis of DNA did not increase in stretched cells, as compared to merely agitated cells, but 2- to 4-fold increases in the rates of synthesis of collagen, hyaluronate, and chondroitin 6-sulfate were found. It was speculated that these results might help to explain the accumulation of collagen in the hypertensive arterial wall. In experiments performed by Buck,[107] subcultured rat aortic smooth muscle cells were plated at subconfluent density on a silicone rubber membrane. After allowing one day for attachment, the membrane was stretched by 20 to 35% at a rate of 4 c/min. Faster rates in this case caused cell detachment. Microscopic study of the cells showed orientation at right angles to the applied stretch, a result that agrees with the usual longitudinal orientation of intimal (but not medial) smooth muscle cells in vivo.

## B. Effects Observed with Nonmuscle Cells

The attempt to explain contact inhibition of cell growth as due to physical phenomena has led to some interesting and germane experimental results. The release of contact inhibition by scraping, or "wounding," a confluent cell monolayer to create a free edge causes, first, migration of cells from the wound edge into the free space and, second, DNA synthesis and mitosis in these and neighboring cells. Stoker[108] considered that a concentration gradient of some substance, taking the form of a diffusion boundary layer over a cell monolayer, might be disturbed by wounding and thereby lead to the migratory and proliferative events. The diffusion boundary layer could affect the availability of a substance critical for growth. As evidence to support this hypothesis, Stoker found that fluid flow or shaking, both maneuvers that can decrease the diffusion boundary layer, could cause a contact-inhibited monolayer of mouse 3T3 fibroblasts to renew proliferation and reach a higher cell density. Interestingly, shaking of cell cultures also caused increased cell locomotion within the confluent monolayer.[109] Although Stoker's postulate of a diffusion boundary layer is plausible in qualitative terms, the appropriate mathematical model must assume a rapid uptake (or secretion) by cells of a putative growth controlling substance. Without such rapid processing at the cell surface, diffusion would not be limiting for growth. Strong evidence against diffusion-limited control of cell growth has been provided by Whittenberger and Glaser,[110] who added Dextran, Ficoll, or methylcellulose to cell cultures to increase medium viscosity up to 25-fold. This maneuver decreases diffusion proportionately, but no change in the saturation density of 3T3 cells was found. An alternative and probably more accurate interpretation of Stoker's results has been suggested by Curtis and Seehar,[111] who pointed out that mechanical stress is induced in cell monolayers by shear forces during fluid flow or shaking. These investigators found that cyclic stretching of chick embryo fibroblasts grown on a fabric mesh increased the numbers of mitoses and thymidine-labeled cells. It should be recalled that Leung and co-workers[50] found no increase in DNA uptake in rabbit aortic smooth muscle cells grown on a cyclically stretched matrix of bovine elastin. Nevertheless, the notion of a mechanical effect on cell growth remains appealing and can be supported by several other general observations. The observations that cell locomotion precedes cell division in a wounding experiment and that shaking also induces cell locomotion could point toward mechanical stress associated with cell motility, as a precursor to mitosis. As cultures approach confluency, locomotion and cell division slow down in concert. Cytochalasin B treatment of cultured endothelial cells sufficient to inhibit migration across a wound edge will also inhibit cell proliferation.[112] However, a contrary finding has been reported with 3T3 cells.[109]

Adhesivity and flattening of cells on standard tissue culture surfaces appear necessary for optimal growth of most nontransformed cell types. By interposing a neutral hydrophilic film of varying thickness (a polymer of 2-hydroxyethyl methacrylate) between the standard polystyrene surface and seeded cells, Folkman and Moscona[113] were able to vary the degree of cell flattening. Cells which remained round on this surface grew poorly, while those which spread and flattened grew at a much faster rate. The results were corroborated by experiments with bacteriologic plastic, to which animal cells are poorly adherent, and by observations on the flattening of cells at a wound edge. Cell shape was proposed as a critical factor associated with proliferation, with the flattened shape associated with the highest degree of proliferation. Work done in other laboratories appears consistent with this proposal.[114-116] However, the designation of cell shape as the critical parameter really begs a question. While cell shape is easy for the investigator to discern, how does a cell know what shape it has? Folkman and Moscona comment that "normal cell growth in vivo may possibly be regulated by a complex balance of *adhesive forces* between cells and between cell and substratum . . . that combine to modulate cell shape, and permit or prevent DNA synthesis" (emphasis added). The adhesive forces postulated here could certainly be related to tensile stress at the cellular level.

The issues of force development, spreading, and other morphogenetic processes at the cellular level are complex, and the reader is referred to a worthwhile review by Vasiliev and Gelfand.[117] Harris[118] made observations on tension development in nonmuscle cells by culturing them on plasma clots and silicone fluids, which serve as mobile substrates. Plasma clots exhibit elastic behavior and are subject to immediate effects on cell contraction or relaxation. The ability of different cell types to spread and flatten on silicone substrates of differing viscosity is taken to indicate degrees of cellular tensile stress during the spreading process. James and Taylor[119] took advantage of the outgrowth of fibroblastic cells from fragments of chick embryo frontal bone to measure tension in a cell monolayer. When placed in proximity to each other, the bone fragments became connected by a sheet of cells and were gradually drawn together. The force necessary to oppose this movement was measured by a glass fiber microbalance. The tensile stress developed by the sheet of cells amounted to $3.4 \times 10^4$ dynes/cm², a value remarkably similar to the tensile stress developed in a contracting skin wound.

## C. Effects Observed with Striated Muscle

Hypertrophy of striated muscle in response to increased work load has been studied extensively. Worthwhile reviews[120,121] and recent publications[122] can be referred to. The discussion here will be limited to in vitro methodology and concepts applicable to the study of smooth muscle. Increased uptake of certain amino acids seems to be an important early event in skeletal muscle hypertrophy.[120] This process is readily studied isotopically, using the nonmetabolized amino acid analog, alphaaminoisobutyric acid (AIB). Excised rat diaphragm can show increased AIB uptake within minutes after the onset of rapid electrical stimulation. The effect does not require protein or RNA synthesis. Interestingly, hemidiaphragms that were fastened rigidly and contracted isometrically consistently concentrated AIB more rapidly than hemidiaphragms that were allowed to shorten against no load. Decreased rates of proteolysis in muscle can be demonstrated in excised skeletal muscle after as little as an hour of electrical stimulation. This phenomenon can also be induced by simple passive stretch of the muscle to 10 to 20% beyond the resting length. In some, but not all studies on excised muscle in vitro, it has been possible to show increased incorporation of labeled amino acids into muscle protein.

Vandenburgh and Kaufman[12] have described an in vitro model of stretch-induced hypertrophy of embryonic chick skeletal myotubes. Myoblasts are plated on the surface of a silicone membrane mounted in a simple stretching frame. After 72 hr of culture, when the myoblasts have fused to form contractile myotubes, the diameter of the silicone membrane is enlarged by 10%. The stretching of the cells is not cyclic, but represents a step change in strain. A number of changes consistent with muscle hypertrophy were found in stretched cultures, including increases demonstrated for AIB uptake, incorporation of amino acids into total and myofibrillar protein, and accumulation of total and myofibrillar protein.[12] By measuring ouabain-sensitive uptake of rubidium-86, activation of membrane Na pumps ($Na^+K^+ATPase$) by stretch was indicated, and inhibition of Na pump activity blocked the early increase in AIB uptake.[123,124] Further work demonstrated that the early biochemical events in the response to stretch can occur in the absence of serum, but that subsequent stretch-induced growth requires serum factors.[125] Using a different, complementary approach on cultured chick myotubes, Brevet and co-workers[126] studied the effect of electrical stimulation on contractile protein synthesis. Stimulation sufficient to produce contraction of the myotubes enhanced the synthesis and accumulation of total protein, but not DNA. Contractile proteins, extractable with pyrophosphate, and myosin heavy chain, quantified on SDS-polyacrylamide gels, were selectively increased in comparison to total protein accumulation.

The role of stretch vs. tension in causing hypertrophy of skeletal muscle has been commented upon by Goldberg and colleagues,[120] who favored the role of tension, either passive or active, as the primary determinant of long-term physiologic muscle hypertrophy. Referring to our previous diagram (Figure 2), we might hypothesize that growth is coupled to a series elastic element in the skeletal muscle cell. Goldberg et al.[120] noted the greater proclivity for hypertrophy due to isometric exercise (emphasizing cyclic tension) as opposed to endurance exercise (emphasizing cyclic stretch). Furthermore, it was noted that stretch in a muscle is limited anatomically by its bony origin and insertion.

## V. SUMMARY

Smooth muscle growth appears to respond physiologically to mechanical factors, the most important of which are tension and stretch. The roles of steady vs. cyclic mechanical loads, and active vs. passive tension in initiating the growth response are largely undetermined at present and may be different for various types of smooth muscle. In interpreting in vivo observations and experiments, one must distinguish between factors that increase thickness of the smooth muscle sheet and factors that cause lumen expansion. Smooth muscle growth in the latter case may not be stimulated directly, but instead may respond to the increased wall tension that regularly accompanies organ dilatation. In conduit organs such as blood vessels and ureters, special mechanisms lead to lumen dilatation in response to increased intraluminal flow, but these mechanisms are largely undetermined at present.

One can hope that the more widespread application of in vitro culture techniques, using apparatus able to deliver precise mechanical stimuli to cells, will increase knowledge of mechanical influences on growth. Careful selection of the type of stimulus and the indicator of a hypertrophic or hyperplastic response is necessary, attempting to mimic situations well characterized in vivo. Such studies with cultured cells face considerable obstacles with regard to (1) anchorage of cells and simultaneous measurement of forces on the cells, (2) the inexpensive provision of numerous experimental replicates, and (3) the organization of cells into tissues for coordinated tension development and growth. Increasing knowledge of the general mechanisms of cell growth, particularly the earliest biochemical manifestations, will aid the task.

## REFERENCES


1. Berry, C. L., Looker, T., and Germain, J., The growth and development of the rat aorta. I. Morphological aspects, *J. Anat.,* 113, 1, 1972.
2. Owens, G. K., Rabinovitch, P. S., and Schwartz, S. M., Smooth muscle hypertrophy versus hyperplasia in hypertension, *Proc. Nat. Acad. Sci. U.S.A.,* 78, 7759, 1981.
3. Laudau, L. D. and Lifshitz, E. M., *Theory of Elasticity,* Pergamon Press, Oxford, 1970.
4. Saada, A. S., *Elasticity Theory and Applications,* Pergamon Press, Elmsford, N.Y., 1974.
5. Frederickson, A. G., *Principles and Applications of Rheology,* Prentice-Hall, Englewood Cliffs, N.J., 1964.
6. Dobrin, P. B., Mechanical properties of arteries, *Physiol. Rev.,* 58, 397, 1978.
7. Milnor, W. R., *Hemodynamics,* Williams & Wilkins, Baltimore, 1982.
8. Robertson, J. H., The influence of mechanical factors on the structure of the peripheral arteries and the localization of atherosclerosis, *J. Clin. Pathol.,* 13, 199, 1960.

9. Carew, T. E., Vaishnev, R. N., and Patel, D. J., Compressibility of the arterial wall, *Circ. Res.*, 23, 61, 1968.
10. Ekstrom, J. and Uvelius, B., Length-tension relations of smooth muscle from normal and denervated rat urinary bladders, *Acta Physiol. Scand.*, 112, 443, 1981.
11. Hume, W. R. and Bevan, J. A., Amino acid uptake in rabbit blood vessels two weeks after induction of hypertension by coarctation of the abdominal aorta, *Cardiovasc. Res.*, 12, 106, 1978.
12. Vandenburgh, H. H. and Kaufman, S., In vitro model for stretch-induced hypertrophy of skeletal muscle, *Science*, 203, 265, 1979.
13. Ruegg, J. C., Smooth muscle tone, *Physiol. Rev.*, 51, 201, 1971.
14. Gabella, G. and Yamey, A., Synthesis of collagen by smooth muscle in the hypertrophic intestine, *Q. J. Exp. Physiol.*, 62, 257, 1977.
15. Clark, J. M. and Glagov, S., Structural integration of the arterial wall. I. Relationships and attachments of medial smooth muscle cells in normal distended and hyperdistended aortas, *Lab Invest.*, 40, 587, 1979.
16. Azuma, T. and Oka, S., Mechanical equilibrium of blood vessel walls, *Am. J. Physiol.*, 221, 1310, 1971.
17. Gow, B. S., Circulatory correlates: vascular impedance, resistance, and capacity, in *Handbook of Physiology, Section 2*, Vol. 2, Bohr, D. F., Somlyo, A. P., and Sparks, H. V., Jr., Eds., American Physiological Society, Bethesda, Md., 1980, 353.
18. Wolinsky, H. and Glagov, S., A lamellar unit of aortic medial structure and function in mammals, *Circ. Res.*, 20, 99, 1967.
19. Wolinsky, H. and Glagov, S., Structural basis for the static mechanical properties of the aortic media, *Circ. Res.*, 14, 400, 1964.
20. Gabella, G., The structural apparatus for transmission of force in smooth muscles, *Physiol. Rev.*, 64, 455, 1984.
21. Reynolds, S. R. M., *Physiology of the Uterus*, 2nd ed., reprinted, Hafner, New York, 1965, 183.
22. De Mattos, C. E. R., Kempson, R. L., Erdos, T., and Csapo, A., Stretch-induced myometrial hypertrophy, *Fertil. Steril.*, 18, 545, 1967.
23. Csapo, A., Erdos, T., De Mattos, C. R., Gramss, E., and Moscowitz, C., Stretch-induced uterine growth, protein synthesis and function, *Nature*, 207, 1378, 1965.
24. Csapo, A., and Sauvage, J., The evolution of uterine activity during human pregnancy, *Acta Obstet. Gynecol. Scand.*, 47, 181, 1968.
25. Anderson, A. B., Turnball, A. C., and Murray, A. M., The relationship between amniotic fluid pressure and uterine wall tension in pregnancy, *Am. J. Obstet. Gynecol.*, 97, 992, 1967.
26. Finn, C. A. and Porter, D. G., *Reproductive Biology Handbooks*, Vol. 1, *The Uterus*, Publishing Sciences Group, Acton, Mass., 1975, 133.
27. Michael, C. A. and Schofield, B. M., The influence of the ovarian hormones on the actomyosin content and the development of tension in uterine muscle, *J. Endocrinol.*, 44, 501, 1969.
28. Harkness, M. L. R. and Harkness, R. D., The collagen content of the reproductive tract of the rat during pregnancy and lactation, *J. Physiol.*, 123, 492, 1954.
29. Csapo, A. I. and Takeda, H., Effect of progesterone on the electric activity and intrauterine pressure of pregnant and parturient rabbits, *Am. J. Obstet. Gynecol.*, 91, 221, 1965.
30. Carsten, M. E., Regulation of myometrial composition, growth, and activity, in *Biology of Gestation*, Vol. 1, Assali, N. S., Ed., Academic Press, New York, 1968, 356.
31. Martin, L., Finn, C. A., and Trinder, G., Hypertrophy and hyperplasia in the mouse uterus after oestrogen treatment: an autoradiographic study, *J. Endocrinol.*, 56, 133, 1973.
32. Cullen, B. M. and Harkness, R. D., Collagen formation and changes in cell population in the rat's uterus after distension with wax, *Q. J. Exp. Physiol.*, 53, 33, 1968.
33. Bakuntz, S. A. and Vantsian, V. T., The electrophysiologic characteristics of the smooth muscles of the ureter, in *Physiology of Smooth Muscle*, Bulbring, E. and Shuba, M. F., Eds., Raven Press, New York, 1976, 121.
34. Rose, J. G., Gillenwater, J. Y., Attinger, F., Sim, P., and Wyker, A. T., Ureteral wall tension: a new means of evaluating ureteral function, *Invest. Urol.*, 10, 480, 1973.
35. Rose, J. G. and Gillenwater, J. Y., Pathophysiology of ureteral obstruction, *Am. J. Physiol.*, 225, 830, 1973.
36. Gee, W. F. and Kiviat, M. D., Ureteral response to partial obstruction: smooth muscle hyperplasia and connective tissue proliferation, *Invest. Urol.*, 12, 309, 1975.
37. Crooks, K. K., Daly, J. J., and Prout, G. R., Jr., Experimental ureteral lengthening in dogs, *Invest. Urol.*, 18, 66, 1980.
38. Gosling, J., The structure of the bladder and urethra in relation to function, *Urol. Clin. N. Am.*, 6, 31, 1979.

39. Smith, D. R., *General Urology,* Lange, Los Altos, Calif., 1972, 114.
40. Mattiasson, A. and Uvelius, B., Changes in contractile properties in hypertrophic rat urinary bladder, *J. Urol.,* 128, 1340, 1982.
41. Peterson, C. M., Goss, R. J., and Atryzek, V., Hypertrophy of the rat urinary bladder following reduction of its functional volume, *J. Exp. Zool.,* 187, 121, 1974.
42. Goss, R. J., Liang, M. D., Weisholtz, S. J., and Peltzer, T. J., The physiological basis of urinary bladder hypertrophy, *Proc. Soc. Exp. Biol. Med.,* 142, 1332, 1973.
43. Langworthy, O. R. and Kolb, L. C., Histological changes in the vesical muscle following injury of the peripheral innervation, *Anat. Rec.,* 71, 249, 1938.
44. Jacobsen, C. E., Jr., Neurogenic vesical dysfunction: an experimental study, *J. Urol.,* 53, 670, 1945.
45. Carpenter, F. G., Histological changes in parasympathetically denervated feline bladder, *Am. J. Physiol.,* 166, 692, 1951.
46. Fry, D. L., Certain hemorheologic considerations concerning the blood vascular interface with particular reference to coronary artery disease, *Circulation,* Suppl. 4, 38, 1969.
47. Meairs, S., Weihe, E., Mittmann, U., Vetter, H., Kohler, U., and Forssmann, W. G., Morphologic investigation of coronary arteries subjected to hypertension by experimental supravalvular aortic stenosis in dogs, *Lab. Invest.,* 50, 469, 1984.
48. Ross, R., Atherosclerosis: a problem of the biology of arterial wall cells and their interaction with blood components, *Arteriosclerosis,* 1, 293, 1981.
49. Schwartz, S. M. and Ross, R., Cellular proliferation in atheroslerosis and hypertension, *Prog. Cardiovasc. Dis.,* 26, 355, 1984.
50. Leung, D. Y. M., Glagov, S., and Mathews, M. B., Cyclic stretching stimulates synthesis of matrix components by arterial smooth muscle cells in vitro, *Science,* 191, 475, 1976.
51. Bevan, A. T., Honour, A. J., and Stott, F. H., Direct arterial pressure recording in unrestricted man, *Clin. Sci.,* 36, 329, 1969.
52. Roach, M. R. and Burton, A. C., The reason for the distensibility curves of arteries, *Can. J. Biochem. Physiol.,* 35, 681, 1957.
53. Vatner, S. F., Pagani, M., Manders, W. T., and Pasipoularides, A. D., Alpha adrenergic vasoconstriction and nitroglycerin vasodilation of large coronary arteries in the conscious dog, *J. Clin. Invest.,* 65, 5, 1980.
54. Holtz, J., Giesler, M., and Bassenge, E., Two dilatory mechanisms of anti-anginal drugs on epicardial coronary arteries in vivo: indirect, flow-dependent, endothelium-mediated dilation and direct smooth muscle relaxation, *Z. Kardiologie,* 72(Suppl. 3), 98, 1983.
55. Leung, D. Y. M., Glagov, S., and Mathews, M. B., Elastin and collagen accumulation in rabbit ascending aorta and pulmonary trunk during postnatal growth, *Circ. Res.,* 41, 316, 1977.
56. Geer, J. C. and Haust, M. D., *Smooth Muscle Cells in Atherosclerosis,* Vol. 2, Monographs on Atherosclerosis, S. Karger, Basel, 1972, 39.
57. Movat, H. Z., More, R. H., and Haust, M. D., The diffuse intimal thickening of the human aorta with aging, *Am. J. Pathol.,* 34, 1023, 1958.
58. Wolinsky, H. and Glagov, S., Comparison of abdominal and thoracic aortic medial structure in mammals: deviation of man from the usual pattern, *Circ. Res.,* 25, 677, 1969.
59. Doyle, J. M. and Dobrin, P. B., Stress gradients in the walls of large arteries, *J. Biomech.,* 16, 631, 1973.
60. Dock, W., The predilection of atherosclerosis for the coronary arteries, *JAMA,* 131, 875, 1946.
61. Nerem, R. M., and Cornhill, J. F., The role of fluid mechanics in atherogenesis, *J. Biomech. Eng.,* 102, 181, 1980.
62. Friedman, M. H., Hutchins, G. M., Bargeron, C. B., Deters, C. B., and Mark, F. F., Correlation of human arterial morphology with hemodynamic measurements in arterial casts, *J. Biomech. Eng.,* 103, 204, 1981.
63. Fox, B. and Seed, W. A., Location of early atheroma in the human coronary arteries, *J. Biomech. Eng.,* 103, 208, 1981.
64. Barrett, A. M., Arterial measurements in the interpretation of cardiomegaly at necropsy: cardiac hypertrophy and myocardial infarction, *J. Pathol. Bacteriol.,* 86, 9, 1963.
65. Wolinsky, H., Effects of hypertension and its reversal on the thoracic aorta of male and female rats: morphological and chemical studies, *Circ. Res.,* 28, 622, 1971.
66. Wiener, J., Loud, A. V., Giacomelli, F., and Anversa, P., Morphometric analysis of hypertension-induced hypertrophy of rat thoracic aorta, *Am. J. Pathol.,* 88, 619, 1977.
67. Seidel, C. L., Aortic actomyosin content of maturing normal and spontaneously hypertensive rats, *Am. J. Physiol.,* 237, H34, 1979.
68. Seidel, C. L., Allen, J. C., Bowers, R. L., Mechanical and biochemical alterations of aorta induced by hydralazine hypotension, *J. Pharmacol. Exp. Ther.,* 213, 514, 1980.

69. Olivetti, G., Anversa, P., Melissari, M., and Loud, A. V., Morphometry of medial hypertrophy in the rat thoracic aorta, *Lab. Invest.*, 42, 559, 1980.
70. Wilens, S. L., The nature of diffuse intimal thickening of arteries, *Am. J. Pathol.*, 27, 825, 1951.
71. Still, W. J. S., The early effect of hypertension on the aortic intima of the rat, *Am. J. Pathol.*, 51, 721, 1967.
72. Leibovich, S. J. and Ross, R., A macrophage-dependent factor that stimulates the proliferation of fibroblasts in vitro, *Am. J. Pathol.*, 84, 501, 1976.
73. Gabbiani, G., Elemer, G., Guelpa, C. H., Valloton, M. B., Baddonel, M. C., and Huttner, I., Morphologic and functional changes of the aortic intima during experimental hypertension, *Am. J. Pathol.*, 96, 399, 1979.
74. Gajdusek, C., DiCorleto, P., Ross, R., and Schwartz, S. M., An endothelial cell-derived growth factor, *J. Cell Biol.*, 85, 467, 1980.
75. Castellot, J. J., Jr., Addonizio, M. L., Rosenberg, R., Karnovsky, M. J., Cultured endothelial cells produce a heparinlike inhibitor of smooth muscle cell growth, *J. Cell Biol.*, 90, 372, 1981.
76. Folkow, B., Grimby, G., and Thulesius, O., Adaptive structural changes of the vascular walls in hypertension and their relation to the control of the peripheral resistance, *Acta Physiol. Scand.*, 44, 255, 1958.
77. Folkow, B., Physiological aspects of primary hypertension, *Physiol. Rev.*, 62, 377, 1982.
78. Mulvany, M. J., Hansen, P. K., and Aalkjaer, C., Direct evidence that the greater contractility of resistance vessels in spontaneously hypertensive rats is associated with a narrowed lumen, a thickened media, and an increased number of smooth muscle cell layers, *Circ. Res.*, 43, 854, 1978.
79. Short, D., Morphology of the intestinal arterioles in chronic human hypertension, *Brit. Heart J.*, 28, 184, 1966.
80. Uvelius, B., Anders, A., and Johansson, B., Structural and mechanical alterations in hypertrophic venous smooth muscle, *Acta Physiol. Scand.*, 112, 463, 1981.
81. Brody, W. R., Kosek, J. C., and Angell, W. W., Changes in vein grafts following aorto-coronary bypass induced by pressure and ischemia, *J. Thoracic Cardiovasc. Surg.*, 64, 847, 1972.
82. Seidel, C. L., Lewis, R. M., Bowers, R., Bukoski, R. D., Kim, H. S., Allen, J. C., and Hartley, C., Adaptation of canine saphenous veins to grafting, *Circ. Res.*, 55, 102, 1984.
83. Goss, R. J., *The Physiology of Growth,* Academic Press, New York, 1978, chap. 15.
84. Gabella, G., Hypertrophic smooth muscle. I. Size and shape of cells, occurrence of mitoses, *Cell Tissue Res.*, 201, 63, 1979.
85. Gabella, G., Hypertrophic smooth muscle. IV. Myofilaments, intermediate filaments and some mechanical properties, *Cell Tissue Res.*, 201, 277, 1979.
86. Ellenberg, M., Diabetic neuropathy: clinical aspects, *Metabolism,* 25, 1627, 1976.
87. Thoma, R., *Untersuchungen uber die Histogenese and Histomechanik des Gefassystems,* Enke, Stuttgart, 1893.
88. Holman, E., Problems in the dynamics of blood flow. I. Conditions controlling collateral circulation in the presence of an arteriovenous fistula, following the ligation of an artery, *Surgery,* 26, 889, 1949.
89. Kamiya, A. and Togawa, T., Adaptive regulation of wall shear stress to flow change in the canine carotid artery, *Am. J. Physiol.*, 239, H14, 1980.
90. Clowes, A. W., Reidy, M. A., and Clowes, M. M., Mechanisms of stenosis after arterial injury, *Lab. Invest.*, 49, 208, 1983.
91. Guyton, J. R. and Hartley, C. J., Effects of flow restriction and reduced velocity pulsation on circumferential arterial growth, *Fed. Proc.*, 43, 973, 1984.
92. Langille, B. L. and O'Donnell, F., Role of endothelium in flow-induced changes in arterial diameter, *Fed. Proc.*, 43, 973, 1984.
93. Dewey, C. F., Jr., Bussolari, S. R., Gimbrone, M. A., Jr., and Davies, P. F., The dynamic response of vascular endothelial cells to fluid shear stress, *J. Biomech. Eng.*, 103, 177, 1981.
94. Ives, C. L., Eskin, S. G., McIntire, L. V., and DeBakey, M. E., The importance of cell origin and substrate in the kinetics of endothelial cell alignment in response to steady flow, *Trans. Am. Soc. Artif. Int. Organs,* 29, 269, 1983.
95. Hull, S. S., Romig, G. D., and Sparks, W. V., Flow-induced vasodilation of large arteries is dependent on endothelium, *Fed. Proc.*, 43, 900, 1984.
96. Moncada, S., Prostacyclin and arterial wall biology, *Arteriosclerosis,* 2, 193, 1982.
97. Weksler, B. B., Eldor, A., Falcone, D., Levin, R. I., Jaffe, E. A., and Minick, C. R., Prostaglandins and vascular endothelium, in *Cardiovascular Pharmacology of the Prostaglandins,* Herman, A. G., Vanhoutte, P. M., Denolin, H., and Goossens, A., Eds., Raven Press, New York, 1982, 137.
98. Frangos, J. A., Ives, C. L., Eskin, S. G., and McIntyre, L. V., Effect of pulsatile flow of $PGI_2$ production by endothelial cells, *Circulation,* 68, III-73, 1983.
99. Furchgott, R. F., Role of endothelium in responses of vascular smooth muscle, *Circ. Res.*, 53, 557, 1983.

100. Griffith, T. M., Edwards, D. H., Lewis, M. J., Newby, A. C., and Henderson, A. H., The nature of endothelium-derived vascular relaxant factor, *Nature,* 308, 645, 1984.
101. Cass, A. S., Dewey, A. W., Jr., and Hinman, F., Jr., Disuse and increased function of the dog ureter, *Invest. Urol.,* 6, 393, 1969.
102. Walker, R. M., Brown, R. S., and Stoff, J. S., Role of renal prostaglandins during antidiuresis and water diuresis in man, *Kidney Int.,* 21, 365, 1981.
103. Wright, L. F., Rosenblatt, S. G., and Lifschitz, M. D., High urine flow rate increases prostaglandin E excretion in the conscious dog, *Prostaglandins,* 22, 21, 1981.
104. Fejes-Toth, G., Naray-Fejes-Toth, A., Ritger, B., and Frolich, J. C., Urinary prostaglandin and kallikrein are not flow dependent in the rat, *Prostaglandins,* 25, 99, 1983.
105. Lelkes, Gy. and Karmazsin, L., Development of elastic elements in tissue cultures, *Acta Morphol. Acad. Sci. Hung.,* 5, 149, 1955.
106. Leung, D. Y. M., Glagov, S., and Mathews, M. B., A new in vitro system for studying cell response to mechanical stimulation: different effects of cyclic stretching and agitation on smooth muscle cell biosynthesis, *Exp. Cell Res.,* 109, 285, 1977.
107. Buck, R. C., Behavior of vascular smooth muscle cells during repeated stretching of the substratum in vitro, *Atherosclerosis,* 46, 217, 1983.
108. Stoker, M. G. P., Role of diffusion boundary layer in contact inhibition of growth, *Nature,* 246, 200, 1973.
109. Stoker, M. and Piggott, D., Shaking 3T3 cells: further studies on diffusion boundary effects, *Cell,* 3, 207, 1974.
110. Whittenberger, B. and Glaser, L., Cell saturation density is not determined by a diffusion-limited process, *Nature,* 272, 821, 1978.
111. Curtis, A. S. G. and Seehar, G. M., The control of cell division by tension or diffusion, *Nature,* 274, 52, 1978.
112. Selden, S. C., III. and Schwartz, S. M., Cytochalasin B inhibition of endothelial proliferation at wound edges in vitro, *J. Cell Biol.,* 81, 348, 1979.
113. Folkman, J., and Moscona, A., Role of cell shape in growth control, *Nature,* 273, 345, 1978.
114. Wittelberger, S. C., Kleene, K., and Penman, S., Progressive loss of shape-responsive metabolic controls in cells with increasingly transformed phenotype, *Cell,* 24, 859, 1981.
115. Ben-Ze'ev, A., Farmer, S. R., and Penman, S., Protein synthesis requires cell-surface contact while nuclear events respond to cell shape in anchorage-dependent fibroblasts, *Cell,* 21, 365, 1980.
116. McKeehan, W. L. and Ham, R. G., Stimulation of clonal growth of normal fibroblasts with substrata coated with basic polymers, *J. Cell Biol.,* 71, 727, 1976.
117. Vasiliev, J. M. and Gelfand, I. M., Mechanisms of morphogenesis in cell cultures, *Int. Rev. Cytol.,* 50, 159, 1977.
118. Harris, A. K., Cell surface movements related to cell locomotion, in *Locomotion of Tissue Cells, Ciba Foundation Symp. 14,* Elsevier, Amsterdam, 1973, 3.
119. James, D. W. and Taylor, J. F., The stress developed by sheets of chick fibroblasts in vitro, *Exp. Cell. Res.,* 54, 107, 1969.
120. Goldberg, A. L., Etlinger, J. D., Goldspink, D. F., and Jablecki, C., Mechanism of work-induced hypertrophy of skeletal muscle, *Med. Sci. Sports,* 7, 248, 1975.
121. Zak, R., Contractile function as a determinant of muscle growth, in *Cell and Muscle Motility,* Dowben, R. M. and Shay, J. W., Eds., Plenum Press, New York, 1981, 1.
122. Ashmore, C. R. and Summers, P. J., Stretch-induced growth in chicken wing muscles: myofibrillar proliferation, *Am. J. Physiol.,* 241, C93, 1980.
123. Vandenburgh, H. H. and Kaufman, S., Stretch-induced growth of skeletal myotubes correlates with activation of the sodium pump, *J. Cell. Physiol.,* 109, 205, 1981.
124. Vandenburgh, H. H. and Kaufman, S., Coupling of voltage-sensitive sodium channel activity to stretch-induced amino acid transport in skeletal muscle in vitro, *J. Biol. Chem.,* 257, 13448, 1982.
125. Vandenburgh, H. H., Cell shape and growth regulation in skeletal muscle: exogenous versus endogenous factors, *J. Cell. Physiol.,* 116, 363, 1983.
126. Brevet, A., Pinto, E., Peacock, J., and Stockdale, F. E., Myosin synthesis increased by electrical stimulation of skeletal muscle cell cultures, *Science,* 193, 1152, 1976.

Chapter 8

# CHEMICAL STIMULI OF THE HYPERTROPHIC RESPONSE IN SMOOTH MUSCLE

Julie H. Campbell and Gordon R. Campbell

## TABLE OF CONTENTS

# I. MODULATION OF SMOOTH MUSCLE PHENOTYPE

Various cell types can alter their character when the environment changes. These alterations in character are called modulations of the differentiated state and involve reversible interconversions between phenotypes. Modulations in cell phenotype may occur as a result of cell interactions, alterations of extracellular matrix, or in response to other signals such as hormones.

Smooth muscle cells are capable of expressing a range of phenotypes. At one end of the spectrum of phenotypes is the smooth muscle cell whose function is almost exclusively that of contraction ("contractile state"). This function is reflected structurally with 80 to 90% of the cell volume occupied with contractile apparatus.[1] Organelles such as rough endoplasmic reticulum (ER), Golgi, and free ribosomes are few in number and located in the perinuclear region. Cells of the taenia coli, vas deferens (Figure 1), circular muscle of the ileum, and smaller muscular arteries (Figures 2, 3) express this phenotype. These muscles contain relatively little connective tissue (e.g., volume of extracellular space is about 30% in taenia coli, and 10 to 15% in the circular muscle of the ileum), and in visceral muscle tissues much of this is synthesized by constituent fibroblasts. The smooth muscle cells therefore have minimal synthetic function. At the opposite end of the spectrum of phenotypic expression is the muscle cell whose function is almost exclusively that of synthesis ("synthetic state"). In line with other cell types actively engaged in production of extracellular matrix (such as fibroblasts), the cytoplasm contains few filament bundles but large amounts of rough ER, Golgi, and free ribosomes. Cells in the synthetic state are seen in development and repair[2] (Figure 4).

It must be emphasized that the contractile and synthetic states as described above represent the extreme ends of a continuous spectrum of phenotypes with most smooth muscle expressing intermediate morphology (Figure 5). For instance, the volume of the connective tissue containing extracellular space in large mammalian elastic arteries is 50 to 60% and smooth muscle is the only cell type present.[3-5] This cell must therefore be not only responsible for contraction/relaxation but the normal turnover of extracellular matrix components. Again this is reflected structurally with 60 to 70% of the cell volume occupied with myofilaments in the adult animal[6,7] (Figure 5).

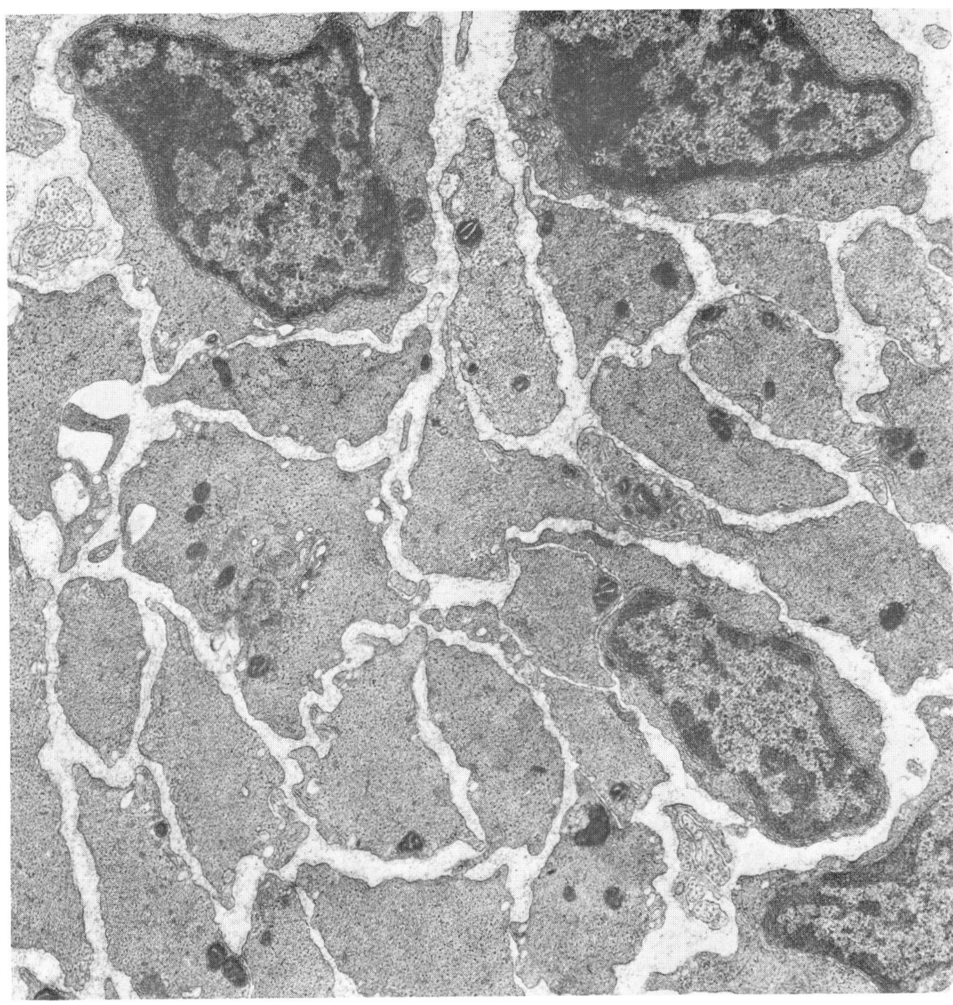

FIGURE 1.   Transverse section of smooth muscle cells from newborn guinea pig vas deferens. Most of the cell cytoplasm is filled with myofilaments, and the extracellular space contains relatively little connective tissue. (× 12,500).

## II. SITUATIONS IN WHICH MODULATION OF PHENOTYPE IS OBSERVED IN VIVO

### A. Development
#### 1. Visceral Smooth Muscle

Initially in the development of most smooth muscle organs, undifferentiated mesenchymal cells surround the newly formed epithelium. These soon develop a typical fibroblast appearance with their cytoplasm containing a large number of organelles, particularly free ribosomes, rough ER, mitochondria, and Golgi.[8-12] With increasing age there is a progressive reduction in the nucleocytoplasmic ratio and the appearance of bundles of thin filaments with associated dark bodies. As time progresses, filament bundles increase in size, thick filaments appear, and there is a concomitant decrease in the number of synthetic organelles. Other characteristic features of smooth muscle such as dense areas along the cell membrane, plasmalemmal vesicles, and a basal lamina appear at about the same time or a little later than thick myofilaments.[10,13]

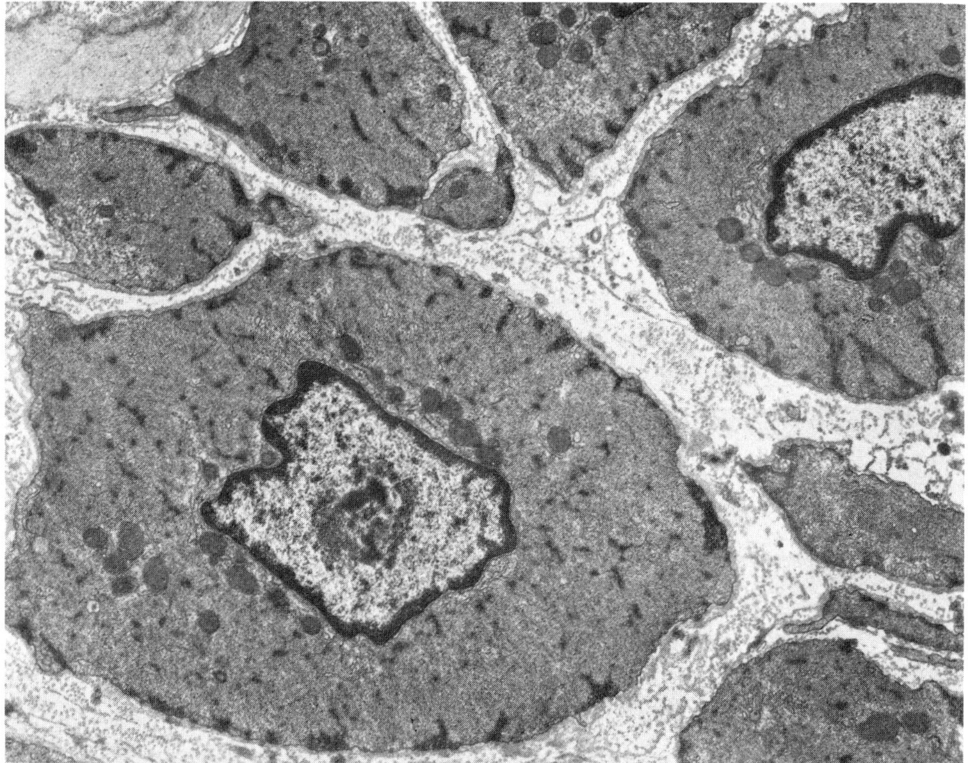

FIGURE 2.    Transverse section through smooth muscle cells of the caudal artery of a 14 week rat, demonstrating organelles in the perinuclear region of a vascular muscle cell whose function is primarily contraction-relaxation. (× 11,300).

## 2. Vascular Smooth Muscle

Blood vessels first develop in the embryo from mesenchyme as a budding network of small endothelial-lined channels.[14] These channels are soon ensheathed by extracellular matrix and later become surrounded by locally derived irregularly shaped mesenchymal cells.[15-17] When elastic lamellae and other extracellular matrix components are first produced, the mesenchymal cells resemble fibroblasts with few filaments but large amounts of rough ER and free ribosomes[16,18] (Figure 6). With time, as the elastic lamellae become more identifiable in large arteries, the muscle cells take on a more characteristic appearance, although they do not do so at the same rate across the wall since with further development new laminae are added onto the adventitial aspect.[4] At parturition most arteries appear to have developed their adult number of lamellar units,[19] for after this stage, the number of layers in the wall does not increase and medial thickening is due to the production of connective tissue. In the newborn rat aorta, myofilaments occupy 7.7% of the volume of smooth muscle cells and are usually confined to the periphery of the cell. Golgi and rough ER constitute about 30% of the cell volume.[6] Over the next 8 to 12 weeks the amount of elastin increases from 6% to 35% and collagen increases from 8% to 26% dry weight of the media, which are the approximate adult values. At about the same time as these adult extracellular matrix values are reached, the volume fraction of smooth muscle cytoplasm occupied by myofilaments reaches the adult level of 70%, with rough ER 7.2% and Golgi 2.4%.[6] Biochemical analysis of actin, tropomyosin, vimentin, and desmin content per cell correlates well with these observed morphological changes. The major type of actin pres-

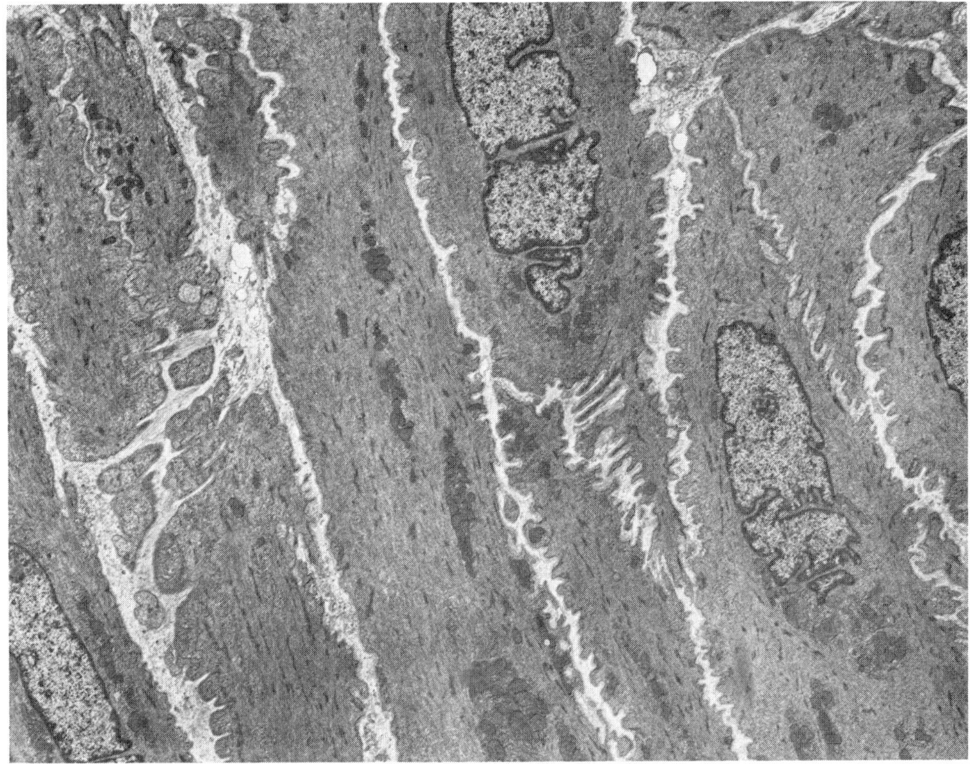

FIGURE 3. Longitudinal section through smooth muscle cells of the caudal artery of a 14 week rat, demonstrating, as in Figure 2, the relative amounts of synthetic organelles vs. myofilaments. (× 8,600).

ent in the fetus and newborn animals is the $\beta$-isoform but this changes to the predominate $\alpha$-isoform with development. Smooth muscle cells of the adult human aortic media contain more than 60% of actin in the $\alpha$-isoform, 31% in the $\beta$ form, and negligible amounts in the $\gamma$ form. Vimentin is found in 87% of medial cells in the fetal rat aorta, 13% contain both vimentin and desmin, and none contain desmin alone. With development there is a gradual decrease in cells containing vimentin alone and an increase in those in which both vimentin and desmin coexist, until adult values of 51% vimentin alone, 48% with both vimentin and desmin, and 1% desmin alone are reached at 12 weeks after birth.[20]

## B. Injury and Repair
### 1. Visceral Smooth Muscle

Synthetic state smooth muscle cells have been widely reported in areas of regeneration and repair after damage to visceral smooth muscle tissues[21-24] (Figure 4), where they appear to be smaller than the contractile state cells from the same tissue.[25] These cells are actively engaged in synthesis of extracellular matrix and/or proliferation. Although several reports have shown that both vascular and visceral smooth muscle are capable of division while contractile,[26-29] the vast majority of smooth muscle cells observed in mitosis are at least partially modulated towards the synthetic state, containing few myofilament bundles and a large number of organelles such as free ribosomes and rough ER[21,30,31] (Figure 7).

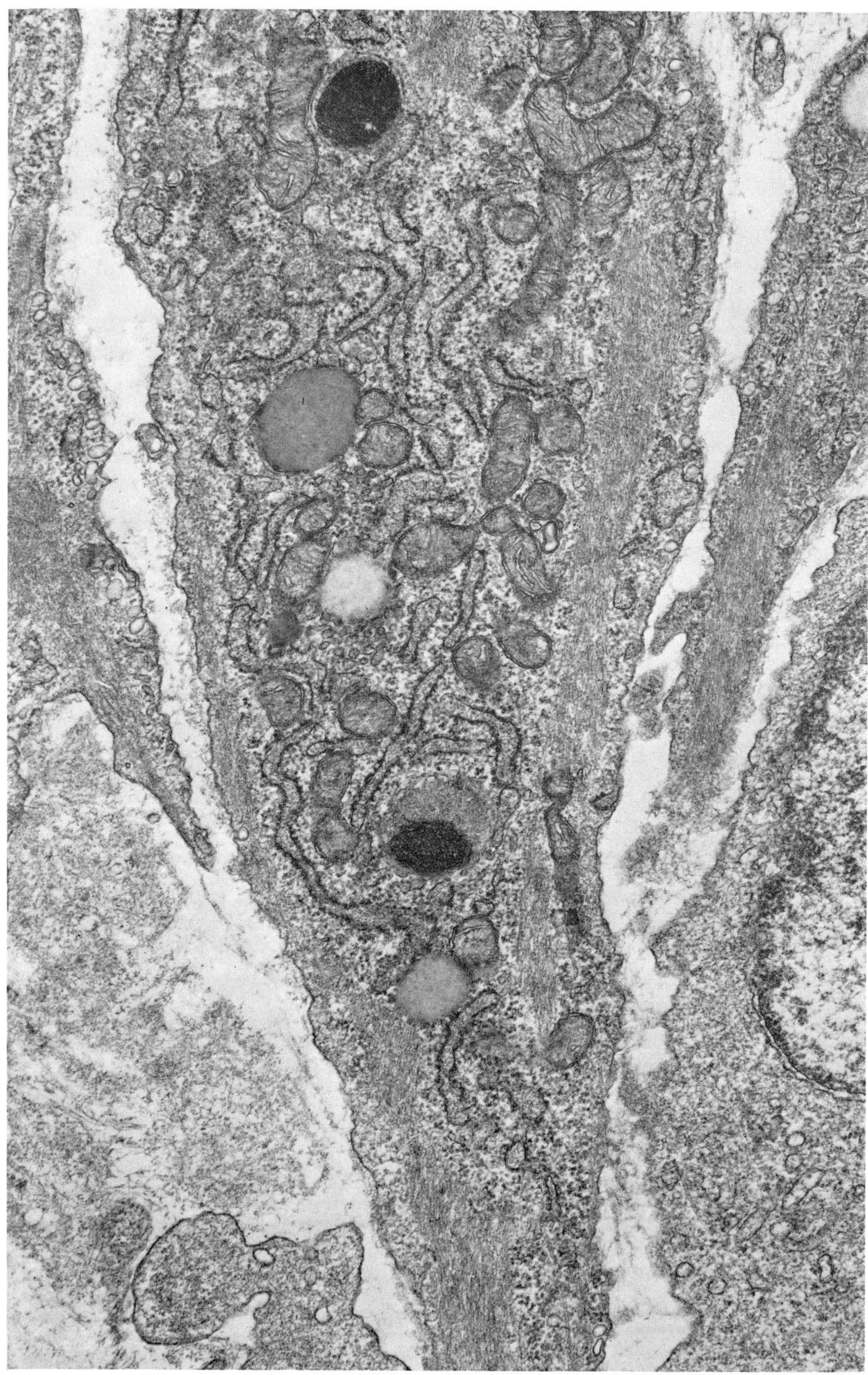

FIGURE 4.    Synthetic state smooth muscle cell from guinea pig vas deferens, one week after transplantation into the anterior eye chamber. Note the large amount of synthetic apparatus present in the cell. (× 25,500). (Reproduced from Campbell, G. R., Uehara, Y., Malmfors, T., and Burnstock, G., *Z. Zellforsch. Mikrosk. Anat.*, 117, 115, 1971. With permission.)

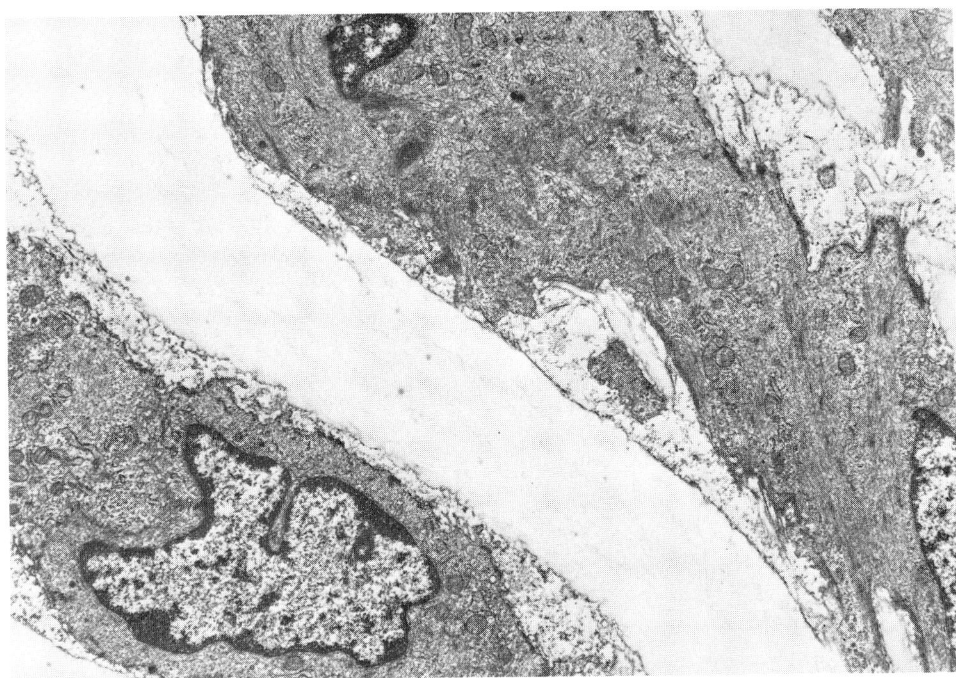

FIGURE 5. Transverse section through aortic media of 4 week old rabbit. Different layers of smooth muscle are separated by elastic laminae. More synthetic organelles are present in these cells than those in Figure 1. (× 9,000).

## 2. Vascular Smooth Muscle

### a. Direct Damage

Studies of direct damage to the wall of a blood vessel show that arterial medial cells respond in a similar manner to damaged visceral smooth muscle[32-40] (Figure 8). Poole et al.[36] wrote in their description of injured arterial media: "On looking at cells in the tunica media further and further away from the site of injury, it was seen that there was a continuous gradient in cell morphology from normal smooth muscle cells of the tunica media (which had presumably not been injured) to the cells showing great lack of differentiation nearer to the silk suture. Mitoses were seen among these cells."

### b. Endothelial Denudation

In recent years many studies have been made of the effects of endothelial denudation, since the fibromusculoelastic intimal thickenings which occur as a result of this procedure have been proposed as an accelerated model for the events leading to the cellular accumulation characteristic of atherosclerotic plaques.[41-43]

Injury to the endothelium has been caused experimentally by a variety of traumata such as painting with acid,[44] balloon catheterization,[45-47] other mechanical methods,[41,48] and air drying.[49,50] If the damage is of sufficient size, smooth muscle cells from the media migrate to the intima where they subsequently proliferate forming a neo-intima of longitudinally orientated smooth muscle cells (Figures 9 and 10). Detailed examination of the phenotypic changes of these proliferating smooth muscle cells has not been made; however, attention has been drawn to the large amounts of rough ER, free ribosomes, and mitochondria in these cells up to 3 weeks after endothelial injury[48,51-53] (Figure 11). These cells have been variously described as modified smooth muscle cells,[47,54,55] undifferentiated cells,[56] immature smooth muscle,[57] smooth muscle-

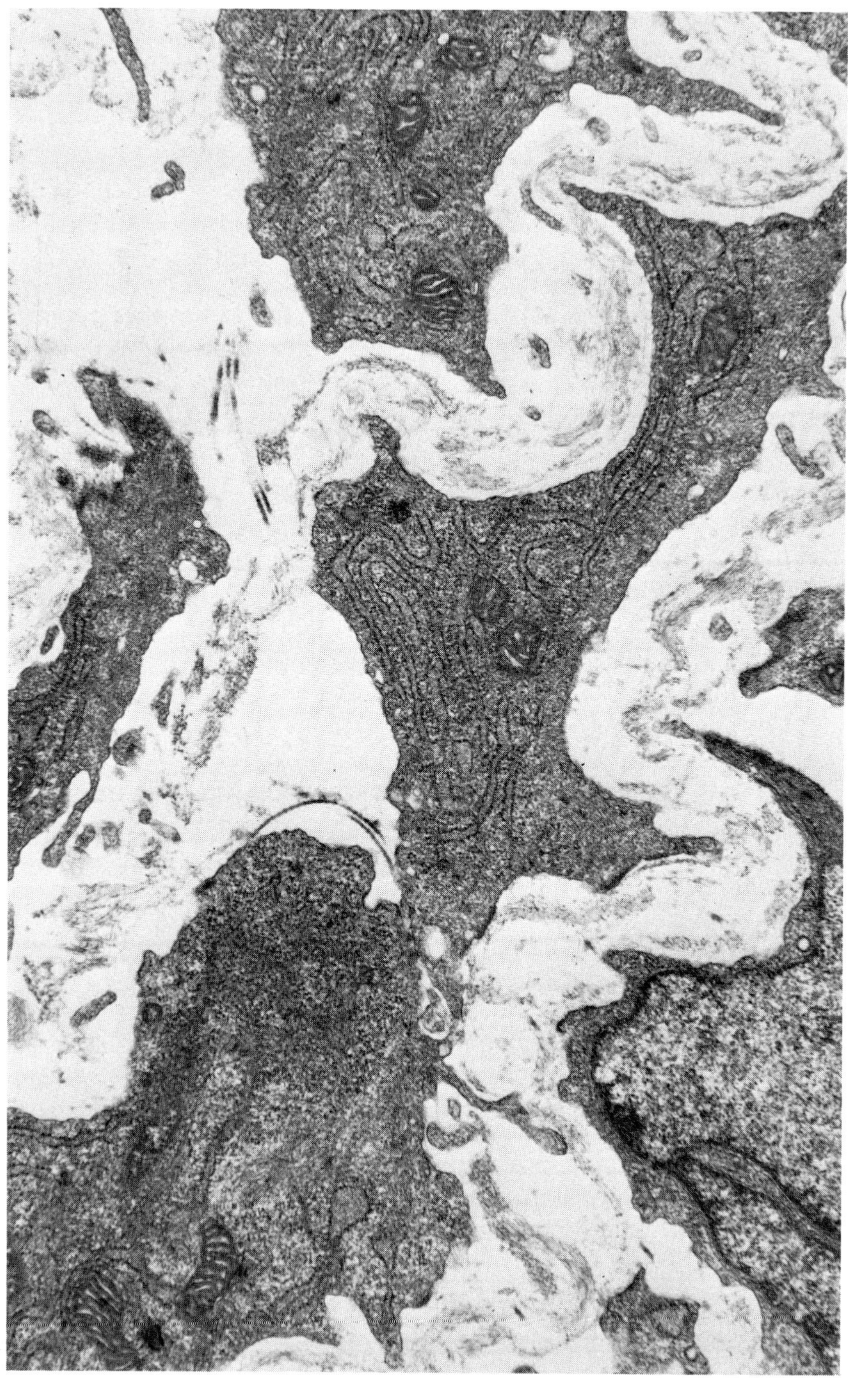

FIGURE 6.    Smooth muscle cells from media of 12 day chick embryo carotid artery. Note the small amounts of collagen and forming elastic lamellae. The muscle cells contain few filaments; most of their cytoplasm is filled with rough ER and free ribosomes. (× 16,000).

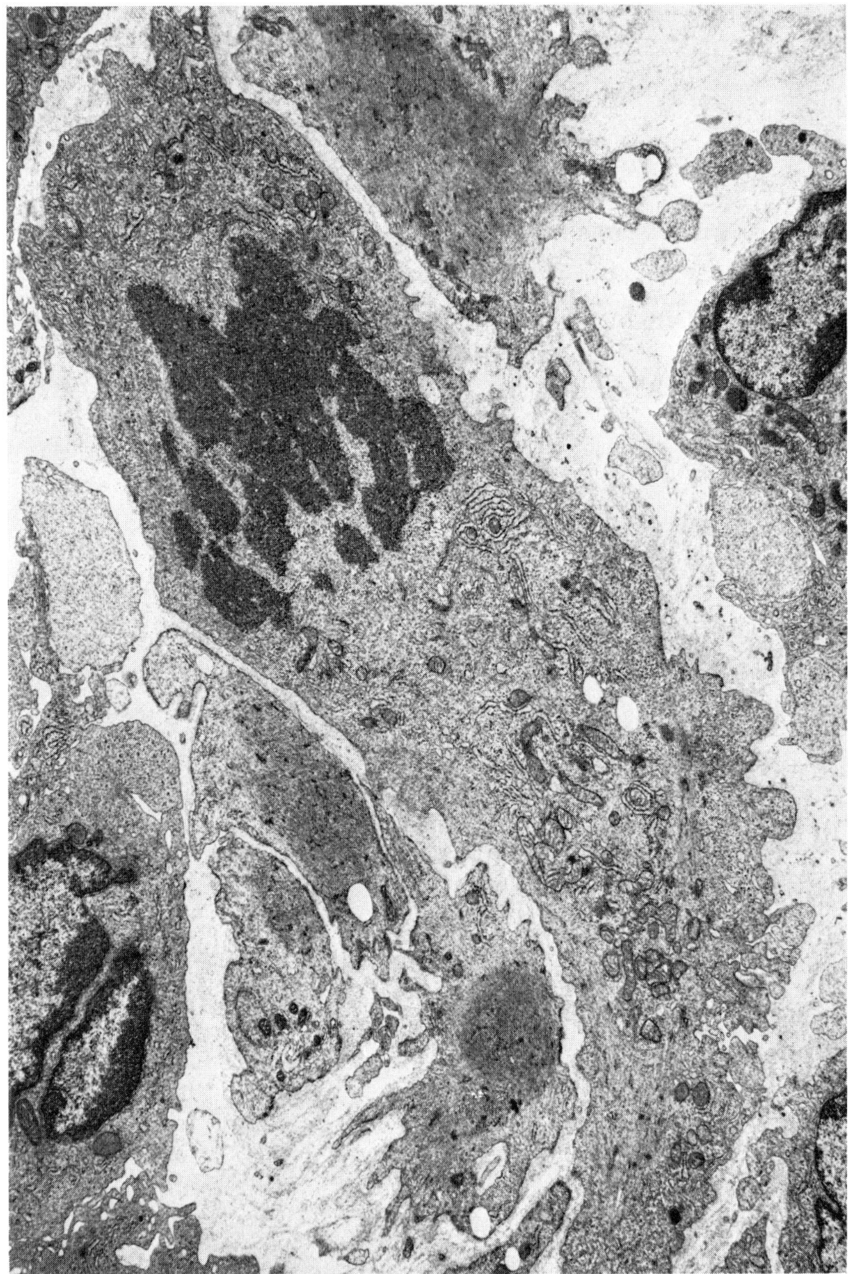

FIGURE 7. Dividing smooth muscle cell from guinea-pig vas deferens, one week after transplantation into the anterior eye chamber. Note that most of the cytoplasm is filled with synthetic organelles. (× 18,000). (Reproduced from Campbell, G. R., Uehara, Y., Malmfors, T., and Burnstock, G., *Z. Zellforsch. Mikrosk. Anat.*, 117, 115, 1971. With permission.)

like,[49] or of synthetic phenotype.[58,59] Upon reestablishment of the overlying endothelial layer the smooth muscle cells of the neo-intima regain an apparently normal complement of myofilaments with a concomitant decrease in the amount of synthetic organelles.[53,59] Autoradiographic studies have shown that thymidine indices reach a maximum in the neointima during the first week following endothelial denudation[49,60-63] and return to the baseline by 8 weeks after the intima is reendothelialized.[53]

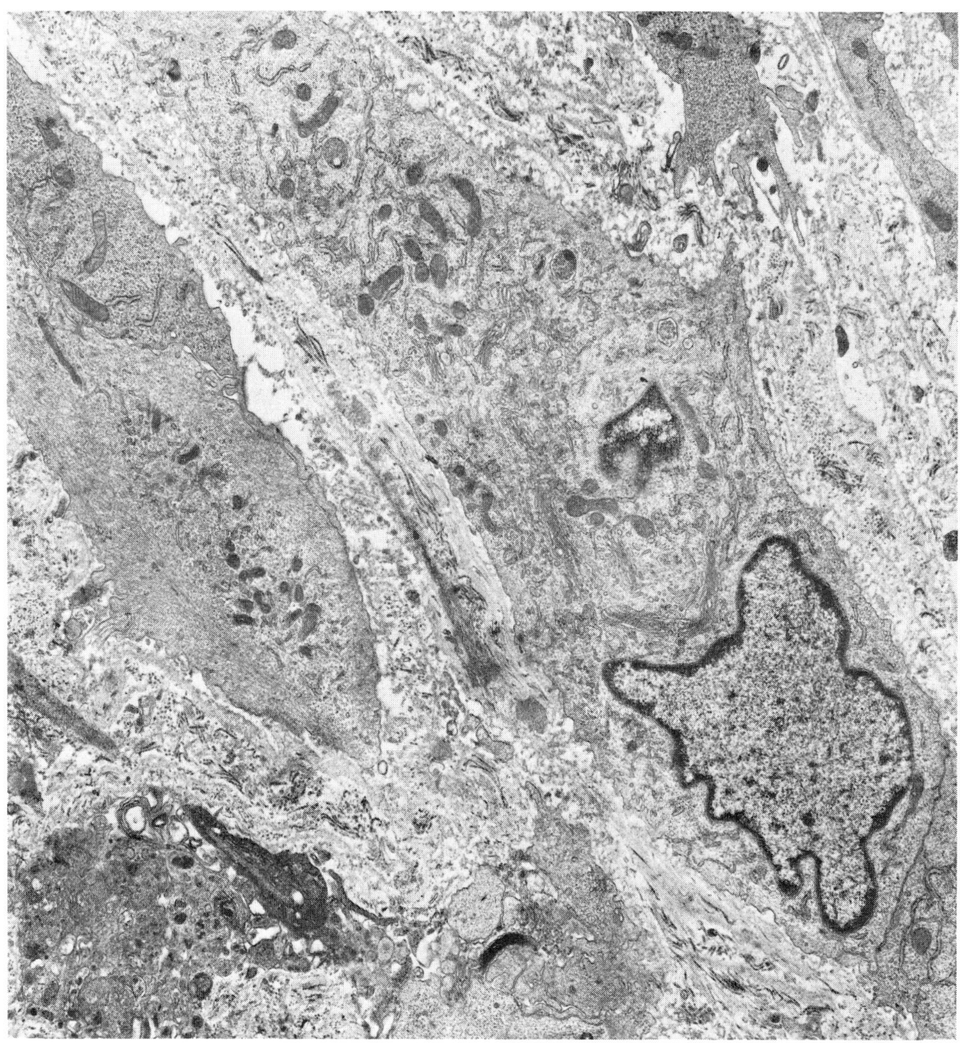

FIGURE 8.    Synthetic state cells in media of renal artery of spontaneously hypertensive rat after mechanical damage caused by passing a needle through the wall. (× 8,000).

Recent studies have shown a different filamentous expression from that of the media in smooth muscle of a neo-intima formed 2 to 3 weeks after endothelial injury. Fifteen days after injury, cells that have migrated into the intima contain decreased amounts of actin and desmin and increased amounts of vimentin. Moreover, $\beta$-actin is the predominant actin isotype in these cells and significant amounts of $\gamma$-actin appear, whereas $\alpha$-actin decreases. Seventy-five days after injury when the endothelium has completely regenerated, the cytoskeletal elements of the neo-intimal smooth muscle cells are similar to those of the media.[52,64,65] Studies using antibodies to the heavy meromyosin end of smooth muscle myosin also demonstrate a loss of myosin in smooth muscle cells of the neo-intima at 14 days after endothelial denudation with a return to nearly normal levels by 6 to 8 weeks.[66]

The smooth muscle cells in the neo-intima formed after endothelial denudation differ in their metabolism and extracellular matrix production, according to whether the

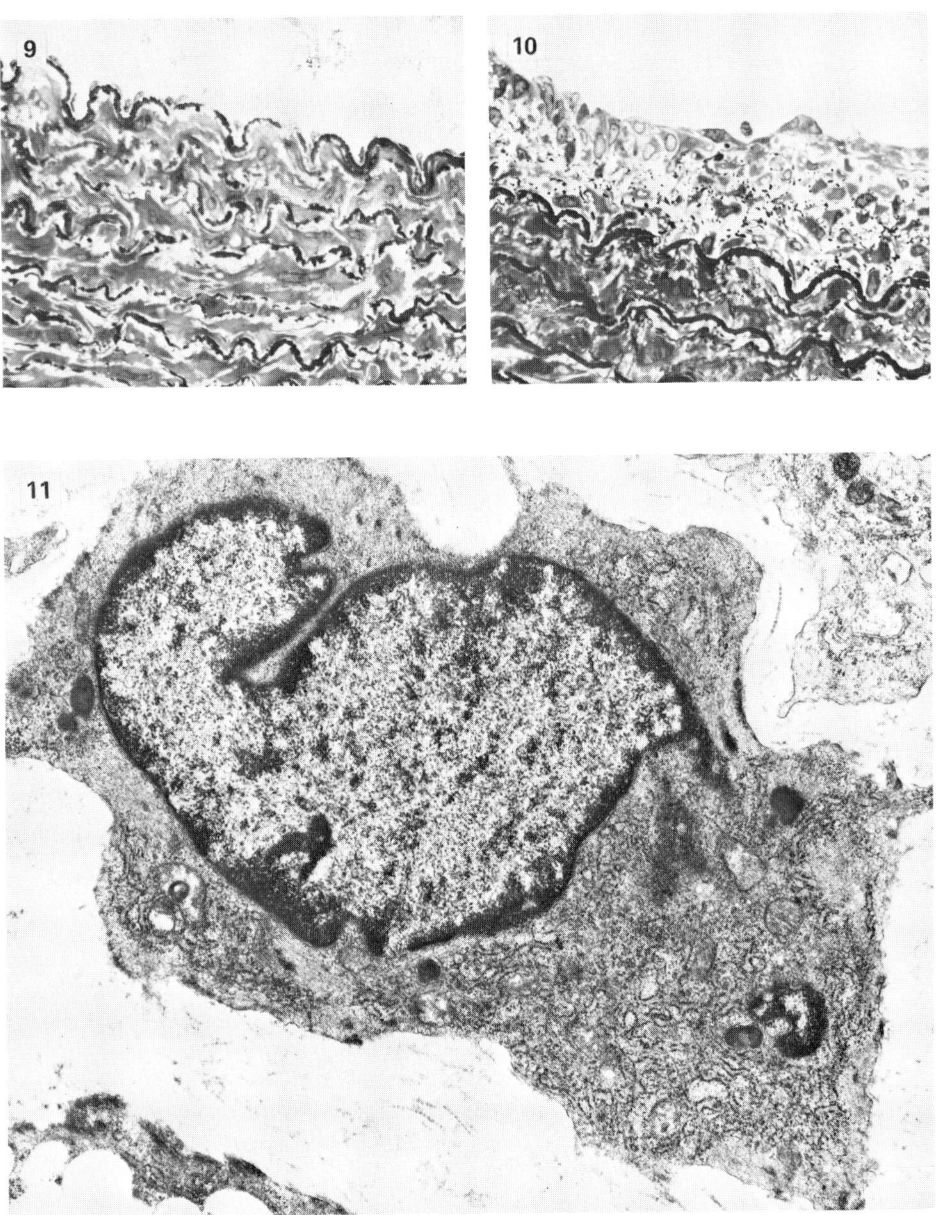

FIGURES 9 to 11.   Figure 9 shows a rabbit left common carotid artery. No cells are present in the intima and the endothelium is closely applied to the internal elastic lamina. (× 280). Figure 10 shows a rabbit right common carotid artery 14 days after endothelial denudation by drying with a jet of air. The lumen is lined by a layer of newly formed endothelial cells and a pseudoendothelium of smooth muscle cells. Beneath these cells is a neo-intima of largely longitudinally disposed smooth muscle cells (× 280). Figure 11 shows smooth muscle cell present in neo-intima of Figure 10. Note the large amount of rough ER and small peripherally located bundles of myofilaments. (× 10,000).

area is still deendothelialized or it has been reendothelialized. Minick et al.[67,68] found in rabbits that lipid accumulates preferentially in the neo-intima of areas covered by regenerated endothelium. Subsequent studies suggested that this accumulation is not simply a result of passive infiltration but due to metabolic differences in the reendothelialized neo-intima.[69,70] The specific activities of the cholesteryl ester metabolizing enzymes acyl CoA: cholesterol acyltransferase (ACAT) and acid cholesteryl esterase and marker enzymes for major cell organelles such as cytochrome C oxidase, N-acetyl-$\beta$-glucosaminidase, and catalase are increased in deendothelialized and reendothelialized areas of injured aortas compared with uninjured controls, but the activity of acid cholesteryl esterase is significantly less in reendothelialized as compared with deendothelialized areas.[71] This altered metabolism of the smooth muscle cells persists for up to one year after a single balloon catheter deendothelialization and reendothelialization.[72] Wight et al.[73] found by morphometric and chemical analysis significant increases in the glycosaminoglycans (GAG) heparan sulfate and chondroitin sulfate in reendothelialized aortas compared to deendothelialized or uninjured aortas. Since GAG and proteoglycans can form insoluble and soluble complexes with plasma lipoproteins,[74-76] this may be another mechanism whereby changes in smooth muscle metabolism contribute to the increased deposition of lipid in reendothelialized regions of injured aorta. Further in vivo[77,78] and in vitro[79] studies demonstrate that aortic GAG content is markedly influenced by the presence or absence of endothelium.

## C. Pregnancy

During pregnancy there is a progressive increase in total collagen (measured as hydroxyproline) in the uterus of mammals, reaching levels at term from 475% (rat) to 810% (man) above those at estrus.[80,81] Smooth muscle cells increase progressively in size during this period, but do not appear to divide.[81] They contain at this time an extensive synthetic apparatus and only small filament bundles[82,83] (see chapter by Sanborn, this volume). A similar modulation of smooth muscle phenotype can be elicited under estrogenic stimulation.[84,85] After a single dose of estradiol diproprionate, smooth muscle of the rat uterus shows an increase in the number of cellular organelles and assumes after 72 hr, a morphology similar to that of embryonic tissue.[86]

Smooth muscle cells in a similar phenotype have also been observed in uterine arteries of guinea pigs,[87] where the cells have been suggested to be involved in the synthesis and degradation of elastic fibers during pregnancy and postpartum involution respectively.[87-89] Estrogen injection produces an increase both in cellular activity and the amount of connective tissue present in aortae.[90-92]

## D. Hypertension

In hypertension, vessels are subjected to an increase in wall stress due to elevated arterial pressure, the response of the vessel being medial hypertrophy. Histological and biochemical analyses demonstrate that this medial thickening is due to increased amounts of smooth muscle, collagen, and proteoglycans.[93-97] Recently, quantitative morphometric studies have added to our understanding of how the vessel wall adapts to increased intraluminal pressure.[7,98] Following surgical constriction, the aorta of young male rats demonstrates a 39% thickening of the wall within 8 days. This is due largely to volume increases of 57% in smooth muscle cells, 30% in elastic laminae, and 136% in collagen. Concomitant with these changes is a 264% increase in the amount of rough ER present within the smooth muscle cells.[7] In long-term hypertension these changes are sustained. In addition, there are increased numbers of smooth muscle cells in the intima (Figure 12), and increased numbers of cells with distinctive synthetic features in the media (Figure 13).

Alterations in aortic metabolism have also been reported in spontaneously hypertensive- and deoxycorticosterone (DOCA) hypertensive rats. In both groups, the lysosomal enzymes N-acetyl-$\beta$-glucosaminidase and acid phosphatase as well as 5'-nucleotidase show enhanced activity compared to age-matched controls.[99,100]

A number of studies have demonstrated significant incorporation of $^3$H-thymidine

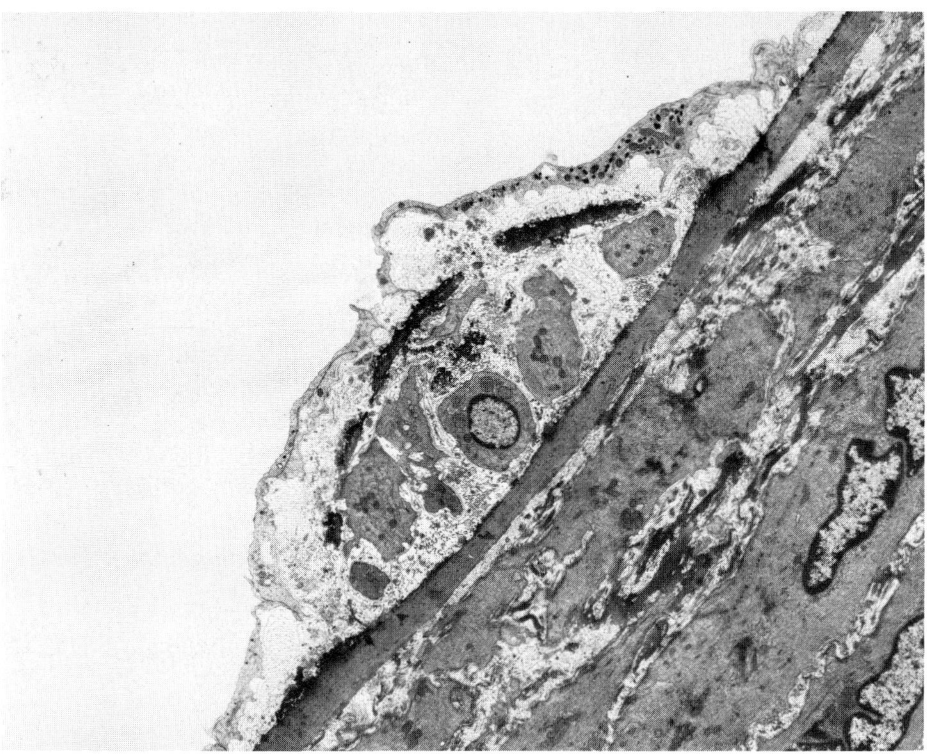

FIGURE 12.   Small neo-intimal thickening from aorta of 8 month old spontaneously hypertensive rat. The intimal smooth muscle cells are much smaller than those of the media and surrounded by laminated basal lamina material. (× 3,200).

into the DNA of smooth muscle cells in early stages of hypertension, followed by an increased DNA content in the vessel wall.[95,101] This implies an increase in smooth muscle cell number in vessels of hypertensive animals. However, more recent morphometric analytical studies[98] have demonstrated that while there is clearly hypertrophy of the walls of large vessels in hypertension, there is no increase in cell number. There is therefore a paradox: increased DNA content in the vessel wall as a result of hypertension, but no increase in cell number. This paradox has apparently been resolved by recent studies of Owens and Schwartz[102-104] (see chapter by Owens, this volume), who found by morphometric analysis that the difference in aortic smooth muscle mass between hypertensive and normotensive rats is due to smooth muscle cell hypertrophy without hyperplasia, accompanied by an increase in the frequency of polyploid cells (i.e., with greater than 2n chromosomes). This is true both of spontaneously hypertensive rats (SHR) and of Sprague-Dawley rats made hypertensive by the Goldblatt procedure. The frequency of polyploidy increases with age, duration of hypertension, and level of blood pressure.[103,104] Polyploid cells therefore represent another phenotypic expression of smooth muscle in response to an altered environment.

Hypertrophy of a vein can also be induced by partial ligation.[105,106] Fourteen days after ligation of the portal-anterior mesenteric vein there is a 60% increase in the v/v of the smooth muscle cells and these cells demonstrate an approximate 3-fold increase in the number of intermediate filaments per cross-sectional area, as compared to controls.[106]

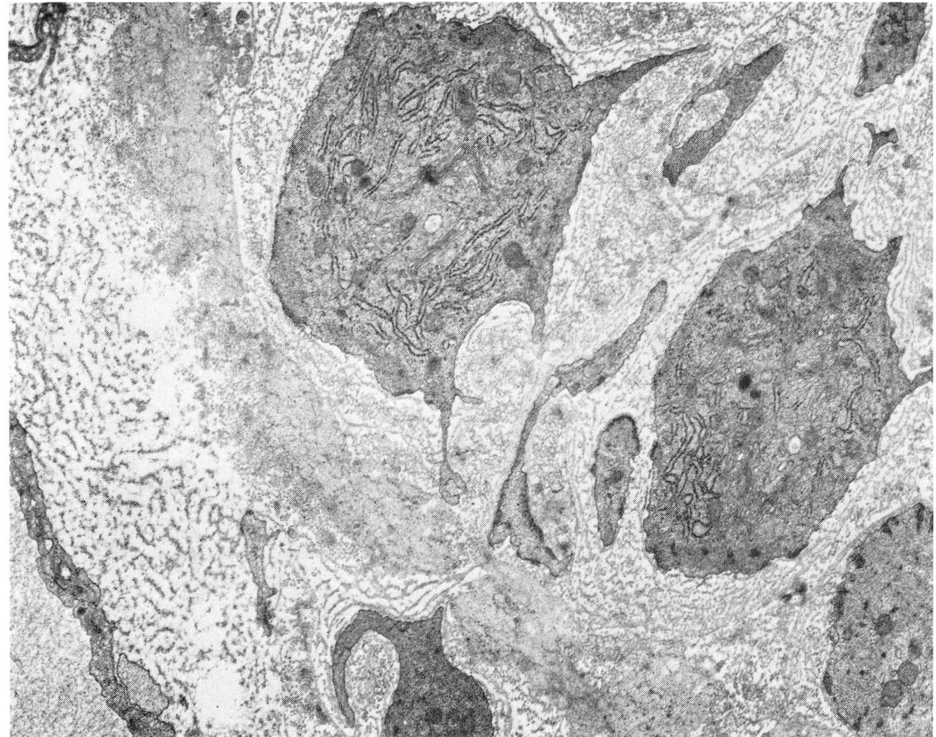

FIGURE 13.    Synthetic state smooth muscle cells in media of caudal artery of 6 month old spontaneously hypertensive rat. (× 8,400).

## E. Intestinal Hypertrophy

There is a marked increase in the lumenal diameter and a compensatory hypertrophy of the wall oral to an experimental partial stenosis of the ileum[107] (see chapter by Gabella, this volume). This hypertrophy is due to cell mitoses, an average increase of 3 to 4 times in cell volume[108] and a significant increase in extracellular matrix.[109,110] Other modulations of smooth muscle phenotype are observed, such as an increase in smooth and rough ER, a reduction in mitochondria,[111] an increase in the size and number of gap junctions between hypertrophic cells,[112] and an increase in the number of intermediate filaments.[113]

## F. Diffuse Intimal Thickenings in Human Arteries

At birth the intima of human arteries is composed of an endothelium separated from the underlying internal elastic laminae by a narrow layer of connective tissue. Over the next three decades there is a gradual increase in the width of this layer to form what is called the diffuse intimal thickening. This development is due to the migration of medial smooth muscle cells through fenestrae or fragmented areas of the internal elastic laminae, followed by their proliferation and elaboration of connective tissue elements.[114] The mechanism by which diffuse intimal thickenings develop remains to be determined. Two mechanisms have been suggested: (1) a response of the vessel wall to repeated injury and (2) a remodelling of the vessel wall due to postnatal growth and increase in blood pressure.[115,116] Regardless of how it is formed, the diffuse intimal thickening is a prominent feature of larger arteries of all large mammals and in human coronary arteries can be up to two times the thickness of the media.

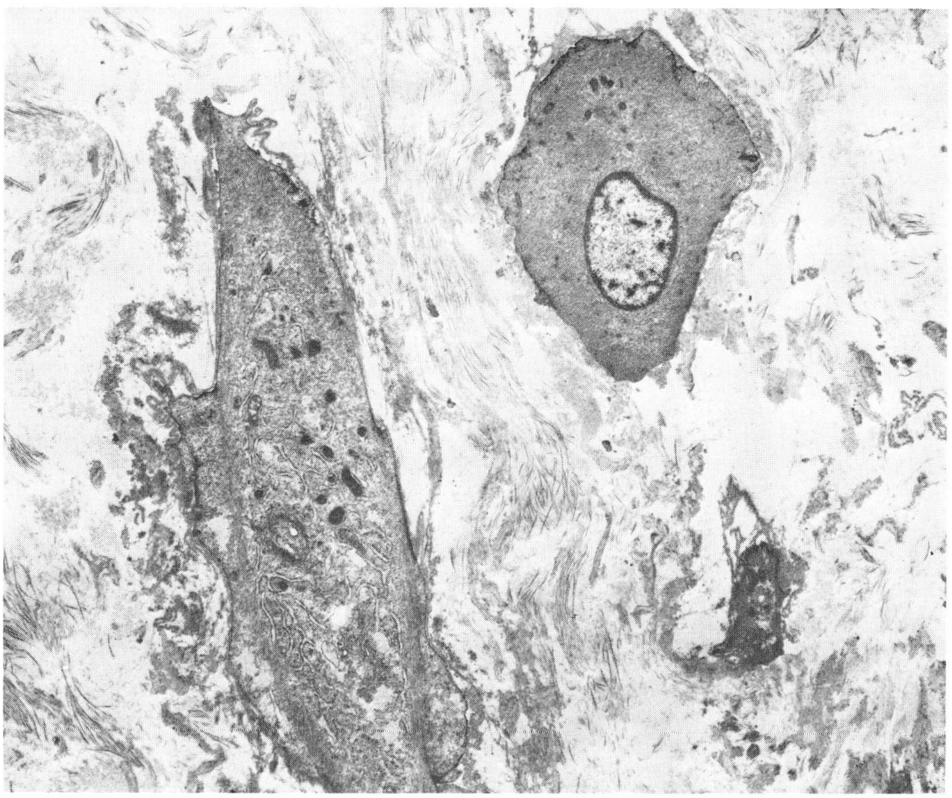

FIGURE 14.   Two smooth muscle cells from a diffuse intimal thickening of human carotid artery. One cell fulfils the criteria of a contractile state smooth muscle cell, the other of a synthetic state cell. (× 5,000).

Diffuse intimal thickenings of human arteries are sites of predilection of atheromatous plaque formation but do not necessarily develop into them.[117,118] Extensive postmortem studies of coronary arteries in subjects whose ages range from fetal to 40 years show that atherosclerotic plaques always develop within preexisting branch pads or intimal thickenings.[119,120] A nodular proliferation of smooth muscle cells and accumulation of foam cells and extracellular lipid in the atherogenic process results in a reorganization of these thickenings.

More than 80% of the cells in human diffuse intimal thickenings can be classified as smooth muscle. These vary widely in size and shape.[117,121] Analysis by morphometric techniques of the volume fraction of synthetic organelles in cells in diffuse intimal thickenings from human carotid endarterectomy material shows an approximate 100% increase compared with cells of the media[122] (Figure 14). This demonstrates that many of these intimal muscle cells exhibit a different phenotypic expression from those of the media. Consistent with this finding is recent data that human diffuse intimal thickening smooth muscle cells contain 15% of their actin in the $\alpha$-isoform, 70% in the $\beta$-isoform and 15% in the $\gamma$-isoform, unlike the aortic media which shows a predominance of actin in the $\alpha$-isoform.[64]

Immunofluorescence staining of both frozen sections and freshly enzyme-isolated cells from human diffuse intimal thickenings, with antibodies to the heavy meryomyosin end of smooth muscle myosin, also demonstrate a range of phenotypic expression in these cells.[66]

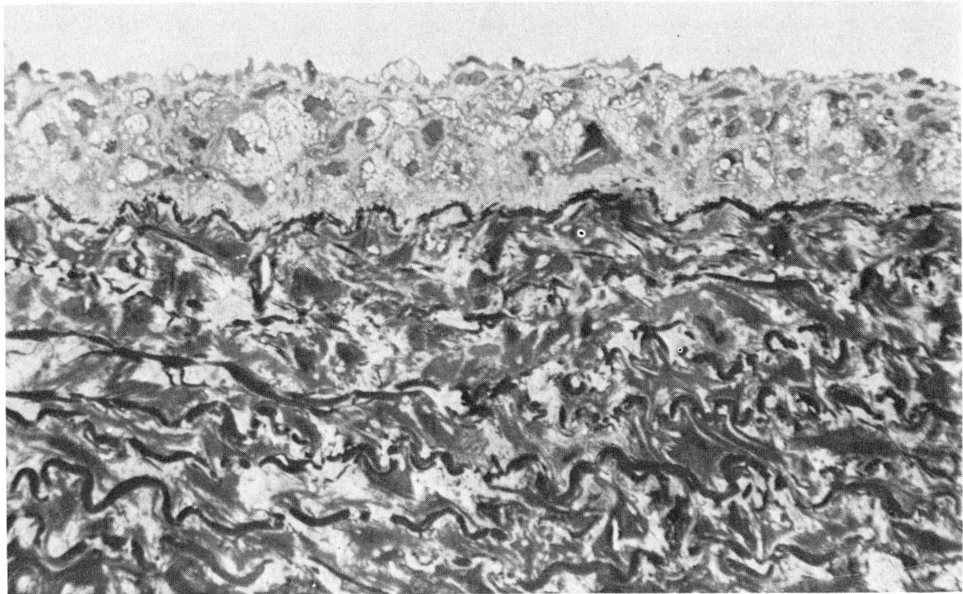

FIGURE 15.    Aorta of rabbit fed a 1% cholesterol supplemented diet for two months. The intima in this area contains many lipid engorged macrophages and smooth muscle cells. (× 500).

## G. Experimental Hypercholesterolemia

Plaques similar to the atheroma observed in man can be produced in rabbits, swine, and certain species of monkey by a diet supplemented with cholesterol (Figure 15). Cholesterol feeding, like endothelial injury, appears to alter a number of metabolic characteristics within the artery wall. Significant increases in synthesis of the extracellular matrix components collagen[123-125] and GAG[73,126] have been demonstrated, although the type of GAG which increases significantly appears to be species dependent. Smooth muscle cells with increased amounts of synthetic apparatus and a concomitant decrease in myofilaments become apparent[114] (Figure 16). These cells have been variously described as "active",[127] rough ER-rich smooth muscle,[128] or "young" smooth muscle cells.[114] In culture, smooth muscle cells derived from atherosclerotic rabbit aortas have increased growth rates and produce more extracellular matrix compared with those from control rabbits.[129,130] Alterations in a variety of enzymes have been observed between cells in the intima of hypercholesterolemic rabbits and those from the media of normal animals.[131]

Both endothelial injury and cholesterol feeding therefore appear to alter metabolic characteristics of the artery wall. As pointed out by Moore,[132] this is probably due in the first instance to a common reaction, since in diet-induced atherosclerosis, the development of an intimal thickening precedes the accumulation of lipid. It might therefore be expected that endothelial injury precedes intimal thickening in hypercholesterolemic animals. In spite of a number of investigations to obtain such data, no clearcut evidence of endothelial desquamation preceding lesion development has been described,[133-136] and other mechanisms must be involved in this process.

## H. Aging

With age, the media of the rat aorta doubles in thickness. This is due to hypertrophy of the smooth muscle cells and an increase in the size of the elastic lamellae and amount of interlaminar connective tissue. The number of cells present, however, is halved.[137]

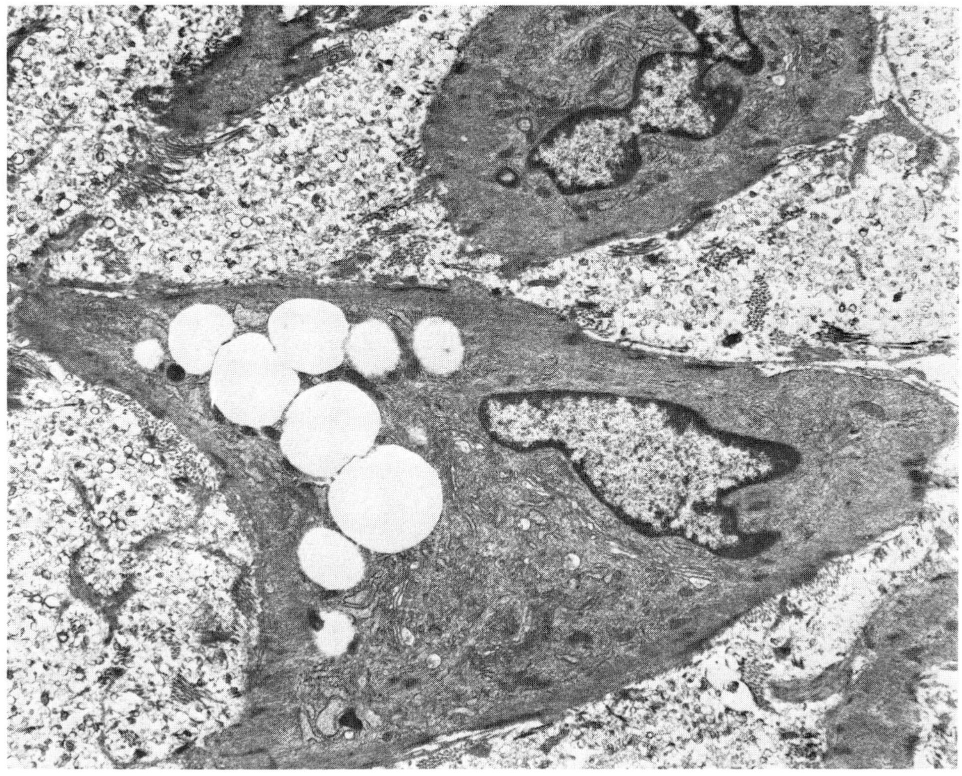

FIGURE 16. Smooth muscle cell from intima of rabbit artery after feeding with a 1% cholesterol supplemented diet for two months. Note the lipid droplets, the large number of synthetic organelles in the perinuclear region, and the small peripherally located filament bundles. (× 9,500).

Associated with this decrease in cell number is an increase in cell necrosis.[137-140] The remaining smooth muscle cells hypertrophy, become very irregular in shape, contain substantial amounts of synthetic organelles such as rough ER, free ribosomes and Golgi, and fewer myofilament bundles confined to the cell periphery.[137,141] They are also surrounded by extensive reduplicated basal lamina condensations similar to that seen in hypertension (see Figure 12). Cliff[137] interpreted the increase in rough ER volume in these hypertrophied medial smooth muscle cells of aged animals as a response of the surviving cells to the decreased cellularity. Recent observations show an increase in the number of polyploid cells present (from <1% at birth to 7% at age 60) in human aorta, and carotid and iliac arteries, with age.[142] It is unknown whether these polyploid cells have increased amounts of synthetic apparatus.

## III. STRUCTURAL AND FUNCTIONAL CHANGES WHICH ALTER WITH SMOOTH MUSCLE PHENOTYPIC EXPRESSION IN VITRO

### A. Morphology

Vascular and visceral smooth muscle enzyme-dispersed into single cells and plated in primary culture, attach and flatten on the culture substrate within 1 to 2 days to closely resemble the same cells in vivo.[31] That is, they express a contractile phenotype, with their cytoplasm containing substantial bundles of thick and thin myofilaments. If they are seeded below a critical cell concentration, the cells undergo a spontaneous change in phenotype to the synthetic state after 6 to 8 days.[143-146] That is they lose most of

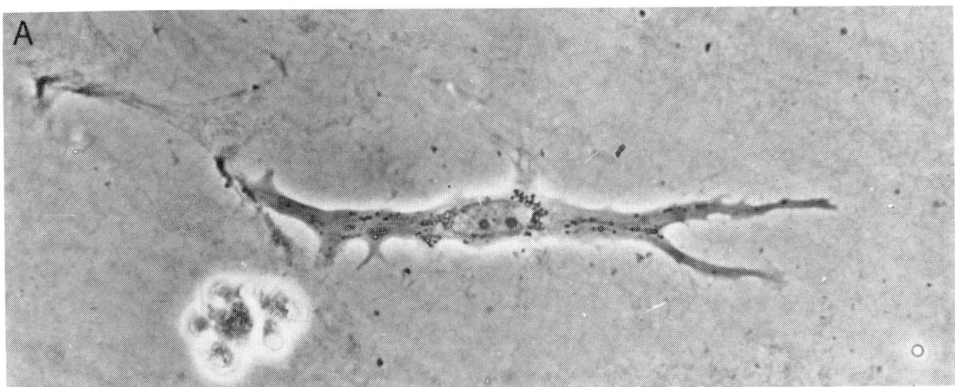

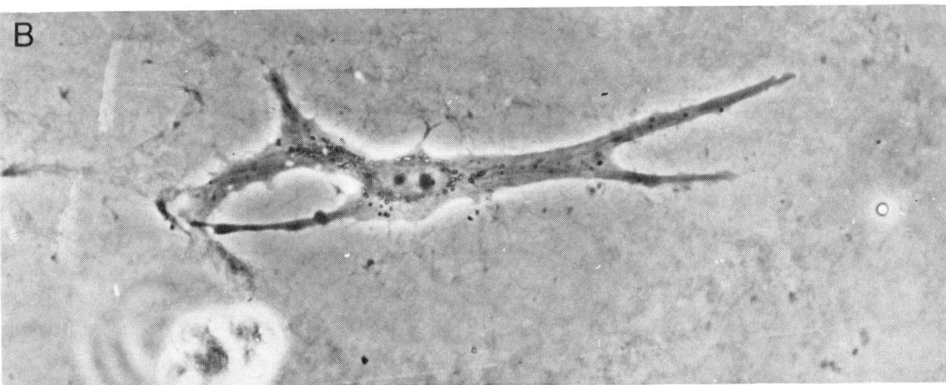

FIGURE 17.    The same smooth muscle cell from the newborn guinea pig vas deferens, shown with increasing time in culture. (A) after 3 days in vitro, (B) after 5 days in vitro, (C) after 7 days in vitro, and (D) after 9 days in vitro. Note that the muscle cell has modulated its phenotype and undergone mitosis. (× 700). (Reproduced from Chamley, J. H., Campbell, G. R., and Burnstock, G., *J. Embryol. Exp. Morph.*, 32, 297, 1974. With permission.)

their myosin-containing thick filaments and gain a substantial amount of organelles involved with synthesis. The number of lysosomes also increase. With phase-contrast microscopy, the cells change from being phase-dense and ribbon or fusiform in shape, to phase-lucent, broad and flat (Figure 17). At the same time there is a decrease or loss in staining reaction with FITC-labelled antibodies to the heavy meromyosin end of smooth muscle myosin. The time of spontaneous phenotypic change is dependent on the donor organ, age, and species. It is not dependent on the presence or absence of serum or serum components in the culture medium.[147,148]

## B. Contraction

In the first few days in primary culture, while the cells morphologically express a contractile phenotype with large numbers of myofilaments, spontaneous contractions at a rate of 1 to 8 per min have been reported in single smooth muscle cells from the guinea pig vas deferens,[29,143,149-152] taenia coli,[152-154] ureter,[145,152] chicken gizzard,[152,155,156] and amnion.[157,158] There are few reports of spontaneous contractions in single vascular smooth muscle cells in culture,[144,159] but the cells can be stimulated to contract with vasoactive agents.[144,160]

Concomitant with the change in morphology to a synthetic state, the cells lose their ability for spontaneous or induced contraction.[31]

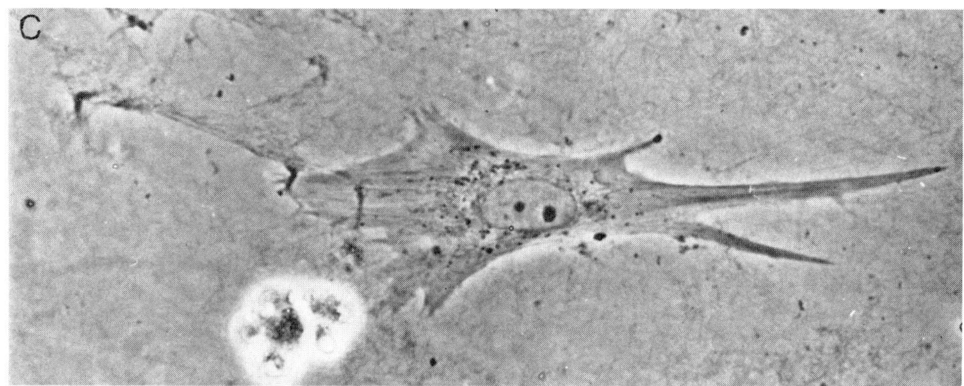

FIGURE 17C

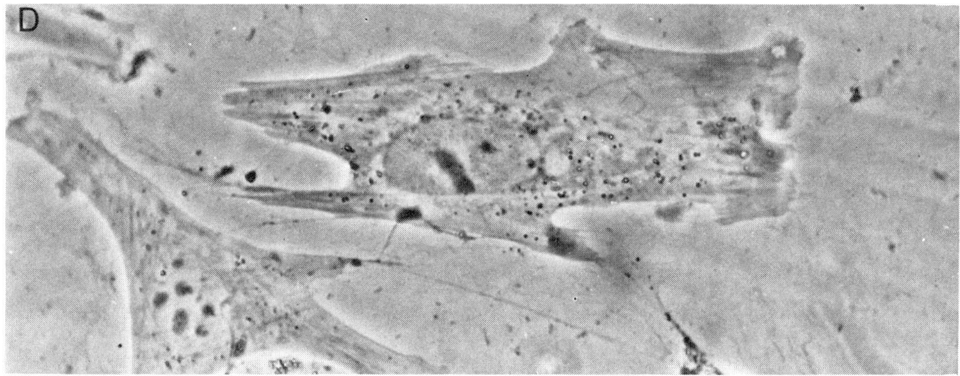

FIGURE 17D

## C. Division

In the first 6 or 7 days of primary culture, only 3 to 5% of aortic smooth muscle cells from the adult pig and monkey incorporate [3]H-thymidine into their nuclei during a 4 hr pulse labelling period in the presence of 5% whole blood serum.[147,161] However, on day 7, 8, or 9, that is, the day after phenotypic modulation to the synthetic state is first evident morphologically, the number of labelled nuclei increases to 10 to 20%. The percentage of cells incorporating [3]H-thymidine into DNA then increases linearly with time until on day 14, 40 to 60% of cell nuclei are labelled during the 4 hr pulse labelling period (Figure 18). The cells begin to proliferate logarithmically in 5% whole blood serum about 24 hr after they show an increased incorporation of [3]H-thymidine (Figure 19). There is no significant increase in cell number while the cells are in the contractile state; however, some contractile cells have been shown to be capable of division.[29]

## D. Synthesis of Extracellular Matrix

Synthetic state aortic smooth muscle cells under basal conditions synthesize 4-fold the amount of collagen as contractile state cells as measured by the incorporation of [3]H-proline into hydroxyproline (Campbell and Campbell, unpublished). The amount of noncollagen protein synthesized doubles under the same conditions. Synthetic state cells also synthesize 5-fold the contractile state levels of sulfated glycosaminoglycans

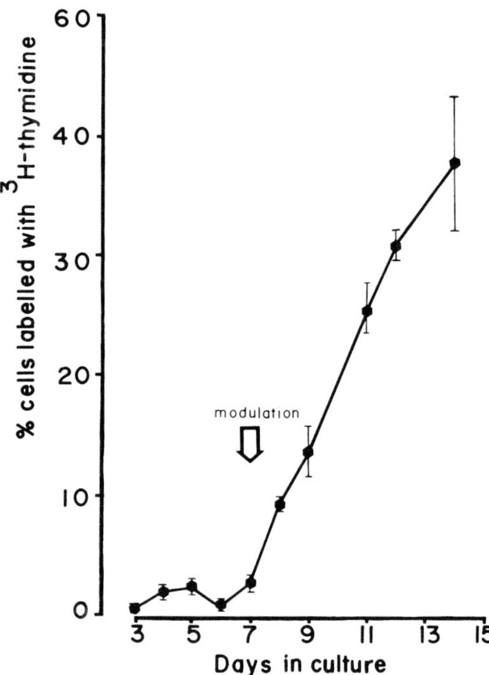


FIGURE 18.   Percentage of cells labelled with ³H-thymidine during a 4 hr pulse period each day in culture in the presence of 5% whole blood serum. Each point represents the mean of 3 dishes ± SD. (Reproduced from Chamley-Campbell, J. H., Nestel, P., and Campbell, G. R., in *Factors in Formation and Regression of the Atherosclerotic Plaque,* Born, G. R. V., Catapano, A. L., and Paoletti, R., Eds., Plenum, New York, 115, 1982. With permission.)

as measured by uptake of inorganic sulfur-35 (Table 1). These increases are not related to cell proliferation since in these experiments the synthetic state cells were maintained in a quiescent growth state.

## E. Enzyme Alterations

Concomitant with the increased amount of synthetic organelles in the synthetic state smooth muscle cells, the specific activity of enzymes associated with mitochondria and rough ER increase.[58] Cytochrome C oxidase and NADPH-dependent cytochrome C reductase increase 2.3- and 12.8-fold, respectively. Increases of 3 to 4-fold in activity of the lysosomal enzymes acid phosphatase and N-acetyl-$\beta$-glucosaminidase occur consistent with the larger numbers of lysosomes present in synthetic state cells. However, acid cholesteryl esterase, which is also found in lysosomes and is responsible for the hydrolysis of cholesteryl esters, decreases slightly but not significantly upon alteration to the synthetic state.

## F. Reversibility of Phenotypic Change

Smooth muscle phenotypic change is reversible or irreversible depending on the initial cell seeding concentration[148] (Figure 20). If the cells are seeded at 10⁶ cells/m*l* or greater, they form a confluent monolayer from day 1 in culture. These cells do not undergo a spontaneous phenotypic change but remain in the contractile phenotype. If they are seeded at less than 10⁶ cells/m*l*, i.e., at 5 × 10⁴ → 1 × 10⁵ cells/m*l*, they undergo the spontaneous phenotypic change on days 6 to 8, then proliferate until they form a confluent monolayer. If the cells undergo fewer than 5 cell doublings in 5 to 7

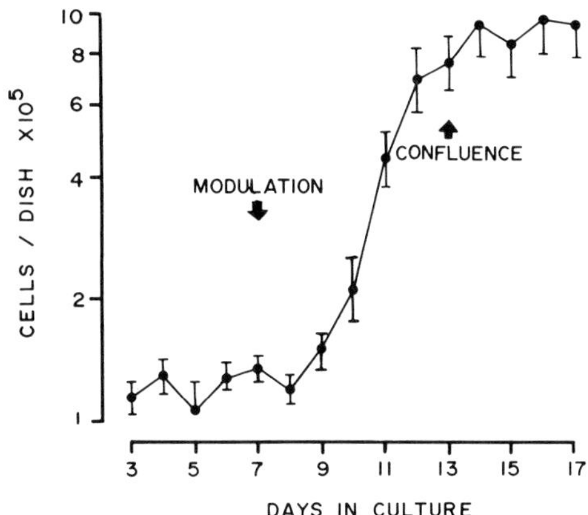


FIGURE 19.   Growth curve in the presence of 5% whole blood serum. Each point represents the mean of 3 dishes ± SD. (Reproduced from Chamley-Campbell, J. H., Nestel, P., and Campbell, G. R., in *Factors in Formation and Regression of the Atherosclerotic Plaque,* Born, G. R. V., Catapano, A. L., and Paoletti, R., Eds., Plenum, New York, 115, 1982. With permission.)

days to form the confluent monolayer, they will revert back to the contractile state 1 to 2 days after confluency is reached.

If, however, they have been seeded sparsely, that is $1 \times 10^3 \rightarrow 5 \times 10^3/m\ell$, the cells undergo the phenotypic change on days 6 to 8, but undergo considerably greater than 5 cell doublings to achieve confluency over 3 weeks. These cells cannot return to the contractile phenotype, but remain permanently in the synthetic phenotype and immediately respond to serum mitogens upon subculture. Smooth muscle cells which are in the synthetic state but are capable of returning to the contractile state are termed "reversible synthetic". Cells which are incapable of returning to the contractile state are termed "irreversible synthetic". The loss of reversion of phenotype is not related to "senescence" in late passage smooth muscle,[162] since the primary cultured cells have undergone many fewer cell doublings and they are still capable of intense proliferation. However, "irreversible synthetic" smooth muscle cells after further cell doublings eventually become senescent and will not proliferate in response to serum mitogens.

## G. Lipoprotein Metabolism

Cultured smooth muscle cells in the synthetic state have an altered ability to metabolize lipoproteins.[58,161,163] Irreversible synthetic state cells bind and degrade considerably more cholesterol ester-rich very low density lipoprotein (CER-VLDL) from hypercholesterolemic rabbit serum than reversible synthetic cells, which in turn have a significantly greater interaction with the lipoprotein than contractile state cells. This is also reflected in their respective rates of cholesteryl ester synthesis and lipid accumulation (Figures 21, 22, and 23).

## IV. CHEMICAL FACTORS WHICH INFLUENCE SMOOTH MUSCLE PHENOTYPE AND THE HYPERTROPHIC RESPONSE

### A. Chalones

It has been known for some time that cell proliferation can be regulated, at least in part, by means of negative feedback, i.e., an inhibitory effect exerted on cell division within a population by the population itself.[164-166] In such a system the basic assump-

Table 1

COLLAGEN AND NONCOLLAGEN PROTEIN SYNTHESIS AND
GAG SYNTHESIS

|                     | Contractile SMC   | Synthetic SMC     | Syn/Con × 100% |
|---------------------|-------------------|-------------------|----------------|
| Collagen            | 18,304 ± 924      | 81,690 ± 2,463    | 446%           |
| Noncollagen protein | 83,058 ± 2,671    | 150,184 ± 16,102  | 181%           |
| Sulfated GAG        | 3,258 ± 397       | 17,240 ± 3,010    | 529%           |

*Note:* Collagen and noncollagen protein synthesis as measured by incorporation of $^3$H-pro-
line (dpm/$10^6$ cells), and GAG synthesis as measured by incorporation of $^{35}$S (inorganic salt) (dpm/$10^6$ cells) during a 4 hr incubation.

tion is that a chemical signal is produced by the cells of the tissue, enabling the mass of the tissue to be gauged and cell proliferation regulated accordingly. The term given to this chemical signal is "chalone".[165] Chalones are tissue-specific, species nonspecific mitotic inhibitors, and have been found in every tissue examined.[167] They are proteins of two distinct molecular weight classes (2,000 to 6,000 daltons and 20,000 to 50,000 daltons) and act at two points in the cell cycle: at the $G_2$ phase inhibiting division, and at the $G_0/G_1$ serving as a switch between the cell cycle and differentiation.

The spontaneous phenotypic change from the contractile to synthetic phenotype of sparsely-seeded aortic smooth muscle cells in primary culture is prevented if the cells are grown in co-culture with a confluent monolayer of contractile state smooth muscle or endothelial cells.[148,168] The two cell layers do not have to be in contact, just bathed by the same nutrient medium as in modified Rose chambers where the test and feeder cell layers are grown on opposite coverslips.[148] Co-culture with confluent adventitial fibroblasts or subcultured synthetic state smooth muscle cells, or subconfluent prolif-erating endothelial cells does not prevent the spontaneous phenotypic change of con-tractile smooth muscle. The approximate molecular weight of the inhibitory substance or substances is 10,000 to 15,000 daltons as determined by inserting Spectrapor wet cellulose dialysis membranes of different molecular weight cutoff between feeder and test layers.[168] Alcohol precipitates of about 12,000 daltons from aqueous extracts of aorta inhibit smooth muscle phenotypic change.[148] Treatment of the aortic extracts with 4 units/ml heparinase from Flavobacterium heparinum destroys the active factor, suggesting that it is heparin or a related GAG (Table 2). Cultured endothelial cells secrete about 50% of their GAG as a highly sulfated (heparinlike) heparan sul-fate,[169-171] and exogenous addition of sodium heparin to culture medium inhibits smooth muscle phenotypic change[148] (Table 2).

In 1936, Simms and Stillman[172] discovered that the aorta contains growth inhibitory substances. Florentin et al.[173] showed that alcohol-precipitable components of an aqueous extract of pig aorta injected intraperitoneally into 8-week-old living swine, in quantities as low as 500 $\mu$g of protein, drastically reduces the rate of entry of arterial smooth muscle cells into mitosis, with no demonstrable effect on epidermal mitosis. Eisenstein et al.[174] reported two classes of molecules in fractions of bovine aortic ex-tracts which inhibit cell growth, one of low molecular weight which inhibits endothelial cells, and the other composed of structural polyanions, including heparin and a pro-teoglycan, which inhibits the growth of both smooth muscle and endothelial cells.

Alcohol precipitates from aqueous extracts of the pig intima plus media can be frac-tionated to produce both stimulators and inhibitors of synthetic state smooth muscle in culture.[148] Feeder layers of confluent endothelial cells or confluent contractile smooth muscle cells inhibit the proliferation of test synthetic state smooth muscle stim-ulated with 5% whole blood serum, while proliferating endothelial cells stimulate the

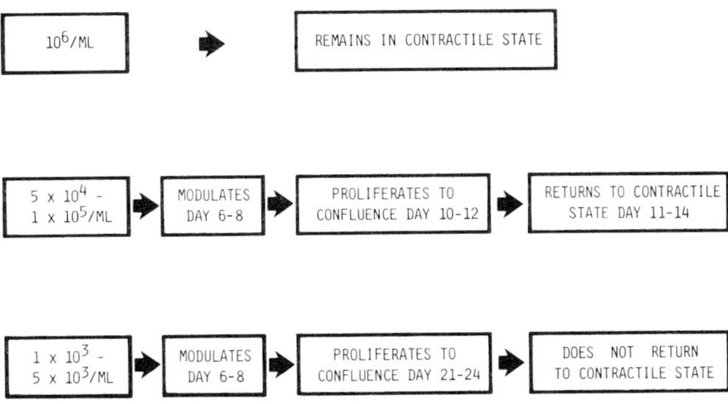



FIGURE 20.  Flow chart representing the cell seeding concentrations under which smooth muscle phenotypic change is reversible or irreversible.

proliferation of synthetic state smooth muscle, and proliferating synthetic state smooth muscle has no effect.[148,168] Similarly, conditioned medium containing 20% whole blood serum from confluent endothelial cells inhibits the proliferation of subcultured smooth muscle cells, while conditioned medium from proliferating endothelial cells and from proliferating or confluent subcultured smooth muscle cells and fibroblasts do not inhibit smooth muscle cell growth.[175] The inhibitory activity is heat stable and not affected by proteases. It is sensitive to Flavobacterium heparinase but not to hyaluronidase or chondroitin sulfate ABC lyase. A crude isolate of GAG (TCA-soluble, alcohol-precipitable) from endothelial cell-conditioned medium reconstituted in 20% serum inhibits smooth muscle cell growth, but no inhibition is seen if the GAG preparation is treated with heparinase. Exogenous heparin at 10 $\mu$g/m$\ell$ inhibits growth, while most other GAG have no effect.[148,175-179] Anticoagulant and nonanticoagulant heparin are equally effective in inhibiting smooth muscle cell growth, as they are in vivo in preventing the formation of myointimal thickening following endothelial denudation.[180,181]

In summary, confluent quiescent monolayers of endothelial cells produce heparinlike substances which appear to have two effects: (1) they maintain smooth muscle cells in the contractile phenotype in which state they are unresponsive to serum and mitogens; and (2) they inhibit the proliferation of serum-stimulated synthetic state smooth muscle cells. In addition, endothelial cells stimulate marked increases in the production of GAG by smooth muscle cells themselves.[79]

On the basis of these cell culture studies it has been suggested that the artery wall itself, by production of a substance (possibly a heparinlike GAG), controls the phenotype of arterial smooth muscle.[115,116,148] That is, in the body of the media the concentration of the substance is presumed to be high, since it is produced by large numbers of contractile state cells, and acts to maintain the cells in this phenotype. In the intima and region of the media near the internal elastic lamina, the smooth muscle cells are maintained in the contractile phenotype by the presence of the substance secreted by the undamaged endothelium. Loss or damage to the endothelium causes the cells to modulate to the synthetic state over a period of several days and thus become susceptible to the action of mitogens, such as the platelet-derived growth factor (PDGF) or the endothelial-derived growth factor (EDGF). This is in line with the studies of Reidy and Schwartz[182] who have shown that there is a critical lesion size of endothelial denudation below which smooth muscle proliferation will not occur. That is, myointimal thickening occurs only in those areas requiring more than 7 days for endothelial regeneration.

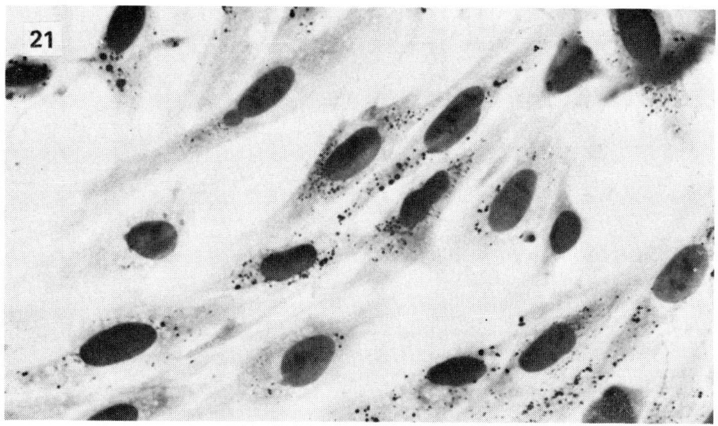

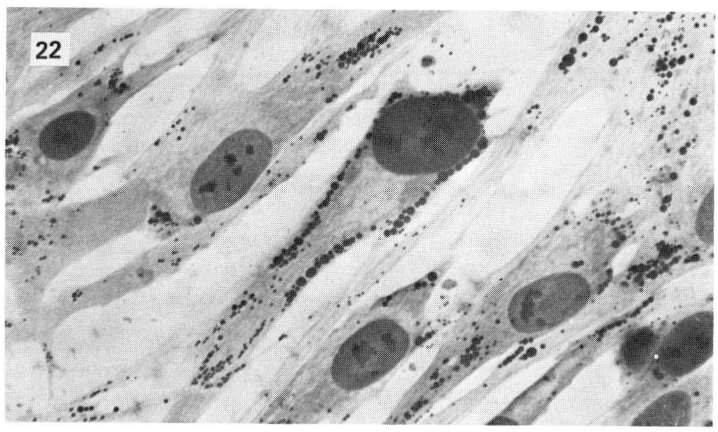

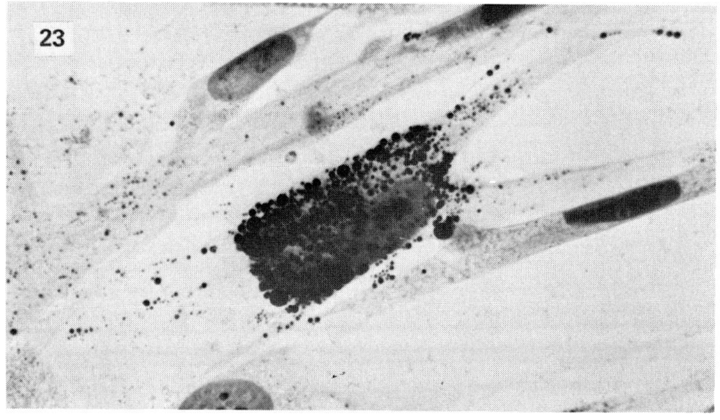

FIGURES 21 to 23.    Figure 21 illustrates contractile state aortic smooth muscle cells. Figure 22 shows reversible synthetic state aortic smooth muscle cells. Figure 23 shows irreversible synthetic state aortic smooth muscle cells. Cells are incubated for 17 hr in 5% lipoprotein deficient serum, then for 24 hr in 50 $\mu$g/m$l$ CER-VLDL, then stained with Fettrot/haematoxylin. The Fettrot stains lipid droplets within the cells. ($\times$ 700).

Table 2
## EFFECT OF GAG EXTRACT, GAG EXTRACT TREATED WITH HEPARINASE, AND SODIUM HEPARIN AT 50 $\mu$g/m$l$ ON SMOOTH MUSCLE PHENOTYPE AND PROLIFERATION

| In culture medium | Influence on phenotypic change | No. SMC $\times 10^5 \pm$ SD |
| --- | --- | --- |
| — (Control) | — | 22.2 ± 4.6 |
| GAG extract | inhibit | 2.9 ± 0.7 |
| GAG extract treated with heparinase | — | 20.6 ± 2.7 |
| Sodium heparin | inhibit | 3.2 ± 0.4 |

*Note:* The smooth muscle cells were seeded at 2.8 ± 0.3 × 10$^5$, and phenotype and cell number determined on day 11.

The demonstration of a heparan sulfate degrading endoglycosidase in platelets led to the suggestion that following endothelial denudation, this enzyme may modify extracellular heparan sulfate within the vascular system.[170,183] Incubation of endothelial cells with a crude platelet enzyme preparation or with a partially purified preparation of heparan sulfate degrading endoglycosidase from the same source, causes a marked release of cell-surface heparan sulfate which appears in the incubation medium as oligosaccharides. The platelet endoglycosidase (heparitinase) scissions heparinlike species at the relatively rare glucuronsyl-glucosamine bonds outside the highly sulfated regions.[184] Castellot et al.[176] proposed that following endothelial denudation or injury, platelets are deposited on the subendothelium releasing heparitinase, platelet factor 4, and PDGF. The heparitinase cleaves heparinlike substances on the endothelium and smooth muscle into fragments which diffuse away, with additional inactivation of the substance by platelet factor 4.[185] With the inhibitory effect of the heparinlike substance removed, PDGF acts on the smooth muscle cells to stimulate their proliferation.

## B. Trophic Factors from Nerves

An apparent trophic influence of sympathetic nerves on functional changes of smooth muscle cells in the caudal artery of the SHR has been reported.[186,187] These experiments suggest that the sympathetic nervous system is capable of causing altered smooth muscle membrane properties in arteries of normotensive rats when translated into the anterior eye chamber of the SHR and reinnervated by nerves supplying the iris. Although it has long been recognized that motor nerves can exert a trophic influence on skeletal muscle, only recently has it been shown that sympathetic nerves can exert a trophic influence on the structure and function of smooth muscle. Denervation of the expansor secundariorum muscle of the 2 week and adult chicken, by sectioning the brachial plexus, results in an approximate 2-fold increase in dry weight over 8 weeks which is due to hyperplasia of the smooth muscle cells.[188] There is also an approximate 20% increase in wall thickness of the radial and ulnar arteries at the same time in the denervated wing[189] (Figures 24 and 25). Denervation of rat and rabbit aortae produces a significant rise in collagen synthesis and an enzymatic shift from aerobic to anaerobic metabolism.[190,191] As well, denervated rabbit aortae exposed to increased serum cholesterol levels accumulate more lipid than control vessels.[190]

These in vivo denervation findings are consistent with cell culture studies where the presence of sympathetic nerve fibers in association with single isolated smooth muscle cells inhibits their phenotypic modulation to a synthetic state and subsequent proliferation in response to serum factors.[143] The addition of sympathetic chain homogenate has a similar inhibitory effect, demonstrating that the trophic effect is elicited by a

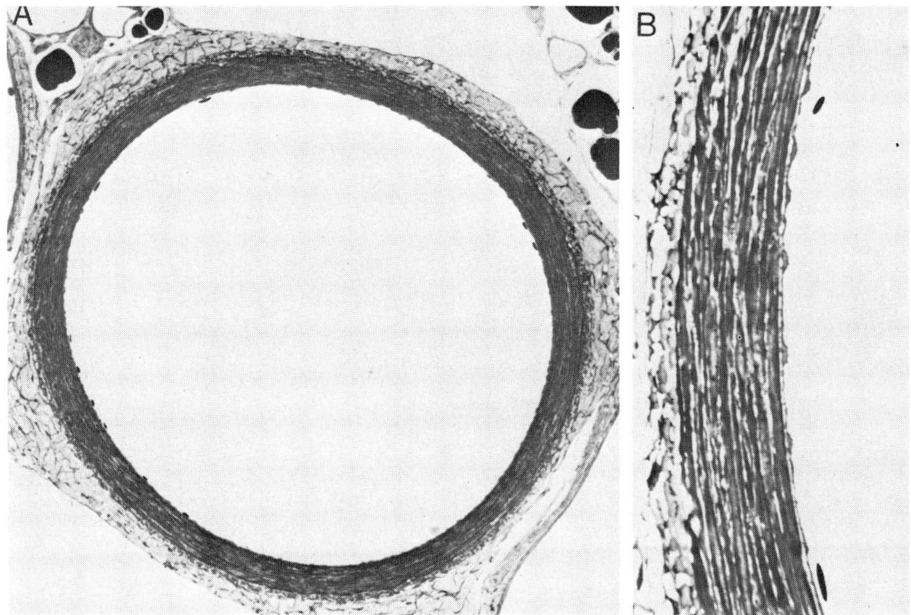

FIGURE 24.   (A) and (B) Transverse plastic section of perfusion fixed radial artery from 10-week-old chicken, stained with methylene blue. (A) Whole transverse section (× 40). (B) Higher magnification of area (× 160) from A. (Micrograph courtesy of Lynne Hartley).

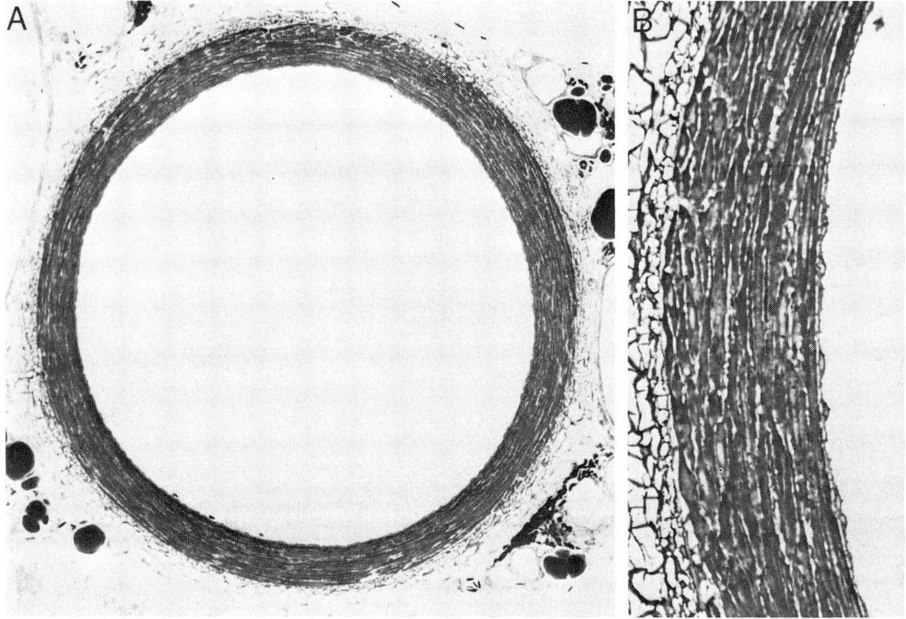

FIGURE 25.   (A) and (B) Transverse section of denervated radial artery from a 10-week-old chicken (denervated by brachial section for 8 weeks). (A) Whole transverse section (× 40). (B) Higher magnification of area (× 160) from A. (Micrograph courtesy of Lynne Hartley).

chemical substance. This effect is mimicked by the addition of dibutyryl cyclic AMP and theophylline to the culture medium,[151] while the addition of the neurotransmitters acetylcholine or noradrenaline has little or no effect. Sympathetic nerves in contact with single, isolated smooth muscle cells in culture also stimulate their association into muscle bundles and their rate of spontaneous contraction.

Denervation of the rabbit ear artery has the opposite effect to that of chick radial and ulnar arteries and expansor secundariorum muscle, and rat and rabbit aortae. Six weeks after unilateral postganglionic sympathetic denervation, the ear artery is lighter, thinner, and stiffer than its control. Although super-sensitive to noradrenaline, the denervated vessel develops a lower maximum contractile force which is reduced proportionately more than tissue weight.[192] Explanations for this heterogeneity of responses to denervation are discussed in Chapter 6 (this volume).

## C. Growth Factors
### 1. Platelet-Derived Growth Factor (PDGF)

PDGF is a basic polypeptide of approximately 30,000 daltons present in the $\alpha$-granules of platelets.[193-197] It is the principal growth factor in whole blood serum and stimulates the proliferation of mesenchymal cells in culture including smooth muscle.[198] Quiescent cells exposed even briefly to PDGF become competent to replicate their DNA and to divide, but do not progress through the $G_0/G_1$ phase of the cell cycle into S unless they are exposed continuously to progression factors in plasma.[199,200] Competence to divide reflects the intracellular accumulation of relatively stable PDGF regulated gene products.[201] The target specificity of PDGF for mesenchymal cells and in vivo studies demonstrating an association between platelet function and the smooth muscle proliferative response to endothelial denudation,[202-204] support the hypothesis that PDGF functions in the maintenance of vascular integrity and in wound repair in general.[205]

Recently it has been shown that there is considerable amino acid sequence similarity between PDGF and the transforming protein $P^{28}$ sis of simian sarcoma virus (SSV), suggesting that expression of growth factor activity may be central for transformation by SSV.[206-210]

### 2. Endothelial-Derived Growth Factor (EDGF)

Endothelial cells produce a mitogen for cultured smooth muscle, the endothelial-derived growth factor (EDGF).[211] It is a polypeptide of molecular weight between 10,000 and 30,000 daltons and is unrelated to PDGF.[212]

### 3. Monocyte/Macrophage-Derived Growth Factor (MDGF)

Mononuclear phagocytes secrete a growth-promoting activity that acts on several nonlymphoid mesenchymal cell types in culture including smooth muscle.[213,214] The growth stimulator, MDGF, is a heat labile, nondialysable protein which contains at least one essential disulfide bond. It is independent of PDGF and serum factors.

### 4. Lipoproteins

Hyperlipemic serum from animals fed a high cholesterol diet stimulates the proliferation of smooth muscle cells in culture to a significantly greater extent than the same concentration of normal serum.[147,215-220] The mitogenic lipoprotein appears to be a large LDL subspecies (molecular weight $> 3.0 \times 10^6$, and buoyant density $< 1.030$ g/m$l$), which occurs both in normolipemic and hyperlipemic serum although in different amounts.[221] Smooth muscle cells in hyperlipemic serum are larger in size and have a 50% increase in protein per cell.[217] Hyperlipemic serum also stimulates the synthesis of macromolecular components of extracellular matrix by the smooth muscle cells.[218-220]

### 5. Catecholamines

Adrenaline, noradrenaline, and to some extent isoproterenol stimulate the proliferation of subcultured vascular smooth muscle cells.[222] The presence of serum is essential for the growth stimulatory effect showing that the catecholamines act synergistically with serum factors. The order of potency of the adrenoceptor agonists and the nearly complete block of adrenaline by phentolamine, suggests that the catecholamines may act through adrenergic receptors.

## D. Hormones
### 1. Insulin

In the presence of 1% whole blood serum there is a linear relationship between the logarithm of insulin dose and growth stimulation of smooth muscle cells in culture.[223-225] The effect is small and can be inhibited by dibutyryl cyclic AMP. In the absence of serum, insulin has no effect, indicating that it acts as a co-factor for growth.[226]

### 2. Growth Hormone

Diabetic rabbit serum stimulates the proliferation of vascular smooth muscle cells in culture.[227-229] The effect is not due to the glucose content of the serum, nor to the amount of lipid, but to elevated levels of growth hormone in diabetic serum.[230]

### 3. Glucocorticoids and Mineralocorticoids

Mineralocorticoids have a direct hypertrophic effect on the artery wall during the prehypertensive and hypertensive states by increasing connective tissue synthesis.[231] In culture, the glucocorticoids cortisol, dexamethasone, and testosterone also act to stimulate the synthesis and secretion of collagen and noncollagen protein.[232] Testosterone inhibits prostacyclin production by these cells while estradiol stimulates prostacyclin synthesis.[233] Dexamethasone, corticosterone, and cortisol inhibit the proliferation of cultured vascular smooth muscle by selectively inhibiting the cells' responsiveness to serum mitogens.[234]

### 4. Steroid Hormones

A number of ultrastructural studies have been made on the effect of estrogen and progesterone on smooth muscle of the urogenital system.[82-86,235,236] After a single dose of estradiol diproprionate, smooth muscle of the rat myometrium shows an increase in the number of cellular organelles and assumes after 72 hr the morphology of synthetic state cells. A similar modification has been observed in uterine arteries of pregnant guinea pigs[87] and estrogen-treated rabbits.[91]

The sensitivity of the myometrium to chemical agents alters during pregnancy and under the influence of hormones. As gestation proceeds, the myometrium of many species becomes more sensitive to oxytocin. In the rat immediately before parturition, oxytocin is 1,000 times more effective than at mid-term.[237] Oxytocin produces depolarization and increased frequency of bursts. The sensitivity is increased by estradiol and decreased by progesterone.

Aortic smooth muscle cells possess specific, high affinity, and limited capacity binding sites for estradiol.[238] Addition of estradiol (0.02 $\mu$g/m$l$) to culture medium inhibits the proliferative effect of hyperlipemic serum on vascular smooth muscle and results in a decrease of stainable collagen and elastin in these cultures.[239] Estradiol increases the degradation of both collagen and elastin in the aorta of rats in vivo[240] and prevents the exaggerated synthesis of collagen and elastin in the aorta of hypertensive rats.[241] Estrogen also decreases the intimal hypertrophy usually complicating vessel transsection and reanastomosis.[242] However, pooled serum from women taking a combined estrogen-progestin oral contraceptive preparation stimulates the proliferation of vascular smooth muscle cells in culture.[243] In vitro addition of varying concentrations of mestranol and norethindone, the two constituents of the preparation, to normal serum does not enhance its mitogenicity, suggesting that the effect is due either with metabolite of these compounds or to the production of other growth-promoting substances during oral contraceptive treatment.

### 5. Vasoactive Hormones

The vasopressor hormones angiotensin II and vasopressin both stimulate the proliferation of arterial smooth muscle cells in culture and induce cellular hypertrophy.[244] This may be of significance in the vessel wall hypertrophy in hypertension.

### 6. Prostaglandins

Prostaglandins are local hormones acting at or near their site of origin. Smooth muscle cells in culture synthesize prostaglandin $E_1$ and $E_2$.[245,246] Their synthesis is inhibited by indomethacin and stimulated by histamine, angiotensin II, and bradykinin. Both $PGE_1$ and $PGE_2$ and their precursor fatty acids 8,11,14-eicosatrienoic acid ($C_{20:3}$) and 5,8,11,14-eicosatetraenoic acid ($C_{20:4}$) have potent inhibitory effects on smooth muscle proliferation.

### 7. Dibutyryl Cyclic AMP

While cyclic AMP is not a hormone, many hormones act through the adenylate cyclase system. Dibutyryl cyclic AMP ($N^6,O^6$-dibutyryl adenosine 3′:5′-cyclic monophosphoric acid) or theophylline (a cyclic AMP phosphodiesterase inhibitor) added to culture medium maintains sparsely-seeded smooth muscle cells in the contractile phenotype.[151] After 8 to 10 days in primary culture, the cytoplasm is almost completely occupied by thick and thin myofilaments with associated dense bodies (Figure 26). Plasmalemmal vesicles are numerous, as are dense areas along the cell. Organelles such as free ribosomes, mitochondria, and Golgi are few and usually are confined to the nuclear region of the cell, characteristic of mature smooth muscle. Cyclic AMP promotes differentiation in many other cell types such as brain cells[247] and skeletal muscle.[248]

## V. CONCLUDING REMARKS

There are at least four different ways for invoking hypertrophy of smooth muscle. These are (1) as a response to injury, (2) as a response to increased pressure, (3) neural, and (4) hormone induced. In this chapter we have stressed that a common feature of all these processes is modulation of smooth muscle phenotype expressed both functionally and structurally. Once in the synthetic phenotype the cells become responsive to chemical factors which stimulate growth and/or extracellular matrix synthesis.

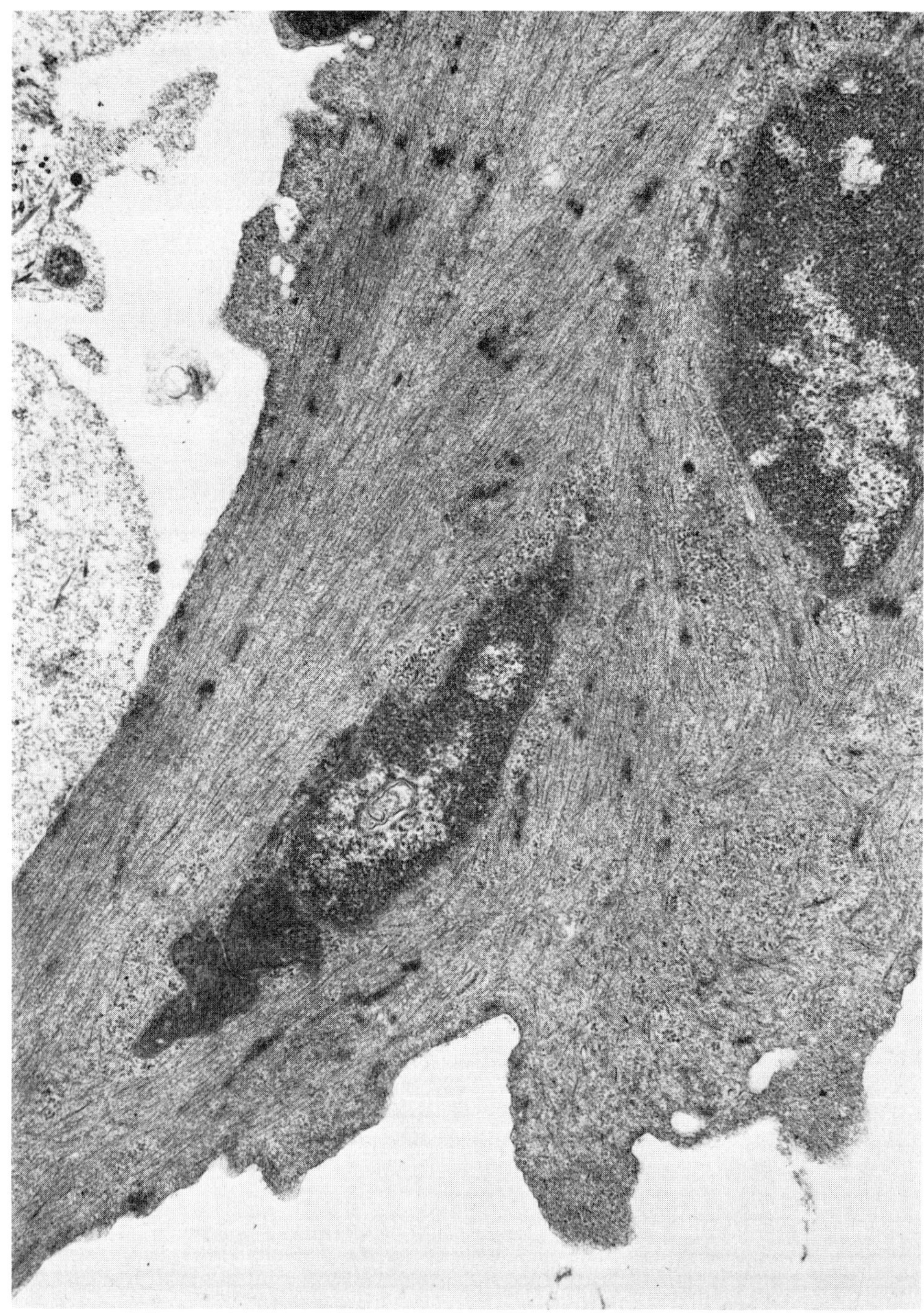

FIGURE 26.   Isolated smooth muscle cell after 8 days in cell culture in the presence of medium with added dibutyryl cyclic AMP + theophylline. The cytoplasm is almost completely filled with myofilaments. (× 15,000). (Reproduced from Chamley, J. H. and Campbell, G. R., *Cell Tiss. Res.,* 161, 497, 1975. With permission.)

# REFERENCES


1. Gabella, G., Structural apparatus for force transmission in smooth muscles, *Physiol. Rev.,* 64, 455, 1984.
2. Campbell, G. R., Chamley-Campbell, J. H., and Burnstock, G., Differentiation and phenotypic modulation of arterial smooth muscle cells, Differentiation and phenotypic modulation of arterial smooth muscle cells, in *Structure and Function of the Circulation,* Vol. 3, Schwartz, C. J., Werthessen, N. T., And Wolf, S., Eds., Plenum, New York, 1981, 357.
3. Pease, D. C. and Paule, W. J., Electron microscopy of elastic arteries. The thoracic aorta of the rat, *J. Ultrastruct. Res.,* 3, 469, 1960.
4. Karrer, H. E., An Electron microscope study of the aorta in young and aging mice, *J. Ultrastruct. Res.,* 5, 1, 1961.
5. Paule, W. J., Electron microscopy of the newborn rat aorta, *J. Ultrastruct. Res.,* 8, 219, 1963.
6. Gerrity, R. G. and Cliff, W. J., The aortic turnica media of the developing rat. I. Quantitative sterologic and biochemical analysis, *Lab. Invest.,* 32, 585, 1975.
7. Olivetti, G., Anversa, P., Melissari, M., and Loud, A. V., Morphometry of medial hypertrophy in the rat thoracic aorta, *Lab. Invest.,* 42, 559, 1980.
8. Yamamoto, I., An electron microscope study on development of uterine smooth muscle, *J. Electron Micros. Tokyo,* 10, 145, 1961.
9. Leeson, T. S. and Leeson, C. R., The rat ureter. Fine structural changes during its development, *Acta Anat.,* 62, 60, 1965.
10. Bennett, T. and Cobb, J. L. S., Studies on the avian gizzard: the development of the gizzard and its innervation, *Z. Zellforsch. Mikrosk. Anat.,* 98, 599, 1969.
11. Imaizumi, M. and Kuwabara, T., Development of the rat iris, *Invest. Ophthalmol.,* 10, 733, 1971.
12. Hoyes, A. D., Ramus, N. I., and Martin, B. G. H., Differentiation of the muscle of the human fetal bladder: an ultrastructural study, *Micron.,* 3, 414, 1972.
13. Yamauchi, A. and Burnstock, G., Post-natal development of smooth muscle cells in the mouse vas deferens. A fine structural study, *J. Anat.,* 104, 1, 1969.
14. Sethi, N., and Brookes, M., Ultrastructure of the blood vessels in the chick allantois and chorioallantois, *J. Anat.,* 109, 1, 1971.
15. Manasek, F. J., The ultrastructure of embryonic myocardial blood vessels, *Dev. Biol.,* 26, 42, 1971.
16. DeSimone-Santoro, I. and Renda, T., Recherches ultrastructurales, histochimiques et histoenzymologiques sur la paroi aortique pendant sa morphagénèse chez l'embryon de poulet, *Ann. Histochim.,* 16, 171, 1971.
17. Blatt, H. J., Über die Entwichlung der Coronararterien bei der Ratte Licht-und electronmikroskopische Untersuchungen, *Z. Anat. Entwicklungsgesch,* 142, 53, 1973.
18. Gonzalez-Crussi, F., Vasculogenesis in the chick embryo. An ultrastructural study, *Am. J. Anat.,* 130, 441, 1971.
19. Wolinsky, H. and Glagov, S., A lamellar unit of aortic medial structure and function in mammals, *Circ. Res.,* 20, 99, 1967.
20. Kocher, O., Skalli, O., Cerutti, D., Gabbiani, F., and Gabbiani, G., Cytoskeletal features of rat aortic cells during development: An electron microscopic, immunohistochemical and biochemical study, *Circ. Res.,* 56, 829, 1985.
21. Campbell, G. R., Uehara, Y., Malmfors, T., and Burnstock, G., Degeneration and regeneration of smooth muscle transplants in the anterior eye chamber. An ultrastructural study, *Z. Zellforsch. Mikrosk. Anat.,* 117, 155, 1971.
22. Kiviat, M. D., Ross, R., and Ansell, J. S., Smooth muscle regeneration in the ureter. Electron microscopic and autoradiographic observations, *Am. J. Pathol.,* 72, 403, 1973.
23. McGeachie, J. K., Smooth muscle regeneration. A review and experimental study, *Monogr. Dev. Biol.,* 9, 1, 1975.
24. Dalley, B. K., Bartone, F. F., and Gardner, P. J., Smooth muscle regeneration in swine ureters. A light and electron microscopic study, *Invest. Urol.,* 14, 104, 1976.
25. Yokota, R., Burnstock, G., Jones, R., and Brown, M. D., An ultrastructural study of regeneration of minced smooth muscle in the vas deferens of the guinea-pig, *Cell Tiss. Res.,* 190, 481, 1978.
26. Florentin, R. A., Nam, S. C., Lee, K. T., and Thomas, W. A., Increased mitotic activity in aortas of swine after 3 days of cholesterol feeding, *Arch. Pathol.,* 88, 463, 1969.
27. Cobb, J. L. S. and Bennett, T., An ultrastructural study of mitotic division in differentiated gastric smooth muscle cells, *Z. Zellforsch. Mikrosk. Anat.,* 108, 177, 1970.
28. Scott, R. F., Jarmolych, J., Fritz, K. E., Imai, H., Kim, D. N., and Morrison, E. S., Reactions of endothelial and smooth muscle cells in the atherosclerotic lesion, in *Proc. 2nd. Int. Symp. Atherosclerosis,* Jones, R. J., Ed., Springer Verlag, Berlin, 1970, 50.

29. Chamley, J. H. and Campbell, G. R., Mitosis of contractile smooth muscle cells in tissue culture, *Exp. Cell Res.*, 84, 105, 1974.
30. Gabella, G., Fine structure of smooth muscle. I. Cellular structures and electrophysiological behavior; fine structure of smooth muscle, *Philos. Trans. Roy. Soc. London Ser. B.*, 265, 7, 1973.
31. Chamley-Campbell, J., Campbell, G. R., and Ross, R., The smooth muscle cell in culture, *Physiol. Rev.*, 59, 1, 1979.
32. Thomas, W. A., Jones, R., Scott, R. F., Morrison, E., Goodale, F., and Imai, H., Production of early atherosclerotic lesions in rats is characterised by proliferation of "modified smooth muscle cells", *Exp. Mol. Pathol.*, 1(Suppl.), 40, 1963.
33. Scott, R. F., Jones, R., Daoud, A. S., Zumbo, O., Coulston, F., and Thomas, W. A., Experimental atherosclerosis in Rhesus monkeys. II. Cellular elements of proliferative lesions and possible role of cytoplasmic degeneration in pathogenesis as studied by electron microscopy, *Exp. Mol. Pathol.*, 7, 34, 1967.
34. Thomas, W. A., Florentin, R. A., Nam, S. C., Kim, D. N., Jones, R. M., and Lee, K. T., Preproliferative phase of atherosclerosis in swine fed cholesterol, *Arch. Pathol.*, 86, 621, 1968.
35. Imai, H., Lee, K. J., Lee, S. K., Lee, K. T., O'Neal, R. M., and Thomas, W. A., Ultrastructural features of aortic cells in mitosis in control and cholesterol-fed swine, *Lab. Invest.*, 23, 401, 1970.
36. Poole, J. C. F., Cromwell, S. B., and Benditt, E. P., Behavior of smooth muscle cells and formation of extracellular structures in the reaction of arterial walls to injury, *Am. J. Pathol.*, 62, 391, 1971.
37. Stary, H. C., Proliferation of arterial cells in atherosclerosis, *Adv. Exp. Med. Biol.*, 43, 59, 1974.
38. Glagov, S. and Ts'ao, C. H., Restitution of aortic wall after sustained necrotizing transmural ligation injury: Role of blood cells and artery cells, *Am. J. Pathol.*, 79, 7, 1975.
39. Kamio, A., Huang, W. Y., Imai, H., and Kummerow, F. A., Mitotic structures of aortic smooth muscle cells in swine and in culture: Paired cisternae, *J. Electron Microsc. (Tokyo)*, 26, 29, 1977.
40. Taura, M., Origin and fate of paired cisternae in mitotic aortic cells of swine, *J. Electron Microsc. (Tokyo)*, 27, 283, 1978.
41. Björkerud, S. and Bondjers, G., Arterial repair and atherosclerosis after mechanical injury. Part 5. Tissue response after induction of a large superficial transverse injury, *Atherosclerosis*, 18, 235, 1973.
42. Stemerman, M. B., Spaet, T. H., Pitlick, F., Cintron, J., Lejnieks, I., and Tiell, M. L., Intimal healing. The pattern of reendothelialization and intimal thickening, *Am. J. Pathol.*, 87, 125, 1977.
43. Ross, R., Smooth muscle cells and atherosclerosis, in *Vascular Injury and Atherosclerosis*, Moore, S., Ed., Marcel Dekker, New York, 1981, 53.
44. Veress, B., Kádár, A., and Jellinek, H., Ultrastructural elements in experimental intimal thickening I. Electron microscopic study of the development and cellular elements of intimal proliferation, *Exp. Mol. Pathol.*, 11, 200, 1969.
45. Baumgartner, H. R. and Studer, A., Consequences of vessel catheterization in normo- and hypercholesterolaemic rabbits, (German), *Pathol. Microbiol.*, 29, 393, 1966.
46. Stemerman, M. B. and Ross, R., Experimental arteriosclerosis I. Fibrous plaque formation in primates, an electron microscope study, *J. Exp. Med.*, 136, 769, 1972.
47. Schwartz, S. M., Stemerman, M. B., and Benditt, E. P., The aortic intima. II. Repair of the aortic lining after mechanical denudation, *Am. J. Pathol.*, 81, 15, 1975.
48. Knieriem, H-J., Bondjers, G., and Björkerud, S., Electron microscopy of intimal plaques following induction of large superficial mechanical injury (transverse injury) in the rabbit aorta, *Virchows Arch. A.*, 359, 267, 1973.
49. Fishman, J. A., Ryan, G. B., and Karnovsky, M. J., Endothelial regeneration in the rat carotid artery and the significance of endothelial denudation in the pathogenesis of myointimal thickening, *Lab. Invest.*, 32, 339, 1975.
50. Van Pelt-Verkuil, E., Van Pelt, W., and Jense, D., Morphometry of air-drying-induced arteriosclerosis in rat carotid artery. Effect of air-flow rate, *Atherosclerosis*, 3, 441, 1983.
51. Haudenschild, C. C. and Schwartz, S. M., Endothelial regeneration. II. Restitution of endothelial continuity, *Lab. Invest.*, 41, 407, 1979.
52. Gabbiani, G., Rungger-Brändle, E., De Chastonay, C., and Franke, W. W., Vimentin-containing smooth muscle cells in aortic intimal thickening after endothelial injury, *Lab. Invest.*, 47, 265, 1982.
53. Clowes, A. W., Reidy, M. A., and Clowes, M. M., Mechanisms of stenosis after arterial injury, *Lab. Invest.*, 49, 208, 1983.
54. Clowes, A. W. and Karnovsky, M. J., Suppression by heparin of smooth muscle cell proliferation in injured arteries, *Nature (London)*, 265, 625, 1977.
55. Taura, S., Taura, M., Imai, H., and Kummerow, F. A., Morphological alteration of aortic wall and mitotic cells after complete endothelial loss induced by repeated balloon denudation of swine aorta, *Tokoku J. Exp. Med.*, 129, 25, 1979.
56. Spaet, T. H., Stemerman, M. B., Veith, F. J., and Lejnieks, I., Intimal injury and regrowth in rabbit aorta. Medical smooth muscle cells as a source of neointima, *Circ. Res.*, 36, 58, 1975.

57. Guyton, J. R. and Karnovsky, M. J., Smooth muscle cell proliferation in the occluded rat carotid artery. Lack of requirement for luminal platelets, *Am. J. Pathol.,* 94, 585, 1979.
58. Campbell, J. H., Popadynec, L., Nestel, P. J., and Campbell, G. R., Lipid accumulation in arterial smooth muscle cells. Influence of phenotype, *Atherosclerosis,* 47, 279, 1983.
59. Larrue, J., Daret, D., Demond-Henri, J., Allieres, C., and Bricaud, H., Prostacyclin synthesis by proliferative aortic smooth muscle cells. A kinetic in vivo and in vitro study, *Atherosclerosis,* 50, 63, 1984.
60. Burns, E. R., Spaet, T. H., and Stemerman, M. B., Response of the arterial wall to endothelial removal: and autoradiographic study, *Proc. Soc. Exp. Biol. Med.,* 159, 473, 1978.
61. Hassler, O., The origin of the cells constituting arterial intima thickening: an experimental autoradiographic study with the use of $H^3$-thymidine, *Lab. Invest.,* 22, 286, 1970.
62. Webster, W. S., Bishop, S. P., and Geer, J. C., Experimental aortic intimal thickening. I. Morphology and source of intimal cells, *Am. J. Pathol.,* 76, 245, 1974.
63. Goldberg, I. D., Stemerman, M. B., Ransil, B. J., and Fuhro, R. L., In vivo aortic muscle cell growth kinetics: differences between thoracic and abdominal segments after intimal injury in the rabbit, *Circ. Res.,* 47, 182, 1980.
64. Gabbiani, G., Kocher, O., Bloom, W. S., Vandekerckhove, J., and Weber, K., Actin expression in smooth muscle cells of rat aortic intimal thickening, human atheromatous plaque, and cultured rat aortic media, *J. Clin. Invest.,* 73, 148, 1984.
65. Kocher, O., Skalli, O., Bloom, W. S., and Gabbiani, G., Cytoskeleton of rat aortic smooth muscle cells. Normal conditions and experimental intimal thickening, *Lab. Invest.,* 50, 645, 1984.
66. Campbell, G. R. and Campbell, J. H., Smooth Muscle Cells, in *The Biology and Clinical Science of Atherosclerosis,* European Atherosclerosis Group, Churchill Livingstone, Edinburgh, 1984.
67. Minick, C. R., Stemerman, M. B., and Insull, W., Jr., Effect of regenerated endothelium on lipid accumulation in the arterial wall, *Proc. Natl. Acad. Sci., U.S.A.,* 74, 1724, 1977.
68. Minick, C. R., Stemerman, M. B., and Insull, W., Jr., Role of endothelium and hypercholesterolaemia in intimal thickening and lipid accumulation, *Am. J. Pathol.,* 95, 131, 1979.
69. Falcone, D. J., Hajjar, D. P., and Minick, C. R., Enhancement of cholesterol and cholesteryl ester accumulation in reendothelialized aorta, *Am. J. Pathol.,* 99, 81, 1980.
70. Falcone, D. J., Hajjar, D. P., and Minick, C. R., Lipoprotein and albumin accumulation in reendothelialized and deendothelialized aorta, *Am. J. Pathol.,* 114, 112, 1984.
71. Hajjar, D. P., Falcone, D. J., Fowler, S., and Minick, C. R., Endothelium modifies the altered metabolism of the injured aortic wall, *Am. J. Pathol.,* 102, 28, 1981.
72. Rosenfeld, R. S., Paul, I., and Spaet, T. H., Long-term effects of deendothelialization of rabbit aorta: In vitro synthesis of DNA, protein and lipid, *Proc. Soc. Exp. Biol. Med.,* 173, 427, 1983.
73. Wight, T. N., Curwen, K. D., Litrenta, M. M., Alonso, D. R., and Minick, C. R., Effect of endothelium on glycosaminoglycan accumulation in injured rabbit aorta, *Am. J. Pathol.,* 113, 156, 1983.
74. Wight, T. N., Vessel prosteorglycans and thrombogenesis, in *Progress in Hemostasis and Thrombosis,* Spaet, T., Ed., Grune & Stratton, New York, 1980, 1.
75. Camejo, G., The interaction of lipids and lipoproteins with the intercellular matrix of arterial tissue: Its possible role in atherogenesis, *Adv. Lipid Res.,* 19, 1, 1982.
76. Camejo, G., Ponce, E., López, F., Starosta, R., Hurt, E., and Romano, M., Partial structure of the active moiety of a lipoprotein complexing proteoglycan from human aorta, *Atherosclerosis,* 49, 241, 1983.
77. Collatz-Christensen, B., Chemnitz, J., Tkocz, I., and Kim, C. M., Repair in arterial tissue: 2. Connective tissue changes following embolectomy catheter lesion: The importance of endothelial cells to repair and regeneation, *Acta Pathol. Microbiol. Scand. Sect. A,* 87, 275, 1979.
78. Richardson, M., Ihnatowycz, I., and Moore, S., Glycosaminoglycan distribution in rabbit aortic wall following balloon catheter deendothelialization. An ultrastructural study, *Lab. Invest.,* 43, 509, 1980.
79. Merrilees, M. J. and Scott, J., Interaction of aortic endothelial and smooth muscle cells in culture. Effect on glycosaminoglycan levels, *Atherosclerosis,* 39, 147, 1981.
80. Harkness, M. L. R. and Harkness, R. D., The collagen content of the reproductive tract of the rat during pregnancy and lactation, *J. Physiol. (London),* 123, 492, 1954.
81. Montfort, I. and Pérez-Tamayo, R., Studies on uterine collagen during pregnancy and puerperium, *Lab. Invest.,* 10, 1240, 1961.
82. Laguens, R. and Lagrutta, J., Fine structure of human uterine muscle in pregnancy, *Am. J. Obstet. Gynecol.,* 89, 1040, 1964.
83. Dessouky, D. A., Electron microscopic studies of the myometrium of the guinea pig. The smooth muscle cell of the myometrium before and during pregnancy, *Am. J. Obstet. Gynecol.,* 100, 30, 1968.
84. Ross, R. and Klebanoff, S. J., Fine structural changes in uterine smooth muscle and fibroblasts in response to estrogen, *J. Cell Biol.,* 32, 155, 1967.

85. Friederici, H. H. R. and De Cloux, R. J., The early response of immature rat myometrium to estrogenic stimulation, *J. Ultrastruct. Res.,* 22, 402, 1968.
86. Bo, W. J., Odor, D. L., and Rothrock, M., The fine structure of uterine smooth muscle of the rat uterus at various time intervals following a single injection of estrogen, *Am. J. Anat.,* 123, 369, 1968.
87. Albert, E. N. and Pease, D. C., An electron microscopic study of uterine arteries during pregnancy, *Am. J. Anat.,* 123, 165, 1968.
88. Albert, E., The effect of pregnancy on the elastic membranes of mesometrial arteries in the guinea-pig, *Am. J. Anat.,* 120, 611, 1967.
89. Albert, E. and Bhussry, B., The effects of multiple pregnancies and age on the elastic tissue of uterine arteries in the guinea-pig, *Am. J. Anat.,* 121, 259, 1967.
90. Gostimirovich, D., Estrogen effects on the aortic wall in young immature rabbits, *Virchows Arch. A.,* 343, 258, 1968.
91. Gostimirovich, D., Effects of different doses of estrogen on the aortic wall of young immature rabbits, *Virchows Arch. A.,* 349, 93, 1970.
92. Fischer, G. M. and Swain, M. L., Influence of contraceptive and other sex steroids on aortic collagen and elastin, *Exp. Mol. Pathol.,* 33, 15, 1980.
93. Ohyama, S., The response of arterial muscular coat to elevated blood pressure. Histometrical studies of renal artery in experimental hypertension of rats, *Tohoku J. Exp. Med.,* 80, 182, 1963.
94. Hollander, W., Kramsch, D. M., Farmelant, M., and Madoff, I. M., Arterial wall metabolism in experimental hypertension of coarctation of the aorta of short duration, *J. Clin. Invest.,* 47, 1221, 1968.
95. Wolinsky, H., Response of rat aortic media to hypertension. Morphological and chemical studies, *Circ. Res.,* 26, 507, 1970.
96. Ooshima, A., Fuller, G. C., Cardinale, G. J., Spector, S., and Udenfriend, S., Increased collagen synthesis in blood vessels of hypertensive rats and its reversal by antihypertensive agents, *Proc. Natl. Acad. Sci., U.S.A.,* 71, 3019, 1974.
97. Nissen, R., Cardinale, G. J., and Udenfriend, S., Increased turnover of arterial collagen in hypertensive rats, *Proc. Nat. Acad. Sci., U.S.A.,* 75, 451, 1978.
98. Wiener, J., Loud, A. V., Giacomelli, F., and Anversa, P., Morphometric analysis of hypertension-induced hypertrophy of rat thoracic aorta, *Am. J. Pathol.,* 88, 619, 1977.
99. Brecher, P., Chan, C. T., Franzblau, C., Faris, B., and Chobanian, A. V., Effects of hypertension and its reversal on aortic metabolism in the rat, *Circ. Res.,* 43, 561, 1978.
100. Tomita, T., Shirasaki, Y., Takiguchi, Y., Okada, T., and Hayashi, E., Aortic cholesterol esterase and other lysosomal enzyme activities in DOCA-salt, renal, and spontaneous hypertension in the rat, *Atherosclerosis,* 39, 453, 1981.
101. Crane, W. A. J. and Dutta, L. P., The utilisation of tritiated thymidine for deoxyribonucleic acid synthesis by the lesions of experimental hypertension in rats, *J. Pathol. Bact.,* 86, 83, 1963.
102. Owens, G. K., Rabinovitch, P. S., and Schwartz, S. M., Smooth muscle cell hypertrophy versus hyperplasmia in hypertension, *Proc. Natl. Acad. Sci., U.S.A.,* 78, 7759, 1981.
103. Owens, G. K. and Schwartz, S. M., Alternatives in vascular smooth muscle mass in the spontaneously hypertensive rat. Role in cellular hypertrophy, hyperploidy, and hyperplasia, *Circ. Res.,* 51, 280, 1982.
104. Owens, G. K. and Schwartz, S. M., Vascular smooth muscle cell hypertrophy and hyperploidy in the Goldblatt hypertensive rat, *Circ. Res.,* 53, 491, 1983.
105. Johansson, B., Structural and functional changes in rat portal veins after experimental portal hypertension, *Acta Physiol. Scand.,* 98, 381, 1976.
106. Berner, P. F., Somlyo, A. V., and Somlyo, A. P., Hypertrophy-induced increase of intermediate filaments in vascular smooth muscle, *J. Cell Biol.,* 88, 96, 1981.
107. Gabella, G., Hypertrophy of intestinal smooth muscle, *Cell Tiss. Res.,* 163, 199, 1975.
108. Gabella, G., Hypertrophic smooth muscle. I. Size and shape of cells, occurrence of mitoses, *Cell Tiss. Res.,* 201, 63, 1979.
109. Gabella, G. and Yamey, A., Synthesis of collagen by smooth muscle in the hypertrophic intestine, *Q. J. Exp. Physiol.,* 62, 257, 1977.
110. Gabella, G., Hypertrophic smooth muscle. V. Collagen and other extracellular materials. Vascularization, *Cell Tiss. Res.,* 235, 275, 1984.
111. Gabella, G., Hypertrophic smooth muscle. II. Sarcoplasmic reticulum, caveolae, and mitochondria, *Cell Tiss. Res.,* 201, 79, 1979.
112. Gabella, G., Hypertrophic smooth muscle. III. Increase in number and size of gap junctions, *Cell Tiss. Res.,* 201, 263, 1979.
113. Gabella, G., Hypertrophic smooth muscle. IV. Myofilaments, intermediate filaments and some mechanical properties, *Cell Tiss. Res.,* 201, 277, 1979.

114. Haust, M. D., Atherosclerosis and smooth muscle cells, in *Biochemistry of Smooth Muscle,* Stephens, N. L., Ed., CRC Press, Boca Raton, Fla., 1983, 189.
115. Campbell, G. R., Chamley-Campbell, J. H., Smooth muscle phenotypic modulation: role in athero-genesis, *Med. Hypoth.,* 7, 729, 1981.
116. Campbell, G. R. and Chamley-Campbell, J. H., Invited review: The cellular pathobiology of ather-osclerosis, *Pathology,* 13, 423, 1981.
117. Geer, J. C. and Haust, M. D., Smooth muscle cells in atherosclerosis, *Monographs on Atheroscle-rosis,* Vol. 2, Pollack, O. J., Simms, H. S., and Kirk, S. E., Eds., S. Karger, Basel, 1972.
118. McGill, H. C., Jr., The lesion, In *Atherosclerosis III, Proc. 3rd Int. Symp. Athersclerosis,* Schettler, G. and Weizel, A., Eds., Springer-Verlag, Berlin, 1973, 27.
119. Velican, C. and Velican, D., Intimal thickening in developing coronary arteries and its relevance to atherosclerotic involvement, *Atherosclerosis,* 23, 345, 1976.
120. Velican, C. and Velican, D., The precursors of coronary atherosclerotic plaques in subjects up to 40 years old, *Atherosclerosis,* 37, 33, 1980.
121. Orekhov, A. N., Karpova, I. I., Tertov, V. V., Rudchenko, S. A., Andreeva, E. R., Krushinsky, A. V., and Smirnov, V. N., Cellular composition of atherosclerotic and uninvolved human aortic subendothelial intima. Light-microscopic study of dissociated aortic cells, *Am. J. Pathol.,* 115, 17, 1984.
122. Mosse, P. R. L., Campbell, G. R., Wang, T. L., and Campbell, J. H., Smooth muscle phenotypic expression in human arteries. I. Comparison of cells from diffuse intimal thickenings of carotid arteries adjacent to atheromatous plaques with those of the media, *Lab. Invest.,* 53, 556, 1985.
123. Ehrhart, L. A. and Holderbaum, D., Stimulation of aortic protein synthesis in experimental rabbit atherosclerosis, *Atherosclerosis,* 27, 477, 1977.
124. Langner, R. O., Gilligan, J. P., and Ehrhart, L. A., The effect of cholesterol feeding on protein synthesis in different regions of the rabbit arterial wall, *Exp. Mol. Pathol.,* 31, 308, 1979.
125. Weigensberg, B. I., Lough, J., and More, R. H., Biochemistry of atherosclerosis produced by cho-lesterol feeding, thrombosis and injury, *Exp. Mol. Pathol.,* 37, 175, 1982.
126. Wangner, W. D. and Salisbury, B. G. J., Aortic total glycosaminoglycan and determata sulfate changes in atherosclerotic rhesus monkeys, *Lab. Invest.,* 39, 322, 1978.
127. Florentin, R. A., Nam, S. C., Daoud, A. S., Jones, R., Scott, R. F., Morrison, E. S., Kim, D. N., Lee, K. T., Thomas, W. A., Dodds, W. J., and Miller, K. D., Dietary-induced atherosclerosis in miniature swine, *Exp. Mol. Pathol.,* 8, 263, 1968.
128. Stary, H. C. and Malinow, M. R., Ultrastructure of experimental coronary artery atherosclerosis in Cynomolgus Macaques. A comparison with the lesions of other primates, *Atherosclerosis,* 43, 151, 1982.
129. Pietilä, K. and Nikkari, T., Enhanced growth of smooth muscle cells from atherosclerotic rabbit aortas in culture, *Atherosclerosis,* 36, 241, 1980.
130. Pietilä, K. and Nikkari, T., Enhanced synthesis of collagen and total protein by smooth muscle cells from athersclerotic rabbit aortas in culture, *Atherosclerosis,* 37, 11, 1980.
131. Doebler, J. A., Anthony, A., Bocan, T. M. A., and Hollis, T. M., Focal alterations in metabolic capabilities associated with development of cholesterol-induced atheromatous lesions: A quantitative cytochemical study, *Exp. Mol. Pathol.,* 36, 227, 1982.
132. Moore, S., Injury mechanisms in atherogensis, in *Vascular Injury and Atherosclerosis,* Moore, S., Ed., Marcel Dekker, New York, 1981, 131.
133. Schwartz, S. M., Vascular Integrity, in *Biologic and Synthetic Vascular Prostheses,* Stanley, J. C., Ed., Grune & Stratton, New York, 1982, 27.
134. Thomas, W. A. and Kim, D. N., Biology of disease. Atherosclerosis as a hyperplastic and/or neo-plastic process, *Lab. Invest.,* 48, 245, 1983.
135. Joris, I., Zand, T., Nunnari, J. J., Krolikowski, F. J., and Majno, G., Studies on the pathogenesis of atherosclerosis. I. Adhesion and emigration of mononuclear cells in the aorta of hypercholestero-lemic rats, *Am. J. Pathol.,* 113, 341, 1983.
136. Schwartz, S. M. and Ross, R., Cellular proliferation in atherosclerosis and hypertension, *Prog. Car-diovasc. Dis.,* 25, 355, 1984.
137. Cliff, W. J., The aortic tunica media in aging rats, *Exp. Mol. Pathol.,* 13, 172, 1970.
138. Kojimahara, M., Sekiya, K., and Ooneda, G., Age-induced changes of cerebral arteries in rats. An electron microscope study, *Virchows Arch. A.,* 361, 11, 1973.
139. Joris, I. and Majno, G., Cellular breakdown within the arterial wall. An ultrastructural study of the coronary artery in young and aging rats, *Virchows Arch. A.,* 364, 111, 1974.
140. Toda, T., Tsuda, N., Nishimori, I., Leszczynski, D. E., and Kummerow, F. A., Morphometrical analysis of the aging process in human arteries and aorta, *Acta Anat.,* 106, 35, 1980.

141. Martin, G., Ogburn, C. E., and Wight, T. N., Comparative rates of decline in the primary cloning efficiencies of smooth muscle cells from the aging thoracic aorta of two murine species of contrasting maximum life span potentials, *Am. J. Pathol.*, 110, 236, 1983.
142. Barrett, T. B., Sampson, P., Owens, G. K., Schwartz, S. M., and Benditt, E. P., Polyploid nuclei in human artery wall smooth muscle cells, *Proc. Natl. Acad. Sci., U.S.A.*, 80, 882, 1983.
143. Chamley, J. H., Campbell, G. R., and Burnstock, G., Dedifferentiation, redifferentiation, and bundle formation of smooth muscle cells in tissue culture: the influence of cell number and nerve fibres, *J. Embryol. Exp. Morphol.*, 32, 297, 1974.
144. Chamley, J. H., Campbell, G. R., McDonnell, J. D., and Gröschel-Stewart, U., Comparison of vascular smooth muscle cells from adult human, monkey, and rabbit in primary culture and in subculture, *Cell Tiss. Res.*, 177, 503, 1977.
145. Chamley, J. H. and Campbell, G. R., Isolated ureteral smooth muscle cells in culture. Including their interaction with intrinsic and extrinsic nerves, *Cytobiology*, 11, 358, 1975.
146. Thyberg, J., Palmberg, L., Nilsson, J., Ksiazek, T., and Sjölund, M., Phenotype modulation in primary cultures of arterial smooth muscle cells. On the role of platelet-derived growth factor, *Differentiation*, 25, 156, 1983.
147. Chamley-Campbell, J. H., Campbell, G. R., and Ross, R., Phenotype-dependent response of cultured aortic smooth muscle to serum mitogens, *J. Cell Biol.*, 89, 379, 1981.
148. Chamley-Campbell, J. H. and Campbell, G. R., What controls smooth muscle phenotype?, *Atherosclerosis*, 40, 347, 1981.
149. Chamley, J. H., Campbell, G. R., and Burnstock, G., An analysis of the interactions between sympathetic nerve fibres and smooth muscle cells in tissue culture, *Dev. Biol.*, 33, 344, 1973.
150. Mark, G. E., Chamley, J. H., and Burnstock, G., Interactions between autonomic nerves and smooth and cardiac muscle cells in tissue culture, *Dev. Biol.*, 32, 194, 1973.
151. Chamley, J. H. and Campbell, G. R., Trophic influences of sympathetic nerves and cyclic AMP on differentiation and proliferation of isolated smooth muscle cells in culture, *Cell Tiss. Res.*, 161, 497, 1975.
152. Gröschel-Stewart, U., Chamley, J. H., Campbell, G. R., and Burnstock, G., Changes in myosin distribution in dedifferentiating and redifferentiating smooth muscle cells in tissue culture, *Cell Tiss. Res.*, 165, 13, 1975.
153. Purves, R. D., Mark, G. E., and Burnstock, G., The electrical activity of single isolated smooth muscle cells, *Plfugers Arch.*, 341, 325, 1973.
154. Purves, R. D., Hill, C. E., Chamley, J. H., Mark, G. E., Fry, D. M., and Burnstock, G., Functional autonomic neuromuscular junctions in tissue culture, *Pflugers Arch.*, 350, 1, 1974.
155. Campbell, G. R., Uehara, Y., Mark, G., and Burnstock, G., Fine structure of smooth muscle cells grown in tissue culture, *J. Cell Biol.*, 49, 21, 1971.
156. Campbell, G. R., Chamley, J. H., and Burnstock, G., Development of smooth muscle cells in tissue culture, *J. Anat.*, 117, 295, 1974.
157. Lewis, M. R. and Lewis, W. H., The contraction of smooth muscle cells in tissue culture, *Am. J. Physiol.*, 44, 67, 1917.
158. Halle, W., Züchtung isolierter und pulsierender Amnionmuskelzellen, *Naturwissenschaften*, 47, 234, 1960.
159. Hermsmeyer, K., De Cino, P., and White, R., Spontaneous contractions of dispersed vascular muscle in cell culture, *In Vitro*, 12, 628, 1976.
160. Mauger, J. P., Worcel, M., Tassin, J., and Courtois, Y., Contractility of smooth muscle cells of rabbit aorta in tissue culture, *Nature*, 255, 337, 1975.
161. Chamley-Campbell, J. H., Nestel, P., and Campbell, G. R., Smooth muscle metabolic reactivity in atherogenesis: LDL metabolism and response to serum mitogens differ according to phenotype, in *Factors in Formation and Regression of the Atheroslerotic Plaque*, Born, G. R. V., Catapano, A. L., and Paoletti, R., Eds., Plenum Press, New York, 1982, 115.
162. Ross, R. and Kariya, B., Morphogenesis of vascular smooth muscle in atherosclerosis and cell culture, *Handbook of Physiology, Section 2*, Vol. 2, Bohr, D., Somlyo, A. P., and Sparks, H. V., Eds., American Physiological Society, Bethesda, Md., 1980, 69.
163. Reardon, M. F., Campbell, J. H., Nestel, P. J., and Campbell, G. R., Metabolism of lipoproteins by rabbit contractile and synthetic state arterial smooth muscle cells in culture, *Arteriosclerosis*, 3, 505a, 1983.
164. Kivilaakso, E. and Rytomaa, T., Erythrocyte chalone, a tissue-specific inhibitor of cell proliferation in the erythron, *Cell Tiss. Kinetics*, 4, 1, 1971.
165. Bullough, W. S., The actions of chalones, *Agents Actions*, 2, 1, 1971.
166. Houck, J. C., Weil, R. L., and Sharma, V. K., Evidence for a fibroblast chalone, *Nature New Biol.*, 240, 210, 1972.

167. Finkler, N. and Acker, P., Chalones: a mini-review, *Mt. Sinai J. Med., N.Y.*, 45, 258, 1978.
168. Campbell, J. H. and Campbell, G. R., Cellular interactions in the artery wall, in *The Peripheral Circulation,* Hunyor, S., Ludbrook, J., McGrath, M., and Shaw, J., Eds., Elsevier, New York, 1984, 33.
169. Wasteson, Å., Glimelius, B., Busch, C., Westermark, B., Heldin, C-H., and Norling, B., Effect of a platelet endoglycosidase on cell surface associated heparin sulfate of human cultured endothelial and glial cells, *Thromb. Res.,* 11, 309, 1977.
170. Gamse, G., Fromme, H. G., and Kresse, H., Metabolism of sulfated glycosaminoglycans in cultured endothelial cells and smooth muscle cells from bovine aorta, *Biochem. Biophys. Acta,* 544, 514, 1978.
171. Oohira, A., Wight, T. N., and Bornstein, P., Sulfated proteoglycans synthesized by vascular endothelial cells in culture, *J. Biol. Chem.,* 258, 2014, 1983.
172. Simms, H. S. and Stillman, N. P., Substances affecting adult tissue in vitro, Part 2, A growth inhibitor in adult tissue, *J. Gen. Physiol.,* 20, 621, 1936.
173. Florentin, R. A., Nam, S. C., Janakidevi, K., Lee, K. T., Reiner, J. M., and Thomas, W. A., Population dynamics of arterial smooth-muscle cells. II. In vivo inhibition of entry into mitosis of swine arterial smooth-muscle cells by aortic tissue extracts, *Arch. Pathol.,* 95, 317, 1973.
174. Eisenstein, R., Harper, E., Kuettner, K. E., Schumacher, B., and Matijevitch, B., Growth regulators in connective tissues. II. Evidence for the presence of several growth inhibitors in aortic extracts, *Pario Arterielle,* 5, 163, 1979.
175. Castellot, J. J., Addonizio, M. L., Rosenberg, R., and Karnovsky, M. J., Cultured endothelial cells produce a heparinlike inhibitor of smooth muscle cell growth, *J. Cell Biol.,* 90, 372, 1981.
176. Castellot, J. J., Jr., Favreau, L. V., Karnovsky, M. J., and Rosenberg, R. D., Inhibition of vascular smooth muscle cell growth by endothelial cell-derived heparin. Possible role of a platelet endoglycosidase, *J. Biol. Chem.,* 257, 11256, 1982.
177. Hoover, R. L., Rosenberg, R., Haering, W., and Karnovsky, M. J., Inhibition of rat arterial smooth muscle proliferation by heparin. II. In vitro studies, *Circ. Res.,* 47, 578, 1980.
178. Hatcher, V. B., Tsien, G., Oberman, M. S., Burk, P. G., Inhibition of cell proliferation and protease activity by cartilage factors and heparin, *J. Supramol. Struct.,* 14, 33, 1980.
179. Nilsson, J., Ksiazek, T., Thyberg, J., and Wasteson, A., Cell surface components and growth regulation in cultivated arterial smooth muscle cells, *J. Cell Sci.,* 64, 107, 1983.
180. Clowes, A. W. and Karnovsky, M. J., Failure of certain antiplatelet drugs to affect myointimal thickening following arterial endothelial injury in the rat, *Lab. Invest.,* 36, 452, 1977.
181. Guyton, J. R., Rosenberg, R. D., Clowes, A. W., and Karnovsky, M. J., Inhibition of rat arterial smooth muscle cell proliferation by heparin. In vivo studies with anticoagulant and nonanticoagulant heparin, *Circ. Res.,* 46, 625, 1980.
182. Reidy, M. A. and Schwartz, S. M., Endothelial regeneration III. Time course of intimal changes after small defined injury to rat aortic endothelium, *Lab. Invest.,* 44, 301, 1981.
183. Wasteson, Å., Höök, M., and Westermark, B., Demonstration of a platelet enzyme, degrading heparin sulphate, *FEBS Lett.,* 64, 218, 1976.
184. Oosta, G. M., Favreau, L. V., Beller, D. L., and Rosenberg, R. D., Purification and properties of human platelet heparitinase, *J. Biol. Chem.,* 257, 11249, 1982.
185. Handin, R. I. and Cohen, H. J., Purification and binding properties of human platelet factor four, *J. Biol. Chem.,* 251, 4273, 1976.
186. Campbell, G. R., Chamley-Campbell, J., Short, N., Robinson, R. B., and Hermsmeyer, K., Effect of cross-transplantation on normotensive and spontaneously hypertensive rat arterial muscle membrane, *Hypertension,* 3, 534, 1981.
187. Abel, P. W. and Hermsmeyer, K., Sympathetic cross-innervation of SHR and genetic controls suggests a trophic influence on vascular muscle membranes, *Circ. Res.,* 49, 1311, 1981.
188. Campbell, G. R., Gibbins, I., Allan, I., and Gannon, B., Effects of long term denervation on smooth muscle of the chicken expansor secundariorum, *Cell Tiss. Res.,* 176, 143, 1977.
189. Hartley, L. and Campbell, G. R., Sympathetic denervation of arteries in the wing of the chicken, *J. Anat.,* 136, 665, 1983.
190. Fronek, K., Trophic influence of the sympathetic nervous system on the arterial wall, *Bibl. Anat.,* 20, 414, 1981.
191. Zemplenyi, T. and Fronek, K., Chemical sympathectomy by 6-hydroxydopamine and arterial enzymes and lactate in the rabbit, *Exp. Mol. Pathol.,* 34, 123, 1981.
192. Bevan, R. D. and Tsura, H., Long-term denervation of vascular smooth muscle causes not only functional but structural change, *Blood Vessels,* 16, 109, 1979.
193. Ross, R. and Vogel, A., The platelet-derived growth factor, *Cell,* 14, 203, 1978.
194. Raines, E. W. and Ross, R., Platelet-derived growth factor. I. High yield purification and evidence for multiple forms, *J. Biol. Chem.,* 257, 5154, 1982.

195. Antoniades, H. N., Scher, C. D., and Stiles, C. D., Purification of the human platelet-derived growth factor, *Proc. Natl. Acad. Sci., USA,* 76, 1809, 1979.
196. Heldin, C-H., Westermark, B., and Wasteson, Å., Platelet-derived growth factor: purification and partial characterization, *Proc. Natl. Acad. Sci., USA,* 76, 3722, 1979.
197. Deuel, T. F., Huang, J. S., Proffitt, R. T., Baenziger, J. U., Chang, D., and Kennedy, B. B., Human platelet-derived growth factor: purification and resolution into two active protein fractions, *J. Biol. Chem.,* 256, 8896, 1981.
198. Ross, R., Glomset, J., Kariya, B., and Harker, L., A platelet-dependent serum factor that stimulates the proliferation of arterial smooth muscle cells in vitro, *Proc. Natl. Acad. Sci., USA,* 71, 1207, 1974.
199. Pledger, W. J., Stiles, C. D., Antoniades, H. N., and Scher, C. D., Induction of DNA synthesis in BALB/c 3T3 cells by serum components: reevaluation of the commitment process, *Proc. Natl. Acad. Sci., USA,* 74, 4481, 1977.
200. Pledger, W. J., Stiles, C. D., Antoniades, H. N., and Scher, C. D., An ordered sequence of events is required before BALB/c-3T3 cells become committed to DNA synthesis, *Proc. Natl. Acad. Sci., USA,* 75, 2839, 1978.
201. Singh, J. P., Chaikin, M. A., Pledger, W. J., Scher, C. D., and Stiles, C. D., Persistence of the mitogenic response to platelet-derived growth factor (competence) does not reflect a long-term interaction between the growth factor and the target cell, *J. Cell. Biol.,* 96, 1497, 1983.
202. Harker, L. A., Ross, R., Slichter, S. J., and Scott, C. R., Homocystine-induced arteriosclerosis: the role of endothelial cell injury and platelet response in its genesis, *J. Clin. Invest.,* 58, 731, 1976.
203. Moore, S., Friedman, R. J., Singal, D. P., Gauldie, J., and Blauchman, M., Inhibition of injury induced thromboatherosclerotic lesions by anti-platelet serum in rabbits, *Throm. Diath. Haemorrh.,* 35, 70, 1976.
204. Fuster, V., Bowie, E. J. W., Lewis, J. C., Fass, D. M., Owen, C. A., Jr., and Brown, A. L., Resistance to arteriosclerosis in pigs with von Willibrand's disease, spontaneous and high cholesterol diet-induced arteriosclerosis, *J. Clin. Invest.,* 61, 722, 1978.
205. Sher, C. D., Shepard, R. C., Antoniades, H. N., and Stiles, C. D., Development gene expression in cancer, *Biochim. Biophys. Acta,* 560, 217, 1979.
206. Doolittle, R. F., Hunkapiller, M. W., Hood, L. E., Devare, S. G., Robbins, K. C., Aaronson, S. A., and Antoniades, H. N., Siminan sarcoma virus *onc* gene, v-*sis,* is derived from the gene (or genes) encoding a platelet-derived growth factor, *Science,* 221, 275, 1983.
207. Waterfield, M. D., Scrace, G. T., Whittle, N., Stroobant, P., Johnsson, A., Wasteson, Å., Westermark, B., Heldin, C-H., Huang, J. S., and Deuel, T. F., Platelet-derived growth factor is structurally related to the putative transforming protein $P^{28}$sis of simian sarcoma virus, *Nature,* 304, 35, 1983.
208. Deuel, T. F., Huang, J. S., Huang, S. S., Stroobant, P., and Waterfield, M. D., Expression of a platelet-derived growth factor-like protein in simian sarcoma virus transformed cells, *Science,* 221, 1348, 1983.
209. Robbins, K. C., Antoniades, H. N., Devare, S. G., Hunkapiller, M. W., and Aaronson, S. A., Structural and immunological similarities between simian sarcoma virus gene product(s) and human platelet-derived growth factor, *Nature,* 305, 605, 1983.
210. Assoian, R. K., Grotendorst, G. R., Miller, D. M., and Sporn, M. B., Cellular transformation by coordinated action of three peptide growth factors from human platelets, *Nature,* 309, 804, 1984.
211. Gajdusek, C., Di Corletto, P., Ross, R., and Schwartz, S. M., An endothelial cell-derived growth factor, *J. Cell Biol.,* 85, 467, 1980.
212. Di Corletto, P. E., Gajdusek, C. M., Schwartz, S. M., and Ross, R., Biochemical properties of the endothelium-derived growth factor: comparison to other growth factors, *J. Cell Physiol.,* 114, 339, 1983.
213. Calderon, J. and Unanue, E. R., Two biological activities regulating cell proliferation found in cultures of peritoneal exudate cells, *Nature,* 253, 359, 1975.
214. Martin, B. M., Gimbrone, M. A., Unanue, E. R., and Cotran, R. S., Stimulation of nonlymphoid mesenchymal cell proliferation by a macrophage-derived growth factor, *J. Immunol.,* 126, 1510, 1981.
215. Fischer-Dzoga, K., Fraser, R., and Wissler, R. W., Stimulation of proliferation in stationary primary cultures of monkey and rabbit aortic smooth muscle cells. I. Effects of lipoprotein fractions of hyperlipemic serum and lymph, *Exp. Mol. Pathol.,* 24, 346, 1976.
216. Fischer-Dzoga, K. and Wissler, R. W., Stimulation of proliferation in stationary primary cultures of monkey aortic smooth muscle cells. Part 2. Effect of varying concentrations of hyperlipemic serum and low density lipoproteins of varying dietary fat origins, *Atherosclerosis,* 24, 515, 1976.
217. Chen, R. M., Getz, G. S., Fischer-Dzoga, K., and Wissler, R. W., The role of hyperlipidemic serum on the proliferation and necrosis of aortic medial cells in vitro, *Exp. Mol. Pathol.,* 26, 359, 1977.
218. Larrue, J., Desgranges, C., Daret, D., and Bricaud, H., Effects of hypercholesterolaemic serum on aortic explants from normal rats, *Cardiovasc. Res.,* 11, 519, 1977.

219. Rönnemaa, T. and Doherty, N. S., Effect of serum and liver extracts from hypercholesterolemic rats on the synthesis of collagen by isolated aortas and cultured aortic smooth muscle cells, *Atherosclerosis,* 26, 261, 1977.

220. Pietilä, K., Long-term effect of hyperlipidemic serum on the synthesis of glycosaminoglycans and on the rate of growth of rabbit aortic smooth muscle cells in culture, *Atheroslerosis,* 42, 67, 1982.

221. Fless, G. M., Kirchhausen, T., Fischer-Dzoda, K., Wissler, R. W., and Scanu, A. M., Serum low density lipoproteins with mitogenic effect on cultured aortic smooth muscle cells, *Atherosclerosis,* 41, 171, 1982.

222. Blaes, N. and Boissel, J-P., Growth-stimulating effect of catecholamines on rat aortic smooth muscle cells in culture, *J. Cell Physiol.,* 116, 167, 1983.

223. Stout, R. W., Bierman, E. L., and Ross, R., Effect of insulin on the proliferation of cultured primate arterial smooth muscle cells, *Circ. Res.,* 36, 319, 1975.

224. Pfeifle, B. and Ditschuneit, H., The effect of insulin-like growth factors on cell proliferation of human smooth muscle cells, *Artery,* 8, 336, 1980.

225. Pfeifle, B. and Ditschuneit, H., Effect of insulin on growth of cultured human arterial smooth muscle cells, *Diabetologia,* 20, 155, 1981.

226. Waymouth, C. and Reed, D. E., A reversible morphological change in mouse cells (stain L, clone NCTC 929) under the influence of insulin, *Texas Rep. Biol. Med.,* 23, Suppl. 1, 413, 1965.

227. Ledet, T., Diabetic macroangiopathy: in vitro study of the growth of aortic smooth muscle cells propagated in human serum from young juvenile diabetics, *Diabetologia,* 11, 358, 1975.

228. Ledet, T., Growth of rabbit aortic smooth muscle cells in serum from patients with juvenile diabetes, *Acta Pathol. Microbiol. Scand.,* 84, 508, 1976.

229. Ledet, T., Fischer-Dzoga, K., and Wissler, R. W., Growth of rabbit aortic smooth-muscle cells cultured in media containing diabetic and hyperlipemic serum, *Diabetes,* 25, 207, 1976.

230. Ledet, T., Growth hormone antiserum suppresses the growth effect of diabetic serum. Studies on rabbit aortic medial cell cultures, *Diabetes,* 26, 798, 1977.

231. Wolinsky, H., Goldfischer, S., Schiller, B., and Kasak, L. E., Modification of the effects of hypertension on lysosomes and connective tissue in the rat aorta, *Circ. Res.,* 34, 233, 1974.

232. Leitman, D. C., Benson, S. C., and Johnson, L. K., Glucocorticoids stimulate collagen and noncollagen protein synthesis in cultured vascular smooth muscle cells, *J. Cell Biol.,* 98, 541, 1984.

233. Nakao, J., Chang, W-C., Murota, S-I., and Orimo, H., Testosterone inhibits prostacyclin production by rat aortic smooth muscle cells in culture, *Atherosclerosis,* 39, 203, 1981a.

234. Longenecker, J. P., Kilty, L. A., and Johnson, L. K., Glucocorticoid inhibition of vascular smooth muscle proliferation: influence of homologous extracellular matrix and serum mitogens, *J. Cell Biol.,* 98, 534, 1984.

235. Bergman, R. A., Uterine smooth muscle fibres in castrate and oestrogen-treated rats, *J. Cell Biol.,* 36, 639, 1968.

236. Bo, W. J., Odor, D. L., and Rothrock, M. L., Ultrastructure of uterine smooth muscle following progesterone or progestersone-estrogen treatment, *Anat. Rec.,* 163, 121, 1968.

237. Kuriyama, H. and Suzuki, H., Effects of prostaglandin $E_2$ and oxytocin on the electrical activity of hormone-treated and pregnant rat myometria, *J. Physiol. (London),* 260, 335, 1976.

238. Nakao, J., Chang, W-C., Murota, S-I., and Orimo, H., Estradiol-binding sites in rat aortic smooth muscle cells in culture, *Atherosclerosis,* 38, 75, 1981.

239. Fischer-Dzoga, K., Wissler, R. W., and Vesselinovitch, D., The effect of estradiol on the proliferation of rabbit aortic medial tissue culture cells induced by hyperlipemic serum, *Exp. Mol. Pathol.,* 39, 355, 1983.

240. Fischer, G. M. and Swain, M. L., In vivo effects of sex hormones on aortic elastin and collagen dynamics in castrated and intact rats, *Endocrinology,* 102, 92, 1978.

241. Wolinsky, H., Effects of estrogen and progesterone treatment on the response of the aorta of male rats to hypertension, *Circ. Res.,* 30, 341, 1972.

242. Rhee, C. Y., Spaet, T. H., Gaynor, E., Lajam, F., Shiang, H. H., Caruso, E., and Litwak, R. W., Suppression of surgically induced vascular intimal hypertrophy by estrogen, *Circulation,* 49(Suppl. III), 92, 1974.

243. Bagdade, J. D. and Subbaiah, P. V., Atherosclerosis and oral contraceptive use. Serum from oral contraceptive users stimulates growth of arterial smooth muscle cells, *Arteriosclerosis,* 2, 170, 1982.

244. Campbell-Boswell, M. and Lazzarini-Robertson, A., Jr., Effects of angiotensin II and vasopressin on human smooth muscle cells in vitro, *Exp. Mol. Pathol.,* 35, 265, 1981.

245. Alexander, R. W. and Gimbrone, M. A., Stimulation of prostaglandin-E synthesis in cultured human umbilical vein smooth muscle cells, *Proc. Natl. Acad. Sci., USA,* 73, 1617, 1976.

246. Huttner, J. J., Gwebu, E. T., Panganamala, R. V., Milo, G. E., Cornwell, D. G., Sharma, H. M., and Geer, J. C., Fatty acids and their prostaglandin derivatives: inhibitors of proliferation in aortic smooth muscle cells, *Science,* 197, 289, 1977.

247. Lim, R., Mitsunobu, K., and Li, W. K. P., Maturation-stimulating effect of brain extract and dibutyryl cyclic AMP on dissociated embryonic brain cells in culture, *Exp. Cell Res.*, 79, 243, 1973.
248. Aw, E. J., Holt, P. G., and Simons, P. J., Myogenesis in vitro. Enhancement by dibutyryl cAMP, *Exp. Cell Res.*, 83, 436, 1974.

# INDEX

## A



## B



## C